Bioprinting

To Make Ourselves Anew

Kenneth Douglas

OXFORD
UNIVERSITY PRESS

OXFORD
UNIVERSITY PRESS

Oxford University Press is a department of the University of Oxford. It furthers
the University's objective of excellence in research, scholarship, and education
by publishing worldwide. Oxford is a registered trade mark of Oxford University
Press in the UK and certain other countries.

Published in the United States of America by Oxford University Press
198 Madison Avenue, New York, NY 10016, United States of America.

Library of Congress Cataloging-in-Publication Data
Names: Douglas, Kenneth, author.
Title: Bioprinting : to make ourselves anew / Kenneth Douglas.
Description: New York : Oxford University Press, 2021. |
Includes bibliographical references and index.
Identifiers: LCCN 2020052586 (print) | LCCN 2020052587 (ebook) |
ISBN 9780190943547 (hardback) | ISBN 9780190943561 (epub)
Subjects: LCSH: Regenerative medicine. | Tissue engineering. |
Three-dimensional printing.
Classification: LCC R857.T55 D68 2021 (print) | LCC R857.T55 (ebook) |
DDC 610.28—dc23
LC record available at https://lccn.loc.gov/2020052586
LC ebook record available at https://lccn.loc.gov/2020052587

DOI: 10.1093/oso/9780190943547.001.0001

1 3 5 7 9 8 6 4 2

Printed by Sheridan Books, Inc., United States of America

Cover art: Used with permission from Helena Sue Martin, David B. Kolesky,
and Jennifer A. Lewis, Wyss Institute for Biologically Inspired Engineering, Harvard University
Cover design: Kenneth Douglas

To my mother, my father, and my sister

and

To my friend Nancy, and to all those who gave her life anew

Contents

A Note to the Reader

This book is titled *Bioprinting*. Just one word. Seems simple enough. But it requires some explanation. In 2016 a group of 19 distinguished scientists published a paper titled "Biofabrication: Reappraising the Definition of an Evolving Field." The paper proposed a "refined working definition" of bioprinting among other clarifications.[1] The proposed definition isn't the one used in this book.

The technology that led to bioprinting is called 3D printing, invented in 1984 by Chuck Hull. 3D printing uses plastic or metal to build a three-dimensional object layer-by-layer. This build process is distinct from machining by removing portions from a block or casting by pouring molten material into a mold. About 15 years after 3D printing was invented, various scientists began to think seriously about using the technique to print living matter such as cells. Bioengineer Thomas Boland, then at Clemson University, tinkered with a commercial inkjet printer to dispense proteins, bacteria, and cell solutions instead of conventional ink. He published his results. Others soon joined in the fun.

As more people tried their hand at it, *bioprinting* seemed like a reasonable way to describe what they were doing. But as so often happens when clever people start to mess about, they came up with alternative ways to do what they had first started out doing. It wasn't long before folks were able to build three-dimensional objects that included living cells—without necessarily doing so layer-by-layer.

If they were using computer-controlled machines and if the three-dimensional result of the method they thought up still resembled the result of layer-by-layer building, it was all too easy for the "popular and journalistic media" to envelop their different practices under the heading of bioprinting.[2]

The publication on biofabrication that I've cited suggested specific definitions for fields such as bioprinting and bioassembly, taking them to be complementary strategies within the larger realm of biofabrication. Bioprinting describes computer-aided transfer processes for patterning and assembling of living and non-living materials by direct spatial placement and arrangement. Bioassembly describes automated assembly of preformed cell-containing building blocks generated by cell-driven self-organization. According to the paper, both bioprinting and bioassembly may involve layer-by-layer deposition of cells and other compatible material, but neither is restricted to this approach. Biofabrication encompasses both bioprinting and bioassembly and also includes subsequent tissue maturation processes.

What it comes down to is that in this book, I've knowingly conflated bioprinting, bioassembly, building biological constructs layer-by-layer, and subsequent tissue maturation processes. My desire was to cut a broad swath through the present-day scientific literature. So I must provide this Note to the Reader for taking liberties with the efforts of experts to tease apart these closely related fields. Mea culpa.

On a similar subject, the majority of the studies I've presented in this book print living cells. These experiments typically print biological or non-biological material in addition to printing cells. But there are cases in which no cells are printed. And sometimes the researchers seeded the cells onto a bioprinted matrix. Seeding cells means the experimenters spread a defined amount of a cell suspension onto the bioprinted matrix, typically using a pipette (a slender tube). On rare occasions, cells aren't added by bioprinting or seeding—for example, an experiment that printed substances encouraging the directed growth of host cells. So it's important to acknowledge upfront the diversity of experimental methods included in this book, all subsumed under the rubric of bioprinting.

One last item: the definition of a bioink. Just as the term *bioprinting* has assumed different meanings in common usage, so too has the notion of a bioink. This will sound familiar, but in 2019 a group of 15 distinguished scientists published a paper titled "A Definition of Bioinks and Their Distinction from Biomaterial Ink."[3]

The authors observe that the "apparent definition of the term bioink" has become increasingly divergent. They propose that bioinks should be defined as "a formulation of cells suitable for processing by an automated biofabrication technology that may also contain biologically active components and biomaterials."[4] They recommend two designations—bioink and biomaterial ink—be used in place of the single word bioink.

In this book we will simply use the term *bioink* rather than the suggested two terms. This is in keeping with the current literature. We will be explicitly clear when a printer is depositing cells or when it is depositing a plastic polymer, for example.

Nevertheless, we still haven't shared with the indulgent reader what constitutes a *biomaterial* or a *biomaterial ink*. The 2019 paper I've just cited does skillfully differentiate between a bioink and a biomaterial ink. In doing so, it also directs you to yet another scholarly paper on the proposed definition of a biomaterial.[5] In thirteen well-crafted pages, the author of that paper puts forth a compelling argument for a radically different definition of a biomaterial than has been used in times past.

And now I must unceremoniously stop.

I felt duty-bound to write this note to the reader. And I started with the best of intentions. But by this point I'm reminded of a story you likely know:

In an oft-told variation of the Hindu myth of cosmology, a young boy asks his father what holds up the Earth. Amused, the father assures his son that the world rests on the back of a very large turtle. "But what holds up the turtle?" the boy asks. After brief reflection, the father says, "A huge elephant." "But," the boy continues, "what is under the elephant?" Sensing that he is rapidly losing control of the conversation [just as I was losing control of A Note to the Reader], the father finally exclaims, "Son, it's elephants all the way down from there!"[6]

Foreword

I am not sure why Ken Douglas invited me to write the Foreword to this book, but here I am. I do not know the author. He sent me an email explaining that he had written what he believes to be (and I agree) the first book on bioprinting written for the lay audience. He also went on to explain that "a good deal of the material in the book is work done at the Wyss Institute. Also, the book's cover will be graced by an illustration arising from work at the Wyss Institute." As the Founding Director of the Wyss Institute for Biologically Inspired Engineering at Harvard University, where many of the examples that are used to highlight the major accomplishments in this field were carried out as part of our 3D Organ Engineering Initiative, this was a pretty good pitch. Furthermore, once I read the book, I realized that Ken highlighted a great deal of my own work that helped to launch the fields of mechanobiology, tissue engineering, and organs-on-chips, from which bioprinting emerged. But what Ken apparently did not know is that I had actually been bitten by the 3D printing bug myself in the late 1990s, even before the field of bioprinting first emerged, and so accepting this offer was an easy decision.

In 1996, I was invited to attend a meeting of the Defense Science Board, a committee of about twenty leading scientists and engineers who advise the Pentagon on scientific and technical matters of special interest to the U.S. Department of Defense. In the past, the scientists who presented updates to the committee on newest advances in their fields had largely been chemists, physicists, material scientists, and engineers because the focus of the military was on nuclear and chemical warfare, advanced materials for armaments, and computer-based communications and control. But everything changed after the first Iraq war, when the U.S. government discovered that Saddam Hussein had amassed a vast amount of biological warfare agents and that he had rockets armed with deadly anthrax. Due to the significance of this threat and the difficulty of the challenge to rapidly develop relevant countermeasures, the military asked the Defense Advanced Research Projects Agency (DARPA) to take on this task. DARPA was founded immediately after Sputnik to ensure that the United States would never be surprised like that again. Its main purpose is to "maintain our nation's technological superiority for the defense of the United States" by focusing on challenges that may take 20–30 years to solve.

I was invited to speak to the Defense Science Board because, due to this new biological threat, they now badly needed input from experts who understood medicine as well as molecular cell biology and infectious disease. The biologists and physicians they sought also needed to be able to communicate and interact effectively with physicists, chemists, engineers, and military experts, and apparently I fit the bill. In my lecture, I explained to them how molecules physically self-assemble to form three-dimensional filamentous cytoskeletal networks that give cells their shape, as well as how cells and molecular scaffolds join together to form tissues and how tissues combine to create organs during embryo formation, using a design principle known as tensegrity. Tensegrities are structural networks composed of opposing tension and compression elements that balance each other, thereby creating isometric tension or a tensile *prestress* that stabilizes the entire structure. Our bodies use the same principle to stabilize themselves as they are composed of multiple stiff, compression-bearing bones interconnected by tensed muscles, tendons, and ligaments, and the tone or prestress in our muscles governs whether we are flexible or rigid. Thus, as Ken describes in this book, my work focused on the central role of mechanics in biology, and I described living cells and molecules as physical materials. This is not what the audience of physicists, chemists, and engineers expected from a biology lecture, and they were reassured to learn that the laws of physics also apply at this size scale, and at this level of complexity.

After my presentation, a DARPA program manager by the name of Shaun Jones came up to me and explained that one of the biggest immediate challenges the military faced was redesigning their uniforms so that they could protect soldiers from biological pathogens and allow them to continue to operate in the battlefield. The existing protective gear for soldiers was essentially thick rubber overalls that zipped tight like deep sea diving outfits. They completely prevented entry of chemicals and infectious agents; the only problem was that the soldier heated up to deadly temperatures in about 30 minutes due to the heat generated by his own body in a closed space, much like the beautiful blond lover of James Bond who died from being painted gold in *Goldfinger*. Then Shaun explained that "there's a huge opportunity out there for anyone who has vision and can come up with a solution to this problem. Do you think you could make synthetic materials that behave like living cells?"

I explained that, in fact, I had helped to found a biotech company called Neomorphics, Inc., with Jay Vacanti and Bob Langer a few years before that engineered artificial skin by combining cells isolated from human foreskin with a felt-like lattice of synthetic polymers made from surgical suture materials. But that tissue engineering approach was too conventional for Shaun, who replied,

> No, no, what I'm thinking is more like an incredibly flexible condom that is so strong and resilient that you can climb into it and cover your entire body. It has to allow heat and moisture to pass, but it must prevent the passage of all biologicals, and it would be great if it also kills any pathogen it contacts.

A few months later, a newly formed company I created called Molecular Geodesics, Inc., was awarded a DARPA contract on "Biomimetic Materials for Pathogen Neutralization." The concept was to build flexible porous fabrics that have great strength and resilience, as well as solid-phase biochemical processing activities, much like the molecular cytoskeleton that gives living cells their shape, but in this case the immobilized enzymes would inactivate pathogens that pass through the network of the fabric while allowing heat to pass. As I had previously shown that living cells stabilize their shape and internal cytoskeleton using tensegrity architecture, I proposed to use the same design principle to build these bioskins. But how could one build structures with this type of complex 3D architecture at the microscale?

In my search for a solution, I discovered the emerging field of solid free-form fabrication that was being used by industry to rapidly prototype parts with complex 3D forms. The first technique I learned about was stereolithography [usually known as 3D printing], in which computer-assisted design software is used to construct an image of a 3D structure of any arbitrary shape, and then it is computationally cut into many horizontal sections like a deck of cards. To build the structure, a thin (a few thousandths of an inch) layer of a solution containing a light-sensitive liquid polymer is poured over the surface of a square chamber. A moveable ultraviolet laser then rasters back and forth across the entire surface line-by-line and emits brief bursts of light at microscopic points that correspond to pixels where the 3D computer image of the structure is sliced by the first horizontal cross section through the image. This solidifies the plastic dot-by-dot, but only in regions that correspond to the material to be built; all of the remaining solution remains liquid. By lowering the platform a few thousandths of an inch at a time, repeating this process layer-by-layer from bottom to top of the computer image, and washing it free of the unpolymerized fluid, a solid form is created that exhibits the exact same 3D shape and internal fine network architecture on the microscale as that originally designed on the computer. Molecular Geodesics used this type of computer-based design and manufacturing strategy to create 3D tensegrity materials, which were first optimized for their material properties in silico using engineering analysis software.

Needless to say, the 3D printing methods in 1997 were limited in resolution, and the materials that could be used were entirely rigid and inorganic (e.g., polymers, ceramics, and metals), and hence not optimal for our application. Thus, we never fabricated bioskins with the necessary fine structure and flexibility required by the military. Molecular Geodesics eventually changed its name to Tensegra, Inc., and moved into the medical device space, using a similar laser-based, layer-by-layer building approach with titanium metal to build medical devices such as artificial lumbar discs. Although these did well when implanted into large animals, the company died along with the fall of the Twin Towers the week of September 11, 2001, just about the same time similar 3D printing methods were being hijacked by a handful of scientists and engineers who developed ways to print biological materials, including living cells, in defined 3D forms, thereby creating the field of bioprinting.

In this book, Ken Douglas takes off where my story ended and tells how bioprinting first emerged at the turn of the 21st century, the enormous potential it offers for transforming the field of organ transplantation, and the real challenges that must be overcome in order for it to succeed and have an impact on the future of health care. He covers a huge range of biology because this is precisely what the field of bioprinting is trying to do. Researchers seek to 3D print living cells, and the extracellular matrix scaffolds in which they live, to recreate living replacements for virtually every type of organ in our bodies: liver, lung, kidney, muscle, bone, skin, brain, and so on. Ken attempts to bring the reader up to speed on the structure–function relations of each of these organs while also introducing them to basic concepts in biology, physiology, medicine, physics, and engineering. It's a challenge, and readers might get confused in parts, but the tale is compelling. It is a story of creativity, drive, perseverance, and passion, and it is told by recounting the personal stories of many of the key scientists who have led the chase, much as I conveyed my own story relating to 3D printing of bioskins.

The field of bioprinting is growing rapidly, and the various elements that contribute to the process must be refined and optimized to make it a success. This involves continuously developing better sources of organ-specific stem cells, designing building materials with enhanced properties and fugitive inks that can be more easily removed selectively, accelerating the speed of fabrication, and discovering how to let the living cells that are printed finish the design process on their own. After all, our bodies form in the embryo entirely through the action and self-assembly of living cells, without any requirement for blueprint, architect, or manufacturing process. In the end, this field will be confronted with real-world challenges, such as scaling up of manufacturing, fidelity of control, reproducibility, and the ultimate killer of breakthrough medical technologies: cost. Only time will tell whether bioprinting can survive this gauntlet of technical and commercial challenges. But the potential is huge, and as always, technology fallout, scientific value, and recruitment of great young minds into science and engineering will likely come from pursuing this new path of investigation, even if it does not lead to the precise products we expected in the beginning.

Donald E. Ingber, M.D., Ph.D.
Judah Folkman Professor of Vascular Biology
Harvard Medical School and Boston Children's Hospital

Professor of Bioengineering
Harvard John A. Paulson School of Engineering and Applied Sciences

Founding Director and Core Faculty
Wyss Institute for Biologically Inspired Engineering
Harvard University

Preface

I was gobsmacked. Absolutely gobsmacked.

For a long moment I was rooted where I stood, like a male Lot's wife turned into a pillar of salt.[1] Then I burst out, "What!"

I'd come in for my weekly physical therapy session. Nancy—office manager, receptionist, insurance billing doyenne, and spiritual anchor of the clinic—wasn't at her desk. She might have been in a back room or out depositing a fistful of checks at the bank. Chris, my physical therapist, was in his treatment room finishing up notes from his previous patient. Then he came out and saw me standing in front of Nancy's desk.

"Nancy will be out for awhile. We're not sure at this point whether it will be six months or maybe less. We've hooked up the computers so she can work from home. She had a kidney transplant on Tuesday. She's doing well."

That's when I blurted out "What!" in my utter astonishment. This was Friday. I had last seen her the previous Friday and she looked in good health. I had no idea that any such thing was supposed to happen. Chris filled me in a little. Nancy had been living with deteriorating kidneys for a long time yet she didn't show any sign of it. As her condition worsened she'd been placed on the recipient list for a donated kidney. She had a knapsack packed so she could go down from Boulder to the Denver hospital at a moment's notice, if and when she got the call that they had a donor match for her.

I had my physical therapy session with Chris. I didn't want to pry and didn't ask him too many questions about Nancy's situation. But I couldn't stop thinking about her. I'd known Nancy for about 10 years. Like me, she was an avid reader. We chatted about many of the books we'd read, from Dickens to Dostoyevsky. And she was an opera lover like me. We talked opera and once or twice when it wasn't too busy at her desk we would watch an opera excerpt on her computer. Now I'd learned that Nancy had been near death. Thank goodness she'd been lucky and they had found a donor match for her.

When I went home, the first thing I did was write her a get well card. Then my mind played connect-the-dots and I remembered the journal articles I'd skimmed on bioprinting. I dimly grasped that bioprinting originated from 3D printing—the technology that's all the rage where folks use a computer-controlled device to build up three-dimensional objects layer-by-layer. But instead of using plastic or metal to make a statuette of Yoda the Jedi Master or a new twist on a socket wrench, a band of adventurous scientists are using living cells to make pieces of human tissue. Skin, cartilage, bone, muscle, nerve, liver, heart, lung, kidney—they're all in play. And the quest is for an organ; even a mini-organ would be a pippin.

In many professional, peer-reviewed articles on bioprinting they mention the shortage of donor organs for people in need of transplants. And of the many on the waiting lists who die while waiting. In the United States, the number of people needing an organ transplant is over 121,000.[2] Kidneys are the organs most in demand: over 100,000. I didn't yet know the details of Nancy's personal story. More on that in a moment. But on the Friday I first learned of her kidney transplant I wondered about the prospects for bioprinted kidneys and I threw myself into the bioprinting literature with passion.

In my reading I was navigating choppy waters. I took my Ph.D. in physics, though I've spent my academic career in highly interdisciplinary work at the junction of physics and biology. By some unaccountable trick of the mind I thought that would be sufficient to fly through the pages to the feast of reason like a hummingbird zipping through the garden to the welcoming nectar of a trumpet vine. However, I wasn't encumbered by previous familiarity with the terms and concepts that came up again and again with the result that when I read the journal articles on bioprinting far into the night I could follow them only to a degree. Words like *extracellular matrix*, *rheology*, *innervation*, *vascularization*, *bioreactors*, *microfluidics*, and *organ-on-a-chip* percolated through the articles and I found their substance elusive. After I dug into these topics and went back to an article, it was as if I was reading it for the first time. And I felt a rush.

That helped to make my reading smoother sailing. And there was a serendipitous bonus. When I started to ferret out the terminology I found the backstory enthralling too. Take innervation as an example. This refers to the distribution of nerves in a tissue or organ. When a nerve is ruptured, the body has a limited capacity to repair the break. A bioprinted nerve graft can bridge the gap and, in animal studies, provides a conduit to support nerve regrowth. Bioprinting avoids the risks due to secondary surgery to harvest a nerve from another part of the body.

I decided to seek out major players in bioprinting and talk to them about their work and their world, to learn more about the backstory. Reading about innervation I found that the work of Gabor Forgacs (pronounced *Forgatch*) on biofabricated nerve grafts was seminal. I was fortunate to be able to speak to him.[3]

Forgacs came to bioprinting circuitously, having started in Hungary as a theoretical physicist who wanted to be a doctor but couldn't stand the sight of blood.[4] As he joked, he did the next best thing and married a doctor.[5] After coming to the United States, he collaborated with biologist Malcolm Steinberg at Princeton University. Steinberg's famous insight into how dissimilar cell types sort themselves out in the developing embryo—once considered a puzzling if not to say inexplicable phenomenon—is called the differential adhesion hypothesis.[6,7,8,9,10] He came up with his idea in the 1960s, melding elements of both biology and physics. His concept is that the embryo employs liquid-like tissue spreading and cell segregation that arise from differences in the adhesiveness (the stickiness) of individual types of tissue.

Steinberg's differential adhesion hypothesis, though conceived to describe embryonic development, informs much of the work Forgacs has contributed to bioprinting.

Not long before he took to bioprinting, Forgacs published detailed studies on the liquid-like properties of a myriad of cell aggregates including heart, liver, and retinal nerve cells.[11] In 2000 he accepted an offer from the University of Missouri, where he began to publish his work in bioprinting.

Previously, he had brought his cellular aggregates near to one another and watched as they fused in the same way that liquid drops coalesce à la Steinberg's hypothesis. Forgacs' group was the first to do this.[12] But creating these cellular aggregates, putting them together, and waiting for them to fuse was tedious and time-consuming. Then he went to a conference and came cheek-to-jowl with a 3D printer for the first time. It was a portentous encounter for both and neither would ever be the same again.

It wasn't long before Forgacs started to alter a printer so it could handle his cellular aggregates. He and a number of collaborators transformed the 3D printer into a bioprinter.* In July of 2004, they used it to produce the first non-trivial biological structure.[13] Forgacs shared a video with me (including audio) of their bespoke bioprinter completing its task—at which point an experimenter erupts in a spontaneous, ecstatic shout of joy that captures the moment.[14]

The science in this volume will be rigorous though absent the minute details that scientific technoweenies thrive on, much as we thrive on micronutrients in our food. There's a lot of serious business connected with bioprinting, the specter of organ donor waiting lists being just one of them. Even so, we can all still cherish and enjoy the science and the scientific backstory, presented with a minimum of technical baggage but without being shallow.

Helmut Ringsdorf is an eminent chemist, well known for successfully bridging the life sciences and materials science. He's also an advocate for going beyond the precise results of research. "We have to try to look behind the curtain of science, look at the acting scientists and the life they had to live and play in."[15] In this spirit, I've spoken to two dozen luminaries in bioprinting. I've had the pleasure and privilege of talking science with them and also hearing stories arising from their lives and work, some of which will appear in the book.

We'll bring this Preface to a close with Nancy's transplantation story.[16]

My friend Nancy ignited my interest in bioprinting. If Nancy lived some time in the future her story might have unfolded very differently. A physician might aspirate

---

* Forgacs was using a bioprinter to precisely position cells. Then he allowed the cells to self-assemble into aggregates by fusion (bioassembly). He was not building a three-dimensional object layer-by-layer. A Note to the Reader elaborates on the convention in this book: the conflation of divergent means of biofabrication in order to cut a broad swath through the present-day scientific literature.

(draw out by suction) cells that were still functioning well from one of her kidneys. Then the clinician would place these cells in a sealed chamber containing a solution with various nutrients and at optimal temperature and pH (optimal acidity or basicity) and perhaps including other support cells to induce a vascular network to form within the growing mass of kidney cells. The physician would take this enlarged aggregation of Nancy's kidney cells and bioprint a sort of cartridge or cassette enclosing them. The kidney has a naturally occurring area, called the Brödel line, which has a paucity of blood vessels. It's here that a surgeon would make an incision in one of Nancy's kidneys to insert the cassette using minimally invasive surgery and with almost no bleeding. Since the cells originated from Nancy's own kidneys, there would be little fear of rejection and minimal need for immunosuppressant drugs.

Some such scenario is for a future Nancy. The reality at present for my friend and countless others isn't so straightforward. In fact, at the present time, restoring kidney function is quite problematic. What follows is how one such story unfolded in the here and now.

Nancy had the first indications of her polycystic kidney disease in her early forties. The disease causes numerous fluid-filled cysts to sprout in the kidneys like so many dandelions in a grassy lot. As more and more cysts develop or as the cysts get too large, they interfere with the ability of your kidneys to keep metabolic waste from building to toxic levels. There's no known treatment for polycystic kidney disease, only treatment for some of the complications it causes, such as high blood pressure.

The disease is an inherited disorder, although Nancy knew of no one in her family who had ever been diagnosed with this malady. About 10 years prior to her kidney transplant she was feeling her usual vibrant self and went in to her primary care physician for a routine physical exam. Part of the exam was a metabolic panel and the blood drawn at her doctor's office showed high levels of creatinine—metabolic waste. Her doctor referred her to a nephrologist (a doctor specializing in kidney care) who ordered an ultrasound of her kidneys. During the ultrasound Nancy looked at the images and saw a large number of black dots in both kidneys. She knew that something was wrong.

The black dots were fluid-filled cysts. She began to see a nephrologist every three months. Slowly her creatinine levels continued to climb. Then high blood pressure became an issue to be treated. The implacable progression of the underlying disease was slow. Her blood began to show elevated levels of potassium and phosphorus. Her kidneys were just not able to do their job. Despite that, she was still feeling healthy, no pain or fatigue. After some eight years the blood abnormalities became a concern. Her nephrologist asked her to attend Kidney Class at a Denver hospital. Shortly before Christmas of 2014, she went to Kidney Class and it so happened that she was the only one to attend that day. She and the doctor sat down and he explained that dialysis would buy her a little time. Time to hope for a kidney transplant.

She was shocked and terrified. She'd been feeling fit. Married and with a child in college, Nancy was working full-time, taking care of her house, and helping to raise her daughter. Now, after her private Kidney Class, she was terribly depressed. This was a big reality check. Dialysis seemed like a death sentence for someone who'd been leading such a busy life. And the doctor told her that with kidney transplantation you're trading one chronic problem (polycystic kidney disease) for another, namely, medications and their side effects for the rest of your life to keep your immune system from rejecting the kidney.

Nancy saw her primary care physician, who recommended that she go to one of the two kidney transplant programs in the state. In March of 2015, she went to a transplant program hospital. It was an all-day meeting with a financial consultant, a social worker, a pharmacist, a medical assistant, a surgeon, and then the doctor who was the head of the kidney transplant program. They were all quite nice and excited. Usually the people that they see in the transplant program are very sick—perhaps diabetic, maybe overweight, smokers, elderly, etc. And here was Nancy, a relatively young person with a positive attitude, working full-time, exercising regularly. Individuals on the transplant team came up to her and said, "You'll do great!" They were convinced of it. The doctor who was to be her surgeon did an exam. Poking around her abdomen, he said with animation, "Oh this is great, you have plenty of room!" A transplanted kidney is placed in your abdomen. And after all, those doing the transplant want success at their job. They were confident that everything would go well.

Then she was dealt a blow. Blood tests showed that she had a plethora of antibodies that probably arose from her pregnancy and made her body more likely to reject a donated kidney. And her blood type was difficult to match with that of a potential kidney donor. She was a poor fit for 84% of the general population. Nancy was unlikely to get a donated kidney anytime soon.

The transplant team put her on a restricted diet to spare her kidneys. Among other yummies, she couldn't eat hummus, her favorite food. She was now in the category of severe kidney disease and consequently she was formally eligible for a kidney transplant. Nancy was given a number—a rank on the list of people in dire need of a kidney transplant. The lower the number the better off you are. Anything under 20 and you're sitting near the top of the list. Her number was 19. She had her presurgical exam to see if there were any comorbidities—any other chronic diseases or conditions. There were not. Nancy, her family, and her doctors played the waiting game, mostly working to keep her blood pressure under control. When she went on the list she prepared her backpack with pajamas, a toothbrush, and a book.

She had been feeling all right. Then fatigue set in. Coming home from a full day's work she collapsed. Even with a good movie on Turner Classic Movies she'd just fall asleep. A week before Memorial Day of 2016, her transplant coordinator called. "There's a remote possibility that you might be a match for a kidney. But don't get your hopes up too high." The potential donor was on life support. The donor was an intravenous drug user. That increases the risk of HIV, hepatitis C, and other diseases. With a mirthless inflection, Nancy said, "Oh, that's OK. I'm good. I'll wait for another possible donor." The transplant coordinator had her speak to the surgeon. He said, "We're really glad that you're feeling fine. But you're a poor match for 84% of the population. There's a very small window to find a match for you." They had tested the donor kidney. "It's young and it appears to be disease-free. And we don't know when another kidney will come along that will be a match for you." They felt that the young woman with the kidney wasn't on the streets using dirty needles. The donor was probably a "clean" drug user. The surgeon advised that Nancy be tested further to see if she was a match and, if so, that she take the kidney. She said OK.

She received a call the next day. The good news: The kidney is a match for you. The bad news: You're fourth on the list for this kidney. So it's probably not going to be your kidney. On the morning of the next day, Tuesday, the transplant coordinator called again. "Yeah, both the kidneys are spoken for. But you'll move up on the list if another one comes along." Nancy took it philosophically. "I went to work and said to myself, 'Great, I don't have to have a surgery today.'" She got a little hungry at work and nibbled on a sandwich.

Later, *on the same Tuesday morning*, another call came through. It was the transplant coordinator once more. "So, have you eaten today?" Nancy said, "Yes, but not much." The coordinator said, "When can you be at the hospital. It's go time." Nancy was exhilarated, scared, and practical. Although she never got to learn the specifics, everything had changed as fast as a somersault. She got her knapsack, and her husband drove her down to Denver to have a kidney transplant.

For three days following her surgery, Nancy lived in the shadowy world of a sleep/wake cycle persistently disrupted. Nurses checking multiple intravenous lines and giving her medications; the beep-beep-beep of machines monitoring her heart rate and temperature and at set intervals the whoosh-whoosh-whoosh of the cuff inflating to measure her blood pressure, clamping down on her arm and then loosening its grip like an apparition in a dream; the cloying hospital smell of disinfectant. She did get to read a little bit. The book was Ayn Rand's *Atlas Shrugged*. She joked, "A nice uplifting, entertaining choice!"

Two days after her surgery, when she went to the bathroom to brush her teeth she looked in the mirror. "I looked like a different person," she said. Over the years, she had become very pale. Being ashen had become the norm. Now she was positively flushed—like she would have looked in her twenties riding her bike up a mountain trail. And her sense of taste and smell returned. They were lost when she was running poor blood through her body for all the years her kidneys were in crisis. I asked her what her feelings were at this point. She said, "Gratitude. Gratitude."

On Friday, she was free to go home. As she was walking out of the transplant department toward the elevator, a member of the transplant team asked her, "Did you ring the bell?" Sure enough, there was a rope attached to a big ship's bell. When every successful transplant patient leaves this hospital, they get to ring the bell. Nancy says she rang the bell "maybe louder and harder than anybody ever had!" Bong, bong, bong! On the drive home she kept saying to her husband, "Everything is so gorgeous, the sunshine streaming in, the trees, the mountains." She was seeing the world for the first time, again.

Acknowledgments

It is a pleasure to thank the people who brought this book into being. I must begin with my friend Nancy, who inspired the writing. The surgeon and his colleagues were right in believing that she would do great with her donated kidney. They surely knew how wonderful that success would be for her husband and daughter. But I don't know if they realized how many other lives it would touch. I thank Nancy for sharing her story with me.

Bioprinting scientists seek to touch many lives, too. And as I've learned, they receive letters and e-mails from strangers asking if they can help a loved one in need. Despite the demands on their time, they were gracious in speaking to me and in responding to follow-up questions. I thank all of them, including those whose work does not appear in the final edit of the book.

Some of these scientists were especially kind. Gabor Forgacs somehow made the time to carefully critique a draft of the manuscript. His incisive suggestions pulled no punches and led to a flood of revisions. Tony Atala also read a draft, as did Dong-Woo Cho, Chuck Hull, Don Ingber, Mike McAlpine, and Koichi Nakayama, and I'm indebted to all of them.

I thank Sam Buchanan in Japan for his generous help in arranging my interview with Koichi Nakayama, and I thank Sangbok Kim for his diplomacy in facilitating my interview with Dong-Woo Cho.

Author and editor Louisa Gilder read my book proposal and gave me direction and encouragement when I needed it, as did author and editor Jane Friedman. I thank them both. Editor Susan Matheson provided innumerable, valuable suggestions in her meticulous reading of the manuscript, and I thank her for her conscientious scrutiny.

Bryan Kelley was in the trenches with me, nominally as a computer graphics and information technology expert, but he contributed so much more. He repeatedly persuaded me to amend bad writing habits. As a former army sergeant, he was steadfast in not letting me get off easy. He filtered not only the figures but also the text, and that was a tremendous bonus. I can't thank him enough.

Jude Gomez of Gomez Graphics did a wonderful job of vectorizing raster images when they didn't have sufficient resolution. Her turnaround time was terrific, and as an Adobe Illustrator expert she was very considerate with questions that Bryan and I had for her.

Justin Konrad delivered expert legal services with a deft touch. I thank him for that and for being so available for questions and unanticipated situations.

Erika Bagley and Kayla McLaughlin are two superb Customer Account Specialists at the Copyright Clearance Center. They were exceptionally good-hearted and knowledgeable in helping me to navigate several challenging paths to copyright permissions.

I thank my acquisitions editor Jeremy Lewis for giving me the opportunity to become an author with Oxford University Press (OUP) and for his support when I had requests to make of my publisher. I also thank the OUP editorial board, which hospitably approved Jeremy's recommendation to take on this manuscript. I'm grateful to OUP editorial assistant Bronwyn Geyer for her many kindnesses and her expert troubleshooting throughout the process of bringing the book to print. I thank Suganya Elango and Felshiya Samuel of Newgen Knowledge Works for their efforts in overseeing production of the book.

I thank Connie Binder for her accomplished touch in creating a high-quality index for the book.

Introduction

Frailty, Thy Name Is Human[1]

Those who are developing the discipline of bioprinting are kaleidoscopic in their training and talents. They need to be. Shakespeare spoke of "the thousand natural shocks that flesh is heir to."[2] Bioprinting aspires to help those afflicted with maladies that cover the spectrum that the Bard alluded to, whether ailments plainly visible such as charred skin or bony malformations or those less obvious to view such as complaints of the liver or kidney.

It's not only that we are heir to so many bodily misfortunes. Even if we look to ameliorate only a single condition, we immediately find the problem calls for a concatenation of skills and perspectives. Clearly biology is at the core, but to mimic the biology that nature has devised we must catch hold of its mechanical needs and constraints, its material substance, and its need for nourishment. We must try to deconstruct how what we see came to be—its temporal and spatial evolution—and discover how best to recapture the essence of its passage through time and space to its finished form. Healthy, functioning tissue is at once Shakespeare's "too, too solid flesh" and also a conjuring act, a phantasm.[3]

Over time, innumerable contributors have gifted us with their insights into how individual cells discover their identity and establish their relationship to other like-minded cells to form tissues and organs. Our understanding of bioprinting will be enriched by the stories of those who, in effect, bore witness to its genesis with their seminal revelations that made the field possible. So we will overlay numerous confluent themes and investigations to get ahold of the state-of-the-art of bioprinting.

We will learn about 3D printing, which gave birth to bioprinting. We'll discover stem cells that have the potential to develop into different bodily cell types and progenitor cells—the early descendants of stem cells. Other items on the menu are bioreactors used to coax newly bioprinted constructs to mature and also bioinks—soft biomaterials containing living cells. Three additional entrées are extracellular matrices, hydrogels, and decellularized extracellular matrices. Extracellular matrices are the cells' living quarters. Hydrogels are water-based substances that can stand in for extracellular matrices. Decellularized extracellular matrices are the cells' living quarters sans cells and can contribute to making a hydrogel.

Researchers hope that bioprinted parts will be transplantable into our bodies. To appreciate the challenges to doing this safely and effectively, we will examine articular cartilage such as found in your knees and hips and elastic cartilage such as found in your ears (Chapter 3). Our critique will devote an entire chapter (Chapter 4) to our vasculature, the 100,000 miles of conduit supplying the oxygenated blood essential to every cell's survival. We will make use of another chapter (Chapter 5) to discuss innervation—our body's means of allowing us to feel the grass under our bare feet, signal for a cab, or smile at the antics of a couple of kittens. Next, we'll familiarize ourselves with bioprinted stand-ins for our skin that envelops us and allows us to dissipate heat, bones that support us, and the skeletal muscle that allows us to move our bones (Chapter 6).

Bioprinting is in hot pursuit of solutions for damaged or failing internal organs. To understand unhealthy states, we'd first like to become versed in what constitutes healthy organs and how various injuries or disease affect these organs. We'll explore our liver (Chapter 7), an organ with over 500 vital functions; our heart (Chapter 8) that shuttles 2,000 gallons of oxygen-rich blood through our body each day; and our kidneys (Chapter 10) that cleanse a half-cup of blood every minute so we don't die in a morass of our own cellular waste products.

At this moment in time, bioprinting is at a proof-of-concept stage. That means there are many demonstrations of bioprinting constructs that seem to work as desired outside a living organism or in a laboratory animal, though not yet in humans. That's a good start. But it's not good enough for those with burned skin all over their face and body or for those on organ donor waiting lists, facing dialysis or death. Nor is it good enough for those whose heart attacks have left their hearts scarred and weakened, for those whose body has attacked their own liver owing to an autoimmune disease, for those disfigured by bone removal during surgery to access a malignant tumor, for those whose limbs are paralyzed, or for those whose every movement is misery because their cartilage is all worn out.

Dr. Anthony Atala is a urologic surgeon and bioprinting researcher. Though research takes up a lot of his time he still continues to see patients and perform surgeries. Atala said, "That's important to me because it's a reminder of who you're serving—who you're doing this for. The aim of this technology is to make patients' lives better. Full stop."[4]

Welcome to bioprinting.

1
Printing Paradigms

In a monastery in Europe around the year 1000 AD, in an ill-lit space heated indirectly by its proximity to the kitchen, a scribe silently works at copying a manuscript. Bent over the stand of his wooden desk he sits in a room set aside for this purpose, a writing room known as a scriptorium. He measures and outlines the page layouts and then carefully transcribes the text from another book, leaving appropriate space for headings or illustrations. His desk is home to ink pots, knives, and quills. The scribe's labor is utterly fatiguing. "Only try to do it yourself and you will learn how arduous is the writer's task. It dims your eyes, makes your back ache, and knits your chest and belly together. It is a terrible ordeal for the whole body."[1] Through such endeavors scribes preserved a portion of the literary history of the Western world for century after century, prior and subsequent to the turn of the first millennium. That was the paradigm.

Automated Printing

Then the paradigm changed, triggered by plague and paper. In mid-fourteenth-century Europe, *Yersinia pestis* bacteria—the pathogen responsible for bubonic plague—is widely thought to have caused the horror known as the Black Death.* From 1347 to 1351, the bacteria killed approximately 25 million people, estimated to be 30–50% of the European population at that time.[2]

Many of Europe's scribes perished. The scarcity of scribes meant that those who survived were in high demand. And non-monastic, commercial scribes commanded astronomically high prices. This was at a time when papermaking was a growth industry: paper made from linen rag was cheap compared to parchment made from animal skins. As James Burke captured the unfolding history in his book *Connections*, scribes were too costly and paper was so cheap you could use it to cover the walls; there was a crying need for some form of automated writing.[3]

In the early fifteenth century, a German goldsmith, gem cutter, and metallurgist started work on a technique for printing.[4] History credits the man we know as Johann Gutenberg, born sometime between 1394 and 1404 in Mainz, Germany, as the person who developed automated printing.[5]

---

* The name Black Death comes from the dark patches on victims' skin caused by subcutaneous bleeding.

Gutenberg set his printing presses in motion in the early to mid-1450s and soon scribes were redundant. The Gutenberg Bible appeared in 1455, his quintessential contribution and the first major book in Europe using automated printing with movable type.[†]

The advent of automated printing was one of the most critical events in the history of mankind, revolutionizing the spread of knowledge. Science historian John Man did the math.[6] In 1455 you could pile all of Europe's printed books in a single wagon, if you were looking for something to pass the time. By the year 1500, titles ran to tens of thousands and individual volumes to millions. Jump half a millennium to today and books gush from the presses to the tune of 10 billion per year—50 million tons of paper. Add to that daily newspapers and magazines and you shoot up to 130 million tons of paper per year. Pile this up and you have a heap four times the height of the Great Pyramid. Multiply by the 500 years since Gutenberg launched it all and you have a small mountain range.

3D Printing—The Foundation of Bioprinting

The transition from scribes in scriptoria to printing presses with movable type draws attention to the enormity of a paradigm shift. These seismic changes typically arise from a cauldron that's been seething and moiling for quite awhile. If bioprinting becomes a paradigm shift in medicine, it will take some doing.

But before we turn to bioprinting, we must look at one more shift in the paradigm of printing: bioprinting's parent, 3D printing.

"You can work on this, but do it on your own time. Do it nights and weekends. You can use one of the labs in the company, but do it on your own time."[7] That's what the president of a small ultraviolet light company told engineer and inventor Chuck Hull after his employee approached him with an original idea. Hull wanted to find a way to rapidly make prototype plastic parts. He thought that he might be able to use ultraviolet light to cure—that is, bond or fuse together—sheets of plastic to make quick models.

As Hull recalled,

One evening in 1983, I called my wife and I told her, "Get on down to the lab right away." That didn't seem like a good idea to her; she already had her pajamas on and she was going to go to bed to watch TV. But I insisted, so she got in the car and drove down to the lab. While she was driving, she was thinking, "This had better be good." It was.[8]

[†] *Movable type* means that individual type pieces with a single letter or character can be easily assembled and moved around to print any combination you want.

Hull had invented 3D printing. When he filed a patent for the technique in 1984, he called his invention stereolithography, a name that would undergo a metamorphosis to 3D printing.[9],[‡] 3D printing has come to be the accepted name for the process of additive manufacturing in which objects are created one layer at a time. 3D printing is a new paradigm in manufacturing. Traditional manufacturing methods use either a forming process (a cast or a mold) or a subtractive process such as milling or turning to remove material.

3D printing begins with an image of the object you want to create. The images are usually made using computer-aided design (CAD) software. A computer program takes the three-dimensional design, cuts it up into a bunch of very thin two-dimensional horizontal slices, and saves these image slices as a file. You send this file to a 3D printer and the printer uses the information to print the slices, one at a time on top of each other, onto a platform, like a tot building in three dimensions with alphabet blocks in a playpen.

Initially, the printer builds the object by moving the print head(s) side-to-side as well as toward you and away from you in a horizontal plane while laying down material (typically plastic or metal) as the computer tells it to. The instructions to the computer are the thin image slices. This enables you to precisely position the droplets or ribbons of material anywhere you want on a horizontal plane. Call this surface the build platform. When that layer is finished, the computer changes the separation distance between the print head and the build platform—either by instructing the platform to move down or instructing the print head to move up. The next batch of building is done on top of the first batch, although the droplets or ribbon may form a different pattern than the first one. That's why it's called additive manufacturing; you add one layer at a time.

When you watch a 3D printer in action, the print head (the business end where material flows out of the nozzle on its way to making structures) appears to go frantically through its movements in a whirl of feverish activity. You get the impression that you're looking at a movie in fast-forward mode—but without the merriment of people's voices that sound like chipmunks.

3D Printing Techniques

Chuck Hull's original invention of 3D printing used a laser. But that was only the beginning. Nowadays, there are many 3D printing techniques that fall under the heading of laser-assisted. Moreover, there are at least 10 distinctive 3D printing techniques and subtle variations within. Fortunately for us, only a small number come into play in bioprinting, our ultimate destination.

[‡] The patent was granted in 1986.

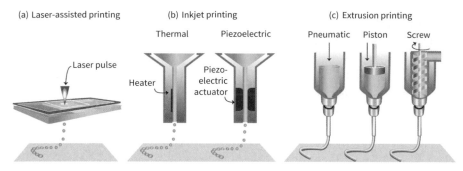

Figure 1.1 Common light- and ink-based 3D printing methods. (a) Laser-assisted printers use lasers focused to generate pressures that propel materials onto a build platform. (b) Thermal inkjet printers electrically heat the print head to vaporize small volumes of the printing fluid. This produces air-pressure pulses that force droplets from the nozzle. Piezoelectric inkjet printers use an actuator (a device providing movement) to apply a mechanical pulse to the fluid. This causes a shock wave that forces the bioink through the nozzle. (c) Extrusion printers use pneumatic, piston, or screw dispensing systems to extrude continuous beads of material.

Source: Reproduced by permission from Jos Malda et al., "25th Anniversary Article: Engineering Hydrogels for Biofabrication," *Advanced Materials* 25, no. 36 (August 23, 2013): 5011–28, https://doi.org/10.1002/adma.201302042.

Figure 1.1 gives you a few of the 3D printing techniques most relevant to our interests. The majority of 3D printers are of the sort called extrusion printers shown in panel (c) of Figure 1.1. You've probably seen a concrete mixer truck—a truck that mixes the components that make up concrete and then ejects the mixture down a short chute to the site you choose. In the examples shown in Figure 1.1 (b) and (c), think of the topmost portion as a very sophisticated incarnation of a concrete mixer and call it a print head.[§] The material that comes out as droplets or ribbons from the print head is the final concrete mixture.

All this movement requires special electromagnetic motors. The typical choice is called a stepper motor. All motors involve the interaction between electric and magnetic fields, but stepper motors (steppers for short) use cleverly designed electromagnets. Steppers turn a shaft through steps—in other words, well-defined angles—rather than through a smooth, continuous motion as a conventional motor does. These specialized motors are the physical innards, the engines that drive 3D printers. Computers control the whole shebang: the movements of the stepper motors, the switching on the fly from one print head to another, and the position of the build platform. Everything is automated. The rattling noise from the energetic

---

[§] Some 3D printers don't use such print heads—for example, projection stereolithography and other laser-based setups. Panel (a) is a cartoon of a laser-based 3D printer.

movement of the print head(s) isn't excessive, and many printers are encased in a housing for noise dampening.

Initially, engineers designed 3D printers for non-biological applications. They built three-dimensional objects using metals, ceramics, and plastics. But in the early 2000s, researchers began to use living cells and other biological materials to build three-dimensional objects: 3D printing begat bioprinting. A 3D printer became a bioprinter.[10,11,12]

Once you've transitioned from 3D printing to bioprinting, then instead of beginning with CAD software, you start with a medical image from a magnetic resonance imaging (MRI) scan or a computed tomography (CT) scan.** The CAD program takes the medical image and modifies it to be read by the bioprinter. A lot of computer programming goes into making the medical scans readable by the bioprinter's computer, but in the end the print head places the cells and other materials with fastidious accuracy.

Trailblazers of Bioprinting

From those heady days of the new millennium, several names are prominent. Bioengineer Thomas Boland, then at Clemson University, took an off-the-shelf inkjet printer and put the bio in bioprinting. He and one of his students modified a commercially available inkjet printer to dispense proteins, bacteria, or cell solutions instead of conventional ink.[13] It was fun to speak to Boland, the person who got bioprinting underway.[14] He told me that he still has a picture on his computer with a date stamp of 10:35 a.m., November 1, 2000, when he pulled off the feat—a picture of the first-ever bioprinted pattern made with his modified commercial inkjet printer.

Following his time at Clemson, Boland joined the faculty at the University of Texas at El Paso, where his graduate students and postdocs coined the term the *Boland effect*. They claimed that every time Boland came into the lab, anything that had been working promptly ceased to work. He shared this while laughing and took their needling good-naturedly.

Then, physician and scientist Vladimir Mironov teamed with Boland.[15] They and their colleagues showed how to place cell aggregates—cells that are only loosely grouped with one another—close enough together in sequential layers of a three-dimensional gel so that the cells fused. Perhaps they got a little ahead of themselves and named the technique "organ printing."[16,17] Nonetheless, bioprinting was off and running.

Mironov is quite a character. When I spoke to him, he followed up by sending me a link to "educational comics for kids" that he created during a stay in Brazil. He said he made them "specially for my granddaughter and grandson."[18]

** CT is also called computerized axial tomography (CAT). Both MRI and CT are medical imaging techniques.

In addition to Mironov and Boland, another innovator and early adopter was the-oretical physicist turned bioengineer Gabor Forgacs, whom we met in the Preface. Of his many contributions, Forgacs' most lasting gift to the field was the development of extrusion bioprinting (panel (c), Figure 1.1). His group was also the first to use the term bioprinting in a published scientific research paper, using the extrusion printer designed by his group.[19] Later, they made the first bioprinter-assisted vascular graft that became a blood vessel after post-printing maturation.[20]

Mechanical engineer Wei Sun, with dual appointments at Tsinghua University in Beijing and Drexel University in Philadelphia, was also at the starting gate.[21,22] He published an early article that explained to the scientific community at large how to design multiple print head (also called multi-nozzle) machines for bioprinting, al-though a few labs had been using more than one print head but hadn't shared the nuts and bolts with everyone.[23] Sun made another generous contribution. In 2009 he founded a journal called *Biofabrication* that extended open arms to the toddler that was bioprinting, and he remains editor-in-chief to this day. What's more, he was the founding president of the International Society of Biofabrication (2010–2014). Sun wrote the bylaws and designed the logo for what has become a vibrant international society, meeting in venues around the world every year.

But Sun was not all business, and it was great fun to speak with him and continue our conversation with multiple e-mails. He shared funny stories with me, like the time when he was trying to catch some sleep and his wife used markers to draw a bunch of little mice all over his face and neck. When he washed his face in the morning, he missed his neck. As the day went by, he noticed that his colleagues were staring at his neck but said nothing. It was only when he got home that he realized there were mice all over his neck.[24]

Materials scientist Brian Derby of the University of Manchester hosted the First International Workshop on Bioprinting and Biopatterning at Manchester in September 2004.[25,26] Derby has been a master craftsman in the inkjet form of bio-printing (panel (b) of Figure 1.1) from the beginning. And biomedical engineer and materials scientist Dietmar Hutmacher of Queensland University of Technology in Australia made seminal studies of scaffold materials and methods.[27]

There were others in attendance at bioprinting's opening ceremonies. For ex-ample, chemist Glenn Prestwich and his student Aleksander Skardal at the University of Utah were part of the National Science Foundation consortium, of which Gabor Forgacs was principal investigator. This was known informally as the organ printing consortium. Prestwich and Skardal produced many of the first successful hydrogel formulations.

The scientists who got bioprinting going with a whoop and a holler are still at it, joined by a legion of others. But multiple hurdles remain in transitioning from 3D printing to bioprinting, and we'll discuss these shortly. To better appreciate the challenges, we first consider applications of 3D printing that started to appear as the new invention entered the arena of medicine, even before bioprinting made its debut.

Medical Successes of 3D Printing

In 1996, a vexing problem confronted surgeons at the Wilford Hall Medical Center at San Antonio's Lackland Air Force Base. They were working to separate a pair of conjoined twins who shared a common leg. The surgeons hoped to split the shared leg lengthwise. But based on what they saw on the CAT scan, the doctors didn't know if this would be possible. The alternative would be that one girl would have the entire shared leg while the other girl would have nothing but her one leg. How to proceed? The surgical team was inspired to send the girls' CAT scan to 3D Systems, the company engineer/inventor Chuck Hull had founded. The company sent back a full-scale model of the girls' bone structure made by a 3D printer. The doctors were surprised to see that the bone was bigger than they expected based on the CAT scan data alone. With the aid of the preoperative model, the surgeons designed a procedure to split the one leg bone lengthwise. The operation took 23 hours. An orthopedic surgeon removed the bone, split it, and replaced the two parts, giving both girls the opportunity to walk on their own legs.

Hull said,

> This is extremely difficult for surgeons, to figure out how they are going to separate the two twins so that you will have two separately living people. When some of those surgeries were first done using the help of our technology, it was really touching for me.[28]

Since that time, there have been many other such cases. One singular instance of conjoined twins occurred in 2016 in New York City. In spite of medical advances independent of the use of 3D printing, the surgery can be exceptionally difficult and perilous depending on where the infants' bodies are fused. For example, when twins are joined at the head. If they are not separated, they typically don't live past their second birthday. The chief surgeon of the New York operation said, "We knew they shared an area of fused brain, but we did not know how complicated it would be until we looked inside." The surgical team used a combination of 3D printed models and virtual planning software to study the connected tissue and plan their course of action. Finally, after four surgeries over the course of seven months the twins were able to recover in their own *individual* hospital beds for the first time. The outcome may have been different without 3D printing.[29]

As word spread, 3D printing found medical uses within the body. A condition known as tracheobronchomalacia typically develops during infancy or early childhood. It presents the problem of airway collapse and is difficult to treat. An infant in Michigan had a collapsed bronchus that blocked the flow of air to his lungs.†† Doctors

---

†† A bronchus is a passageway in the respiratory tract that carries air to the lungs.

told his parents there was a significant chance he would not leave the hospital alive.[30] A team of doctors and a biomedical engineer obtained emergency clearance from the U.S. Food and Drug Administration to create and implant a tracheal splint using a biopolymer made by 3D printing. The splint was sewn around the infant's airway to expand the bronchus and give it a skeleton to assist in proper growth. Because of the choice of biopolymer material, his body was expected to fully resorb the splint in three years. The result was published in the esteemed *New England Journal of Medicine*.[31]

There are some circumstances in which surgeons can implant a non-bioresorbable 3D printed form—and save a life. This occurred not with an infant but, rather, a fourteen-year-old teen from Texas named Hannah. For Hannah, each breath was a struggle. She had been mistakenly treated her entire life for what doctors took to be progressive asthma. Finally, as symptoms continued to worsen, doctors diagnosed her in 2014 with life-threatening tracheobronchomalacia. "They told us her airway was so narrow that it was like breathing through a straw," her father said. "It was shocking to everyone that she had survived it for so long."[32]

The Texas doctors had one last option. They found the work of the University of Michigan team that had provided the infant's 3D printed airway splint, and then they performed a similar procedure. Hannah became the first adolescent to receive a custom airway splint. Whereas the infant's splint was made of bioresorbable material designed to grow with the child and eventually dissolve, Hannah's airway splint will last the rest of her life.

Transitioning from 3D Printing to Bioprinting

The previous stories do not yet have their counterparts in bioprinting. Supplying blood to bioprinted tissues—providing vascularization—is one of the major obstacles facing bioprinting that is absent in 3D printing of non-biological matter.[33,34,35] When you're bioprinting a cell, if that cell isn't closer than 100–200 millionths of a meter to a blood vessel of some sort, even one as small as a capillary, the cell will die. It's one of the many wonders of the human body that our vasculature—the arrangement of our blood vessels—provides each of our roughly 30 trillion human cells with the blood it requires to survive.[36,‡‡,§§]

We'll devote a chapter to vascularization because it's such a crucial issue in making functional bioprinted tissue. However, in this introductory chapter we offer a preview. The key is to make branching networks of blood vessels, which is what nature does. For a sense of how nature comes to grips with vascularization, we turn to Dr. Joseph P. Vacanti of the Harvard Medical School.

---

‡‡ According to the cited endnote, "relatively modest quantitative differences apply for women."
§§ If we're speaking about the vasculature supplying blood to cells, we really should use only 16% of the 30 trillion estimate since red blood cells are thought to comprise about 84% of the 30 trillion estimate.

Vacanti framed the question by asking how nature solves the problem of three-dimensional growth.[37] He pointed out that as a mass of cells enlarges, its surface area increases as the square of the radius of the aggregation of cells but the volume of the aggregate increases as the cube of the radius. He asked, "How does nature tackle this mismatch so that the cells on the interior can be nourished?"[38]

If you're rusty on your junior high school geometry, don't stress. Surface area is the outside area of an object, and volume is the amount of space inside the object. Suppose we specify the cell aggregate as a sphere for the purpose of visualization. As the radius of the sphere increases, its volume increases at a greater rate than its surface area. That's what Vacanti was talking about. He was directing our attention to the ratio of the surface area to the volume. The aggregation of cells initially relies on diffusion for getting the nutrients it needs and for ridding itself of metabolic waste products.

Diffusion is a passive, spontaneous way in which oxygen and nutrients move from regions of higher concentration to regions of lower concentration. But as growth continues, there is comparatively less surface area for the diffusion of nutrients because the ratio of the surface area of the cell to the volume of the cell aggregate becomes smaller. This could result in the death of cells at the center of the aggregate—what's called a necrotic core, the bane of any bioprinted construct. *Necrosis* means death of bodily tissue; the term is derived from an ancient Greek word for *dead body* or *corpse*.

Having a decreasing surface area to volume ratio is a worry for bioprinting, but eons before that it was a predicament for nature. Vacanti answered his rhetorical question of how to keep interior cells nourished: "Nature uses branching networks to achieve this goal of matching surface area to volume."[39] It's such a successful solution that nature uses this same innovation in both animal and plant worlds. Think trees: trunks, limbs, twigs, leaves. "All organs are composed of intertwined branching networks and all communication between organs is accomplished by branching systems."[40] The next time you glance at some English ivy, a potted jade plant, or a fiddle-leaf fig tree, reflect on Vacanti's observation:

> Indeed, animals are structurally not much different from plants in this fundamental repeating pattern. To be successful in their niche, animals must think, react, and move; hence, the developed nervous system, the covering of the skin, and the mobility allowed by a musculoskeletal system. But, in essence, we are branching networks, much as plants are.[41]

Innervation is another big hurdle for bioprinting that's also a non-issue in 3D printing non-living matter. Innervation describes the nerve excitation necessary for life and functionality of the various organs throughout the body. Like vascularization, innervation is also a vital bioprinting subject with its own chapter in this book. Researchers have at least two strategies. The first is to induce innervation after transplantation using pharmacologic means or naturally occurring proteins that stimulate cell growth—a technique called growth factor signaling. An alternate strategy is to stimulate nerve formation in a bioreactor prior to transplantation.

A bioreactor is a container that provides an appropriate biological environment for a chosen cell type to thrive and proliferate. A bioreactor controls such matters as temperature, nutrient delivery, gas exchange, mechanical or electrical stimulation, and removal of waste products generated by the cells. If it happened to occur to you that the human body is a bioreactor, you're quite right. That's really the crux of the matter.

Bioreactors are part of the firmament of the bioprinting world, and ideally they provide the bioprinted version of a given tissue with the lovely environment that the native version of that tissue experiences inside your body. That is, the bioreactor environment tries to mimic the body in a tissue-specific manner. After tissue is bioprinted, it's typically placed in a bioreactor to encourage it to be fruitful and multiply, to borrow a phrase from a best-selling book.

Apropos of multiplying, in order to be useful for human transplantation the bioprinting process has to be scalable. That is, beyond proof-of-concept experiments in glassware or small animals, the technique needs to be capable of creating tissues that will have dimensions and cell densities relevant to human needs. Scale-up of tissue constructs has been another stumbling block for bioprinting.

There are additional complications in realizing the potential of bioprinting. One is the fragility of living cells and other biological matter compared to materials routinely used in 3D printing. If you want to 3D print the metal frame of a bicycle built for two or a plastic rack for your golden oldies CD collection, you can get away with using harsh solvents and high temperatures. That's because in 3D printing the inks are metals, ceramics, and thermoplastic polymers, and they can tolerate chemicals and temperatures that are toxic to biomaterials. To bioprint human tissue, you must be much more circumspect in your choice of ink.

The generic term bioink refers to the ink used in bioprinting. Bioinks are biomaterials that may contain living cells, non-cellular biomaterial, or both. They are the raw material used in the bioprinting process.[42] A hitch in bioprinting is that there are a limited number of bioinks that have been successful in creating tissue. The reason is that living cells need a particular biological environment in order to connect with one another and to flourish. The bioink has to provide that opportunity to the cells. The biological environment is called the cell's extracellular matrix, and a successful bioink mimics the cell's extracellular matrix. The extracellular matrix is the non-cellular material present within all tissues and organs, but the composition of the matrix is distinct for each tissue type. It's made up of proteins and other biological molecules. The matrix provides a mechanical framework for cells that is critical for guiding cell shape, orientation, and function to form specialized structures for each tissue type—skin, bladder, lung, and so on.[43]

Rheology is a further reason that bioprinting is not a walk in the park with the fallen autumn leaves swish swishing underfoot with your every stride. Rheology is the science of the flow and deformation of materials: liquids, soft solids, and even solids that flow. In bioprinting, if the rheological properties—the flow properties—of your bioink aren't right, you'll clog the print head or something equally unproductive.

Rheological properties are used to determine a bioink's printability—that is, whether it can be accurately and precisely deposited where and when you want it to be.[44]

So much for putting something where you want it to be. What about putting nothing where you want it to be? What about incorporating the absence of material—for example, cavities or open channels? With 3D printing, it's not hard to design a metal strut or a plastic support piece, or to leach it away after printing. But in bio-printing, if you must design a void of some description, how can you do that and not have the mechanically weak mass of cells collapse during the building process? And if you do need to put in some sort of prop or brace, it had better be something that can be removed easily or be biocompatible if it must remain in the final product—the product that's going to end up in your body.

This Rubik's Cube with hollows is solved either by the printing of biomaterials that act as scaffolds for the bioprinted cells or sometimes by what's called scaffold-free bioprinting. Scaffold-free bioprinting is a catch-all phrase to describe tricks that researchers have come up with to do without scaffolds.

From this brief summary, it's clear that 3D printing has to be tweaked big time to achieve bioprinting. We'll save vascularization and innervation for chapters of their own. But now we'll flesh out other topics in bioprinting that we've lightly touched on. Topics such as bioreactors, bioinks, extracellular matrices, rheology, and scaffolds dot the bioprinting landscape like raisins speckle a good slice of cinnamon raisin bread. It's a little ticklish to choose how to order these bioprinting leitmotifs because they're all intimately related. However, the basic ideas are very simple, and the themes will reappear in context later in the book when we start talking about bioprinting experiments to build tissues of cartilage, bones, heart, liver, and kidney. We'll start with bioreactors.

Bioreactors—A Boarding School Where Bioprinted Constructs Go to Mature

Bioreactors use several maneuvers to coax newly bioprinted constructs to mature. For example, imagine you find a way to introduce a vascular network in your bioprinted tissue. That's good. But the time required for that vasculature to fully assemble and get blood to all the cells in your tissue construct may be longer than the time the cells can survive on their own. That's not good. However, suppose that immediately after the bioprinting process—we'll call it post-processing—you stick your constructed tissue in a bioreactor. This can help to keep the cells alive and buy time necessary for post-processing tissue fusion (remember our chat about fusion in the Preface) and subsequent development.[45] We'll have more to say on development in a moment when we talk about the extracellular matrix.

Cells grow significantly better under dynamic rather than static culture conditions. This is a result of the continuous cycling of nutrients as well as the removal of metabolic wastes. One task in a bioreactor is oxygen transport to all cells, including those in the inner regions of a bioprinted construct. Bioreactors often accomplish this by perfusion: They create a means of circulating oxygen and nutrients continuously without stagnation, and simultaneously they remove the waste products produced by the living cells.*** This is commonly done when the cells have been printed in a porous scaffold that allows oxygen and nutrients to flow through the pore spaces in tortuous paths determined by the architecture of the scaffold.

Remember that bioreactors are tissue-specific. So, for example, if a bioreactor is designed to help bioprinted cartilage mature, then the design may include a means of mechanical stimulation. After all, articular cartilage—cartilage at your joints where there's a meetup between bone and bone—is providing a shock absorber for you. It helps you to bear up under static or dynamic loads: you're standing around chatting or you're jumping for joy for some reason best known to yourself. The bioreactor design subjects the bioprinted cartilage substitute to mechanical loads, and this actually helps the cartilage to learn these wonderful load-bearing skills.

Bioinks—The Write Stuff

Bioinks are soft biomaterials containing living cells. Saying a bioink is a biomaterial means that it can be safely introduced into your body. With this understanding, a metal hip implant is a biomaterial but obviously it isn't soft. Bioinks are most commonly scaffold-based, meaning that live cells are mixed with materials that will make up a scaffold and then the scaffold substance plus cells are bioprinted. A classic example of a scaffold substance is a solution of long molecules made up of hundreds or sometimes thousands of identical clusters of atoms. Each atomic cluster forms a small molecule, and the long molecule is a repeating chain of these small molecules stuck together. Scientists call these long molecules polymers: *poly* means "many" from an ancient Greek word, and *mer* is from an ancient Greek word for "a share"—hence, having many parts.

One popular polymer substance is called a hydrogel because it's able to absorb large amounts of water and it can easily be made to form a soft solid if the polymers are cross-linked. Cross-linking indicates that the polymer strands are joined in various places by chemical bonds—the polymer strands are holding hands. For bioprinting purposes, some of the most common hydrogels are derived from vertebrate animals. Examples are collagen and gelatin. Because they come from animals, they have built-in signaling molecules that help produce cell adhesion. We'll have much more to say about cell signaling later in this book. Other hydrogels, like alginate (from the cell wall of brown algae, a type of seaweed) or agarose (from red seaweed), are also popular in forming scaffolds to harbor cells, though they lack the inherent signaling molecules.[46]

*** Perfusion is the passage of fluid (often blood) through an organ or tissue.

Extracellular Matrices—The Places Cells Call Home

Biologists have a name for a cell's home within a tissue. It's called the extracellular matrix. The extracellular matrix is the non-cellular material present within all tissues and organs. The extracellular matrix provides essential physical scaffolding for the cellular constituents. It also initiates critical biochemical and biomechanical cues necessary for tissue form and structure, cellular differentiation into one of the individual cell types in the human body, and cellular stability.

When he was a student, University of Virginia cell biologist Douglas W. DeSimone was at a meeting where a prominent biologist questioned him as to why anyone would want to work on "that stuff." At the time many viewed the extracellular matrix and its surrounding connective tissues as having a necessary but largely passive structural role while filling the spaces between cells. As the eminent cell biologist said of the extracellular matrix, it was "the Styrofoam packing material" of cells and tissues.[47] That view has changed radically during the past 40 years.

A big takeaway point is that the cells themselves make the extracellular matrix in which they live.[48] A typical bioprinting approach is to load a hydrogel with cells and substances that mimic the native extracellular matrix, bioprint this combination, and place it in a bioreactor. Alternatively, it's feasible to first print the cells and, after that, place the cells plus the substances that mimic the matrix in the bioreactor.[49] Then you set the thermostat in the reactor so the cells are comfortable in their home away from home, and by thus nurturing the cells you give them a chance to make their own extracellular matrix.

We'll have more to say about an interesting word combination—decellularized extracellular matrix—later on in our narrative. It's the extracellular matrix minus the cells. Panel (a) of Figure 1.2 is a cartoon of a cell within a hydrogel, with substances added to mimic the cell's natural extracellular matrix. The large blob in the center is the cell, and the darker circular object within it is the nucleus of the cell where the cell's DNA resides. The other ornaments are various proteins. Panel (b) is a teaser. It's a scanning electron micrograph showing a portion of the decellularized extracellular membrane from a pig's urinary bladder. We're looking at the convoluted fibrous structure that remains after all the cells are detached by a combination of chemical means and physical agitation.[50] It gives the impression of a bird's nest, doesn't it?

Rheology—It's Flow Time

Bioinks must be devised so that they don't clog the print heads in the bioprinter but deliver the goods to make printed tissue structures robust. Bioinks must behave like a liquid during printing but be self-supporting after printing. To pull this off, you must tune the rheology of the bioink.

(a)

(b)

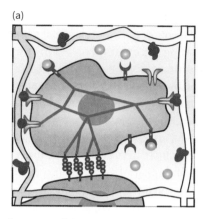

Figure 1.2   (a) Cartoon of a hydrogel environment to accommodate cells. This three-dimensional environment is designed to mimic the intricacy of the cell's natural extracellular matrix. (b) The natural extracellular matrix of a porcine urinary bladder shown in a scanning electron micrograph.

Source: Reproduced by permission from Chelsea M. Magin, Daniel L. Alge, and Kristi S. Anseth, "Bio-Inspired 3D Microenvironments: A New Dimension in Tissue Engineering," *Biomedical Materials* 11, no. 2 (March 4, 2016): 022001, https://doi.org/10.1088/1748-6041/11/2/022001.

Here's the headache: To get precise positioning, especially for the popular extrusion-based bioprinters (panel (c) of Figure 1.1), it's best to concoct a bioink that will have two properties. First, it will become less viscous (thinner) when forced through the nozzle of a bioprinter. An everyday example of this is how ketchup flows when you whack the bottom of the bottle and the ketchup goes from a thick paste to a runny liquid when it exits the neck of the bottle, akin to the bioink exiting the nozzle of the bioprinter. Second, you want the bioink to gel fast after printing so you can build layer after layer and still keep the printed shape, just like you want the ketchup to settle down and stay in place when it gets to your French fries and not run onto your lap.

But getting the best combination of the two properties is tricky. If you max out these bioink properties, you'll also find you get a lot more force on the cells during the printing process, and this causes cell death and injury. So when you prepare your bioink, you have to find a compromise, an optimal window of the bioink's flow properties. You have to optimize the rheology of the bioink.

For the record, really understanding the thinning phenomenon, or shear thinning as scientists refer to it, is a headache for theoretical physicists. So if the innocuous ketchup example seems puzzling when you start to think about what's really going on, you have lots of company. We meet this phenomenon in other ways too and barely give it a thought—paint, for instance. Paint is another case of this weird shear thinning business where the viscosity seems to change so capriciously. How can paint be thin enough so it flows from a stroked brush but a split second later it's thick enough not to drip down the wall? Shear thinning again.[51] Or how about toothpaste? When you squeeze the tube, you're putting a force on its contents. When the paste exits the tube, the viscosity quickly lowers as it flows, but then once it's on your brush it sets up and calmly stays put on the bristles. So in this morning and evening habit of brushing your teeth, you're enacting a shear thinning ritual.

Rheological happenings are common events, but we'll soon be reading about them in the uncommon context of bioprinting.

Scaffolds

The use of scaffolds is widespread in bioprinting. But it's possible to do without a scaffold, and some researchers embrace what's called a scaffold-free technique. They believe that since cells generate tissue, the direct manipulation and control of cells is paramount. In the absence of a scaffold there are no concerns about the scaffold or its resulting decomposition products being compatible with cells. And this technique allows the cells to be grown in conditions more approximating their environment in the body. It's also biomimetic of phenomena in the growing embryo where tissues and organs are formed without any solid scaffolds.[52]

When I spoke to Gabor Forgacs, he mentioned a new angle on scaffold-free printing of tissue spheroids. Tissue spheroids are sphere-shaped aggregates of cells that form spontaneously under a variety of conditions in the laboratory.[†††] A common method to form spheroids is to deposit the cells of interest on a surface of plastic that has been coated so that the cells cannot adhere to it.[53]

Surgeon and biomedical engineer Koichi Nakayama of Saga University in Japan invented the new angle on scaffold-free printing, so I talked to Nakayama about his unusual approach. He named it the Kenzan method after the traditional Japanese art Ikebana, in which, for floral arrangements, the stalks are impaled in a metal needle array called "Kenzan." In Japanese, *ken* = "sword" and *zan* = "mountain."[54] The Kenzan method is essentially a spheroid-assembling method. It uses spheroids that contain around 20,000 cells and are about 0.5 mm in diameter. The medical-grade metal needles robotically impale these preformed spheroids, providing them with spatial organization and a controlled opportunity to interact, fuse, and secrete extracellular matrix.

In one experiment, Nakayama and his colleagues aimed to print a vascular prosthesis (an artificial body part), a tube for repair of blood vessel failures (i.e., a bioprinted vascular graft). They wanted to make prostheses with a diameter less than four millimeters and lined with the right type of cells to inhibit clotting. The gold standard for tubes with such a small diameter is to use the patient's own blood vessels from undamaged areas for the necessary repairs, but issues remain.[55] So Nakayama's team printed vertical columns of skewered tissue spheroids and programmed their

---

[†††] The formation of aggregates is driven by tissue fluidity/liquidity, as we discussed in the Preface.

(a) (b) (c)

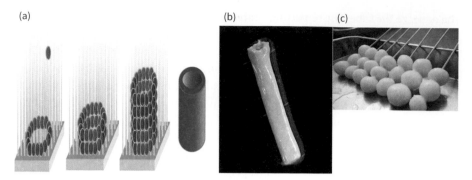

Figure 1.3 (a) A printer (not shown) skewers preformed tissue spheroids onto selected needles. Spheroid fusion forms a tube. (b) A scaffold-free, tubular vascular graft made by the process shown in (a). The tube is 7 millimeters long with an inner diameter of 1.5 millimeters. (c) *Dango* is a Japanese dumpling and sweet that's served skewered, a delicacy that resembles preformed tissue spheroids on needles.

Sources: a, Reproduced by permission from Manabu Itoh et al., "Scaffold-Free Tubular Tissues Created by a Bio-3D Printer Undergo Remodeling and Endothelialization When Implanted in Rat Aortae," ed. Donghui Zhu, *PLoS One* 10, no. 9 (September 1, 2015): e0136681, https://doi.org/10.1371/journal. pone.0136681; b, reproduced by permission from Koichi Nakayama; c, reproduced from 123rf.com, image ID 82325743.

home-built printer to select needles from the array that formed a circular pattern (circular when looking down on them). Then when the spheroids fused, they formed a tube 7 millimeters long with an inner diameter of 1.5 millimeters. After perfusion (passage of appropriate fluid through the tube) in a bioreactor, the tube's inner wall was lined with cells to inhibit clotting (Figure 1.3).

Nakayama shared a story with me. One day, Japanese Prime Minister Shinzō Abe was scheduled to visit Nakayama's lab along with several other labs. Government officials told Nakayama that the Prime Minister only had a short time to stop in at each lab and would he please explain his work in some simple, easy way so as not to embarrass the Prime Minister but that would still impress him and get the point across. Nakayama happened to stop at a convenience store that sold *dango*, a popular dumpling and sweet that's served skewered (panel (c) of Figure 1.3). It was a serendipitous visit. Nakayama realized that this food simile would work well, and his presentation was a smash hit.[56]

A Return to Paradigm Shifts

We began this chapter with a look at paradigm shifts in printing. The introduction of the printing press is widely regarded as the greatest breakthrough in human history.[57] It's fair to say that 3D printing will have an uphill battle to overtake the printing press as the top human innovation. Still, it takes time for the reverberations of a big shake-up to be felt. 3D printing has only been around since the mid-1980s. But it continues to travel imaginative paths, like the phototropic bending of tree branches seeking the sun.

For example, Professor of Biologically Inspired Engineering, Jennifer Lewis, and her team at the Wyss Institute for Biologically Inspired Engineering at Harvard University recently unveiled rotational 3D printing, a new spin on the subject.[58] The key to 3D rotational printing is a pas de deux, a dance for two—think Gelsey Kirkland and Mikhail Baryshnikov—between the speed of the lateral movement and the rate of rotation of the 3D printer print head. Lewis' team designed a print head that rotates at user-controlled, varying, revolutions per second at the same time that it extrudes a fiber-containing composite ink in a predetermined path. The payoff is that the user can program the printer to decide on the directional orientation of the fibers at distinct locations as the printer spits out the ink. Figure 1.4 is an illustration of one possible fiber alignment using rotational 3D printing.

Rotational 3D printing allowed Lewis to create a fiber-reinforced composite material where she tailored properties such as stiffness or flexibility in certain locations by

Figure 1.4 Illustration of the alignment of fibers in the nozzle of a 3D printer and in the printed material.

Source: Reproduced by permission from Brett G. Compton and Jennifer A. Lewis, "3D-Printing of Lightweight Cellular Composites," *Advanced Materials* 26, no. 34 (June 18, 2014): 5930–35, https://doi.org/10.1002/adma.201401804.

introducing variable orientation of the fibers in her printed bioink.[59] Now consider how this might find a place in bioprinting.

Suppose you want to bioprint a substitute for bone or cartilage that can function like the real thing. Naturally occurring bone contains collagen fibers, and it's well known that the fibers are choosy about their orientation—in order to help you bear loads more successfully. Or take cartilage; it's very fussy about orientation. A vexing problem in trying to bioprint a substitute for natural articular cartilage such as you have in your knees and hips is that there are no fewer than three well-defined layers in cartilage, each with its own orientation of cells. This is known as articular cartilage's zonal architecture. And it's one of a handful of reasons why everyone suffering from osteoarthritis doesn't already have bioprinted cartilage available to ease their pain. Down the road, Lewis' rotational printer might better enable the engineering of aligned tissue microarchitectures.[60]

In time, 3D printing may well take its place in the pantheon of great breakthroughs in industrial processes along with Henry Ford's assembly line. If bioprinting has the same impact in changing the medical paradigm, it would stand together with vaccination (1796), anesthesia (1846), pasteurization (1863), and penicillin (1928) as one of the greatest of medical innovations.[61]

But to go from bioprinting proof-of-concept laboratory experiments to bioprinting within the surgical suites of hospitals in Anytown will take time. The majority of scientists leading bioprinting's coming of age will either demur from speculating on if/when bioprinted solid organs will be a reality or say that this vision is at least decades away.[62,63,64,65,66,67,68]

The drama that is the human body has balanced casting. It's an ensemble cast, not a troupe with only one or two principal characters and the rest as supporting performers. The multiple stars of the show are cells, all manner of cells. They're able to share the same stage, dress in the same period costuming, and speak the same language; only their scripts are distinct. Scientists hope to use bioprinting to replace or augment all kinds of cells that have been injured or depleted. Our story next takes us to the ensemble cast of players, the cells.

We propose to round up more than the cells that represent the major tissues and organs of the body—those cells at center stage. To be prepared to thrash out bioprinting, we also need introductions to the players in the wings—those not in the playing space visible to the audience. These more obscure actors, with names like stem cells or progenitor cells, are a big reason that bioprinting scientists stay up late at night. Or toss and turn in their sleep.

2
All About Cells

You've all completed the course in recreational scuba diving and we're almost ready to begin our tour of the living cell. All cells live in a liquid world, so we have to tour the cell by swimming.

Now that I've gone over a brief step-by-step on how to use your aqualung, does anyone have any questions?

No? OK.

We'll start by approaching the cell coming from a blood vessel, specifically an artery. That will give us a chance to check out extracellular structures, the ones on the outside of the cell's membrane. In order to tour the cell we have to shrink ourselves in size. Typical human cells are about 10 millionths of a meter—10 micrometers—in their longest dimension.* We usually refer to a micrometer as a micron, so cells are typically 10 microns long. For comparison, the average width of a strand of human hair is about 100 microns. Of course, our generous host who is allowing us access to his radial artery at the wrist will not shrink himself. Nor will his assistant, who will administer the small pinprick that will provide us with entry to our host's arterial blood vessel.

When we tour the inside of a cell, we're going to be checking out protein molecules, and then we'll need to shrink even more because an average protein is only about one-thousandth the length of the cell.

Mapping out Our Tour

Before diving into the blood vessel, we'll first look at a map so we can get our bearings. After we leave the blood vessel and pass through the multiple tissue layers that wrap around it, we'll be in the extracellular matrix. As we approach the cell through the tangle of the extracellular matrix, we'll catch sight of the cell membrane and its knobby structure. Then we take advantage of the membrane's ability to engulf us and transport us into its interior.

The most prominent object inside any animal cell is the nucleus.† The nucleus holds the cell's genetic library, its DNA. Outside the nucleus yet still within the cell's membrane boundary, there's a thick soup of water and proteins. For the word lovers in our

---

* One-millionth of a meter = one micrometer = one micron.
† Red blood cells are a notable exception because they don't have a nucleus.

little group, here's the scoop: *Nucleus* is derived from the Latin word for nut, and the two parts of the cell—the nucleus and the thick soup—are related to each other like the stone and the pulp in a cherry. Like a cherry, the cell is covered by a skin or membrane.[1] On this tour, our air supply limits our time, so unfortunately we won't be able to enter the cell's nucleus; we'll content ourselves with views of other cellular marvels.

For instance, within the cell there's an impressive network of girders and cables, some of them immensely long, in an intricate inner scaffold of crisscrossed hollow microtubular and solid microfilamentous elements. Collectively, they form the cytoskeleton, the skeleton of the cell.[‡] Like our own skeleton, it's not merely a static but also a dynamic structure. The cytoskeleton forms the bones and muscles of our cells.[2] At a given point in time, you can think of the cytoskeleton as a fairly rigid infrastructure that connects different cell parts with each other. It's largely responsible for the shape of the cell and for the cell's attachment to neighboring cells and to other extracellular anchoring points. Many parts of the cytoskeleton are continually disassembled and then reassembled in a new manner. Still others are made to slide along each other by small molecular motors. These changes largely determine movements within the cell and of the cell itself.

Let's return for a moment to the proteins we mentioned. In the cell and in the entire human body, protein molecules are the movers and shakers. Within a cell there are at least several thousand individual proteins, and their main function is to boost the speed of chemical conversions without actually being consumed in these reactions.[3] Proteins also pitch in with other tasks—as we'll see, they act as gatekeepers for traffic to and from the cell. Proteins can be shaped like globs, almost spherical, whereas others have more exotic shapes: bent, twisted, or bulging with lumps and bumps. Still others thin out to slender threads, often coiled into a helix. And these shapes swell, pulsate, elongate, contract, or uncoil.[4]

Last, let's be clear that the thick soup, called the cytoplasm, surrounding the nucleus isn't simply a glop of jelly. Within the cytoplasm there are a bunch of objects that are similar to elfin versions of our organs. In fact, they're called *organelles*, a diminutive form of the word organ. They may be tiny but they have specific functions, and in the diminutive world of the cell they're powerful. We can think of them in two categories: internal affairs and external affairs.[5] The internal affairs department grapples with building complicated molecules and also with energy production. The external affairs department tackles communication and exchanges of matter between the cell and the world around it. This hustle and bustle also engages the membranous skin of the cell.

So that's the map. Based on this, we'll have two itineraries in today's dive. Our first route will take us to the outskirts and skin of the cell. The second part of our dive will take us into the cytoplasm within the cell. So strap on your aqualung and let's plunge in.[6]

[‡] The prefix *cyto-* will pop up a lot. It denotes a *cell* and goes back to the ancient Greek word meaning "hollow receptacle."

Cell Surface Membrane

Now that we're fully immersed in a blood vessel, you can feel for yourself how turbulent the flow is, particularly in the larger arteries such as the one in which we've begun our journey. Disc-like red blood cells are all around us. Every heartbeat creates a powerful surge that rhythmically propels us—along with our companion blood cells—as we're carried along and tossed about by the rapidly moving current of plasma, the fluid part of the blood. As we continue, the artery becomes narrower, and notice how the force of the current is quieting down. This gives us a good opportunity to look at the cells that line the walls of the blood vessels. They only form a single layer and collectively look like an irregular tiling or pavement. Notice how very flat they are over most of their surface so that their nuclei bulge out into the open space of the vessel, creating a wall with a lumpy texture.[7]

Now we've reached a capillary, the smallest of the blood vessels. It's here that most of the exchanges between blood and tissue take place—and where we too should leave if we don't want to get caught up in the return traffic in a vein going back to the heart. So let's wiggle through a couple of these cells that line the capillary wall and get out of the bloodstream.

We find ourselves in a connective tissue jungle. It's called the extracellular matrix. The main fibers we see in the matrix are made of the protein collagen. Collagen is found throughout the body as a structural support—in skin, tendons, ligaments, cartilage, and bone to name a few places. Working our way through the jungle we come to the surface of a cell, though as we can see cells are rarely alone and on their own. Mostly they huddle together into clustered groups with a characteristic shape. Regardless, the most striking features of the cell surface are its unevenness and changeability. As to unevenness, biochemist, Nobel Prize laureate, and cell explorer Christian de Duve described the great variety of cell surfaces:

Some cells are ravined by deep clefts or pitted by craterlike depressions. Others are deformed by protrusions ... or studded with fingerlike projections identified as microvilli ... or cilia. ... Yet others have their surface pleated by filmy veils. The variety is endless.[8]

As to changeability, there's ceaseless animation at the cell's surface:

Cilia [tiny hairs] beat, membranous veils wave and undulate, microvilli [short projections] bend and twist, pseudopods [protrusions] bulge, craters erupt suddenly in the midst of a quiet surface spewing secretory products [newly synthesized proteins], or in contrast, invaginations yawn open [and engulf proteins]."[9]

In summary, a cell's surface is highly irregular and ever-changing: a dynamic indoor rock climbing wall with no charge for admission (Figure 2.1). The wishbone-like,

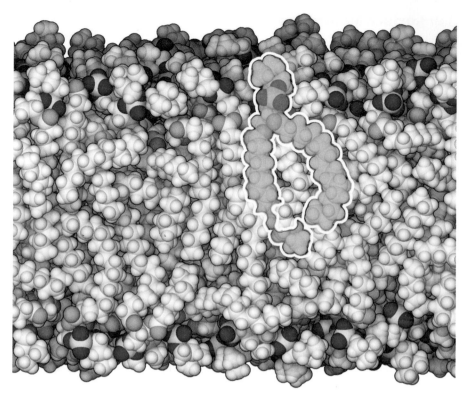

Figure 2.1 A close-up of a cell's surface membrane shown in cross section. The odd-shaped molecule called out in the figure is one lipid molecule in the topmost of the two lipid layers.

Source: Reproduced by permission from David S. Goodsell, *The Machinery of Life* (New York: Copernicus, 2010), 22.

highlighted object in the figure is a single lipid molecule in the upper of the two adjacent lipid layers that make up the cell's membranous surface.[§] Call the wishbone shape the tail of a lipid molecule, and call the bunch of atoms atop of the wishbone's junction the head of a lipid molecule. The two lipid layers are stacked so that the tails face each other.

The thickness of the cell's membrane is usually taken to be 8–10 nanometers. A nanometer is one billionth of a meter. That includes the membrane proteins and sugar chains that are attached to them and stick out from the double lipid layers. We typically refer to these double lipid layers as lipid bilayers. The thickness of the lipid bilayers by themselves is only about 4 nanometers. To get an idea of what this amounts to, consider an airplane. The thickness of a lipid bilayer in comparison to the diameter

---

[§] A lipid is a molecule that does not dissolve in water; fats are one type of lipid.

of the cell is similar to the thickness of an airplane fuselage (the exterior shell of the airplane's body) in comparison with the diameter of the plane's body. The exterior shell of an airplane is about one centimeter thick and the airplane's overall diameter is about five meters—a ratio of 1:500.[10]

Despite the tumultuous contortions of the cell's surface (its membranous skin), the cell remains intact as its membrane adapts with exquisite plasticity. But the membrane is much more elaborate than, for example, a soap bubble or a balloon. Cell membranes resemble medieval cities that protected themselves with a wall and a moat. Even so, bridges and gates are needed to both allow and regulate the two-way traffic that supports life within the city.[11] Swimming alongside the cell's surface, we can see that while the majority of the cell membrane is made up of the lipid bilayer, we also catch sight of proteins known as membrane proteins that play the role of bridges and gates.

Cell Membrane Proteins

Some of these proteins have parts embedded in the lipid bilayer, including transmembrane proteins that span the entire bilayer. Many of the proteins bind long chains of sugar molecules on the outside of the membrane. The chains act as surface receptors, and we can think of them as conduits that enable transfer from the outside of the cell to the inside of the cell, as we'll soon describe.

Figure 2.2 is a rendering of the cell membrane's lipid bilayer replete with embedded proteins and their bound sugar molecules.[12] When a passing molecule on the outside of the cell binds to a surface receptor—presto chango—the surface receptor protein takes on an altered shape than before. It opens up like time-lapse photos of a blossoming yellow lily. Now there is a passageway between the outside of the cell and the inside.

We take the occasion to gape close-up at a few of the proteins. As you can see in Figure 2.3, they're anchored to the cell's membrane, the same skin that we saw before in Figure 2.1 only there it was blown up big to show you its composition. In Figure 2.3, that same membrane is represented by the modest, lightly colored, horizontal crosspiece, interrupted by several proteins. The large protein in panel (a), with much of it reaching into the cell, is called a protein pump. It's in the business of pumping sodium ions out of the cell and potassium ions in. An ion is an atom in which the number of protons and the number of electrons isn't the same; so, an ion has an electrical charge, whereas an atom is electrically neutral.

As a result of pumping ions, the protein in panel (a) creates a difference in electric charge across the membrane and also a difference in concentration of sodium and potassium ions across the membrane. This has pretty significant consequences for all of us, for instance, in the transmission of nerve signals. A part of each of our nerve cells juts out in a long projection called a nerve fiber or axon. What happens in our nerve

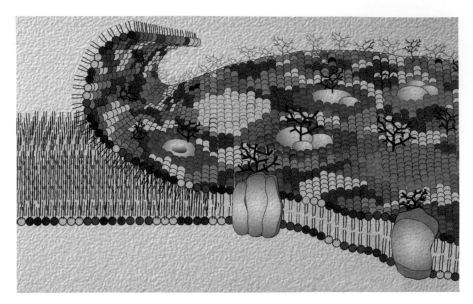

Figure 2.2 An exuberant rendering of the cell membrane's lipid bilayer replete with embedded proteins and their bound sugar molecules.

Source: Reproduced by permission from Michael Edidin, "Lipids on the Frontier: A Century of Cell-Membrane Bilayers," *Nature Reviews Molecular Cell Biology* 4, no. 5 (May 2003): 414–18, https://doi.org/10.1038/nrm1102.

cells is that the axons deplete themselves of sodium ions and then use special sodium channels (not shown in Figure 2.3) to allow the ions to rush back in during a nerve impulse. Our sodium–potassium protein pump has the job of keeping the axon ready, willing, and able for the next signal by maintaining the electric charge difference and the concentration difference across the membrane.[13]

In panel (b) of Figure 2.3, we see another kind of protein machine. It's known as a glucose transporter, and its mission in life is to deliver glucose molecules one-by-one across cell membranes. When we take in food and break it down, we are feeding our cells so they can power themselves. We need to supply each of our cells with glucose, delivered throughout the body by the blood, and each cell gathers what it needs using glucose transporters.[14]

Molecular biologist and scientific artist David Goodsell of the Scripps Research Institute has a name for these proteins that transport glucose: shape shifters. They manage the traffic of glucose across the cell's outer membrane by alternating between two shapes. First, the transporter molecule has an opening facing the outside of the cell where it picks up a molecule of glucose that's passing by in the bloodstream. Then it shifts shape (rightmost image in panel (b) of Figure 2.3). It opens toward the inside of the cell and releases the glucose molecule into the hungry cell.[15]

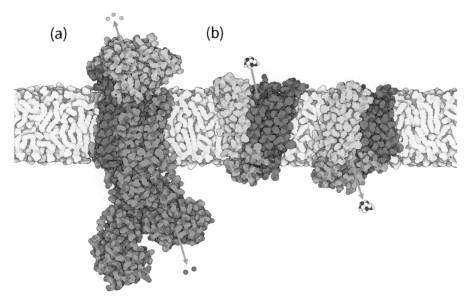

Figure 2.3 (a) A sodium–potassium protein pump with sodium ions pumped out of the cell and potassium ions pumped into the cell. (b) Two views of a protein that functions as a glucose transporter. The leftmost view shows the protein open to the outside of the cell, allowing a glucose molecule to enter; the rightmost view shows the protein changing its shape so it's now open to the inside of the cell, where it releases the glucose into the cell.
Source: Reproduced by permission from David S. Goodsell.

Cell Factories

It's time for us to enter a cell, to swim into the cytoplasm. We take advantage of endocytosis, the cell membrane's ability to engulf us and transport us into its interior.

Ah, at last we're inside a cell. Our first impression is the tumultuous action. There's no shilly-shallying here. The cell runs an export industry and has factories to manufacture the molecules it exports. An example of these factories is the enormous object over there. It has an odd sounding name—the Golgi apparatus—because Italian scientist Camillo Golgi first identified it at the end of the nineteenth century. Figure 2.4 shows a cross-sectional illustration of part of the Golgi, as it's affectionately called. De Duve had this to say about the Golgi: "It is a sprawling complex of membrane-bounded cavities big and small, with the untidy appearance—which actually hides a considerable degree of order—often seen in warehouses and packaging centers."[16] Figure 2.4 conveys that impression. De Duve mentioned "membrane-bounded" and it's true. The organelles within the cells have the same type of lipid bilayer skin as the

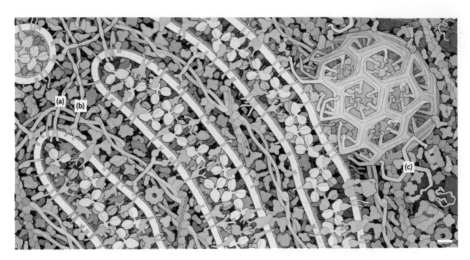

Figure 2.4 A cross section of part of the Golgi apparatus, a molecular export factory that resides in the cytoplasm of a cell. Labels "a" and "b" call out huge tethering proteins that guide small transport vehicles to the right place. Label "c" identifies another protein that assists in encasing the manufactured proteins—the ones that will be exported out of the cell—in the transport vehicles. Scale bar = 10 nm.

Source: Reproduced by permission from David S. Goodsell, *The Machinery of Life* (New York: Copernicus, 2010), 78–79.

cell itself and also similar protein machines to help get materials and supplies in and out between the organelles and the cell's cytoplasm that surrounds each organelle.

As to its function, the Golgi is the processing and sorting plant of the cell. Sugars and fats are attached to the manufactured proteins that need them. When the proteins are properly modified and sorted, they're delivered through the cell in small transport vehicles. In Figure 2.4, you'll see labels a, b, and c. Labels a and b call out huge tethering proteins that guide the transport vehicles to the right place. Label c identifies another protein that assists in encasing the manufactured proteins—the ones that will be exported out of the cell—in the transport vehicles.[17] Figure 2.4 is quite a scene. De Duve referred to journeying through organelles such as the Golgi as like wandering through a surrealistic maze: "Blank walls everywhere; endless spreads of membranes, always tightly sealed, tantalizingly opalescent, hiding from our view vast sets of machinery which, judging from their manifestations, must be of stupendous complexity."[18]

Time restricts the scope and depth of our visit, so we're unable to check out any of the other organelles within the cell. Nor can we take this opportunity to enter the cell's nucleus, since the air supply on our aqualungs is running down. We take our leave of the cell as de Duve did: We imitate a virus and use a bit of membrane to wrap around ourselves. This is the process of exocytosis, the opposite of endocytosis that was our means of cell entry. Mindful of the fluidity, plasticity, and self-sealing properties of the membrane skin of the cell, we push gently yet firmly against its outer layer. It yields and bulges under our pressure and then flows and reshapes around us. The process seals the cell membrane behind us and leaves the cell essentially intact.[19] Now we find our way back through the connective tissue to the bloodstream and head for home.

Becoming a Cell—Stem Cells

Back on dry land after our tour we now have an idea of what makes up a cell. There are many types of cells—for instance, skin cells, bone cells, liver cells, heart cells, and kidney cells. That's not the whole story. Each of these cell groups is composed of distinct specific cells. For example, the most prevalent liver cell is called a hepatocyte, but the liver also contains Kupffer cells, stellate cells, endothelial cells, and more. The human body has over 300 distinguishable cell types, and they all have to begin somewhere. We want to know how a cell becomes a cell.

Stem cells made the headlines starting in the early 1990s.[20] A distinctive characteristic of stem cells is that they're not distinctive—they're unspecialized.[21] Stem cells have the potential to develop into many cell types early in our lives. As we age, they are a reservoir that can replenish other cells during our adult years.

Stem cells are also unlike other kinds of cells in our body. Muscle cells or nerve cells (as examples) don't normally replicate themselves. Stem cells can replicate many times. They're said to be capable of long-term self-renewal.[22]

What also makes stem cells unusual is that under certain conditions they can be encouraged to become cells that are tissue- or organ-specific with the special functions that go along with that tissue or organ.** To say they can be encouraged is speaking politely. The correct word is induced; stem cells can be told to become a specialized cell and they comply.[23]

Until fairly recently, scientists had only two kinds of stem cells to work with: embryonic stem cells and non-embryonic, adult stem cells. An embryo is an early developing organism, and the word refers to the period of development between the fertilized egg and the fetal stage. So an embryonic stem cell is a cell derived from a very early time in development, about five days.[24] Embryonic stem cells can become any of the cell types we possess. Adult stem cells are more limited, restricted to only becoming specialized cells of a particular organ or tissue type. For instance, blood stem cells can produce all the cell types in the blood but not the cells of other organs, for example, the liver or the brain. When we said earlier that stem cells are unspecialized, we need to qualify that. Embryonic stem cells can become any type of cell; adult stem cells are not as unrestricted in their options.

** When we call out tissues and organs we mean: a tissue is a group of cells working together and an organ is several tissues working together.

There has been a contentious debate about the use of embryonic stem cells and the ethical dilemma they present—the moral status of the embryo. There is a question of whether destroying the early embryo to obtain stem cells means destroying a potential human life. Then in 2006, a new scientific discovery came into play and changed the landscape.

Induced Pluripotent Stem Cells

"We have colonies."[25] This announcement startled senior investigator Shinya Yamanaka of Kyoto University. His postdoc insisted it was true, and together they raced into their laboratory. The two scientists saw the evidence through a microscope. Starting with adult mice skin cells—not adult mice stem cells—they had succeeded in transforming these cells to perform like embryonic stem cells capable of developing into bone, muscle, cardiac, or virtually any other cell type. Yamanaka recalled his reaction: "At that moment, I thought, 'This must be some kind of mistake.'" But every repetition of the experiment gave the same result.[26]

Shinya Yamanaka shared the Nobel Prize in Physiology or Medicine in 2012. The prize was awarded for the discovery that mature, specialized cells can be reprogrammed to become pluripotent—that is, scientists can manipulate the mature cells so they revert to immature cells that are able to develop into all types of cells in the body. Embryonic stem cells are also pluripotent.[27] The reprogrammed, specialized, adult cells are called *induced* pluripotent stem cells.

Almost twenty years after Yamanaka's discovery (that he later extended from mice to humans), there are many questions that remain to be answered about induced pluripotent stem cells. Nonetheless, his work has revolutionized our understanding of how cells and organisms develop.

We'll encounter induced pluripotent stem cells repeatedly when we turn to specific bioprinting studies on various tissues and organs. So it's worth our while to present more information on these types of stem cells right up front. There are upsides and downsides, and we'll try to give a balanced picture. One advantage of induced pluripotent stem cells is that they're a good way to make pluripotent stem cell lines that are specific to a disease or even to an individual patient. The phrase *cell line* refers to cells that can be maintained and grown outside of the body, originating from a single cell and thus having a uniform genetic makeup.[28] Disease-specific stem cells are useful ways to study the cause of a disease and find drugs or treatments. Patient-specific stem cells are likely to minimize rejection or the need to suppress the patient's immune system if they're used in transplants.

After Yamanaka's discovery of induced pluripotent stem cells, it seemed possible they would provide an alternative to embryonic stem cells—they might become any of the cell types we possess and wouldn't leave science and society with ethical dilemmas. But it turns out that they aren't fully equivalent to embryonic stem cells; they subtly keep the imprint of their initial specialization and they present a risk of developing a tumor.[29] Currently, these stem cells are a great research tool, and the search continues for ways to make induced pluripotent stem cells that don't pose a risk of developing tumors.[30]

Additional Sources of Stem Cells

So far, our inventory of stem cells includes three sources: embryos, adult tissues, and adult tissue cells induced to return to a state of innocence before they became cartilage, bone, muscle, and what have you. Before we go on, we reemphasize the distinction between adult stem cells and adult tissue cells that have been finagled to become stem cells. In the first case a scientist harvests stem cells that already exist in adult tissues. In the second case, a scientist messes about with ordinary adult tissue—for example, skin— and manipulates skin cells using exotic tricks that cause them to become stem cells.

There are other sources for stem cells that don't get quite as much attention. One of them is known as cord blood, short for umbilical cord blood. Umbilical cord blood remains in the placenta (the organ carrying us in our mother before we emerge to face the world) and in the attached umbilical cord that runs from an opening in the baby's stomach to the placenta in the maternal womb. Umbilical cord blood contains stem cells. Cord blood banks are places that store this blood at ultracold temperatures for safe preservation.

Another source for stem cells is less well known. Amniotic fluid is contained within the amniotic sac (also known as the amnion) that's contained within the placenta. A team led by Dr. Anthony Atala isolated amniotic stem cells in 2007.[31] Atala was the force behind the Amnion Foundation, a non-profit organization that has extended the cord blood banking model to establish a public amniotic stem cell bank.[32]

Progenitor Cells

From a bioprinting perspective, researchers have found that in many situations stem cells are not ideal, regardless of their source, and that progenitor cells work better. This is in spite of the fact that progenitor cells are not pluripotent.

If you think of stem cells as your great-great-great-great-great-grandparents, think of progenitor cells as your great-great-great-grandparents. Progenitor cells are early descendants of stem cells. They can become specialized to form one or more kinds of cells, though they're typically more limited than stem cells in the kinds of cells they can become. And progenitor cells can't keep dividing—and thus reproducing—for as long as stem cells can. Progenitor cells can turn into several types of specialized cells that are restricted to a certain organ or tissue type. Progenitor cells in blood can become the various cell types found in blood, for example, but cannot become any of the cells that comprise the gallbladder.[33] In Chapter 3, we'll see a bioprinting case in which progenitor cells are superior compared to stem cells.

The Different Cells in Our Body

John le Carré wrote the espionage thriller *Tinker Tailor Soldier Spy*. The title comes from an old, old nursery rhyme. The modern English version is:

> Tinker, Tailor,
> Soldier, Spy,
> Rich Man, Poor Man,
> Beggar Man, Thief.

How about:

> Liver, Larynx,
> Eardrum, Eye,
> Skin Cell, Nerve Cell,
> Bladder Cell, Bone.

How do cells manage to become so different? A peek inside the nucleus will answer that question. The nucleus contains molecules called chromosomes that are a set of instructions for how to build—you. Chromosomes are long, long strings of DNA molecules and genes are long stretches of DNA within chromosomes. The Human Genome Project that was completed almost 20 years ago estimated that humans have between 20,000 and 25,000 genes based on how many genes result in the production of proteins. This set of genes is in virtually every cell of your body. Now some scientists contend the number is closer to 50,000 partially based on genes that result in the production of RNA rather than proteins.[34] Many scientists believe that RNA was the first molecule of heredity, preceding DNA.[35]

Regardless of the number of genes, the cells in your body are distinctive because they use the same set of genes differently. Each cell chooses which genes it wants to switch on or keep turned off. This happens because of the elegant orchestration that goes on between DNA aided and abetted by three musketeers of RNA that are present in your cells. The process starts in the nucleus, where DNA makes messenger RNA that exits the nucleus (hence the name messenger) and goes into the cell's cytoplasm.†† Once in the cytoplasm, two other forms of RNA work in concert to synthesize proteins. When cells have dissimilar proteins onboard, the proteins lead to differences in cell structure, shape, and function.

---

†† *Synthetic* messenger RNA is used by several biotech companies as the starting place for their COVID-19 (coronavirus) vaccines. Synthetic RNA is a variation of the natural messenger RNA found in the nucleus of the cells in our bodies.

The previous paragraph doesn't really address how common origins generate distinct human cell types. We'd like to know what determines the genes that are expressed in a given cell at the earliest stages of life and which then leads to bladder cells, pancreas cells, and numerous others. We don't know the answer to this question. Scientists continue to investigate, but there isn't a definitive answer.[36] Figure 2.5 conceptually illustrates cell differentiation.

Nature's developmental tour de force of creating specialized cells is very relevant to bioprinting. Researchers trying to mimic nature are of at least two schools of thought: One is that it's better to print a structure that has a high density of cells and then let biology take over. This is the epitome of the "if you build it they will come" school of thought. Inkjet bioprinting does this well. Another opinion is that if you get the cells positioned accurately at the beginning, this will make the cells more likely to behave the way you want them to behave. Laser-based bioprinting does this well. Extrusion bioprinting is in between these two approaches. This may account for extrusion being the most popular form of bioprinting.[37]

"The big question is biology," says Brian Derby of the University of Manchester. "There is optimism coming from the engineers who say 'we can hack it,' but this reflects a lack of understanding of the biological complexity."[38] Dietmar Hutmacher agrees:

> After printing, we need to make sure that the cells form enough of a natural tissue-like architecture and composition. That's where the biology kicks in and we still need to learn a lot more about how we can direct the biology.[39]

To say that there are dissimilar cells in the body is to speak the obvious. We know that heart cells beat and cells in the eye enable vision. Other distinctive cellular attributes may not be so familiar—size, for example. It may come as a surprise to learn that cells in the human body range in size over more than five orders of magnitude. An order of magnitude means a factor of 10. Five orders of magnitude means $10 \times 10 \times 10 \times 10 \times 10 = 100,000$. A type of small cell found in the blood is about 7 microns in diameter. Large liver cells can be 50–100 microns in diameter. Giant nerve cells including their very long axons can stretch out to over one meter.[40] The ways in which cells are organized in disparate tissues and organs are also variable. For instance, the liver has a roughly hexagonal arrangement of plates of hepatocytes, its most common cell type. A sheet of heart muscle cells has a striped appearance.

It would be wrong to leave the reader with the impression that the cells in the human body have been pigeonholed. Differences in size or shape and, most important, function are certainly critical. Your blood needs to produce 200 billion cells per day. Contrast that with the central nervous system, in which many neurons last a lifetime.[41] As one scientist said, "The concept of 'cell type' is poorly defined and incredibly useful."[42] But perhaps that can be improved upon.

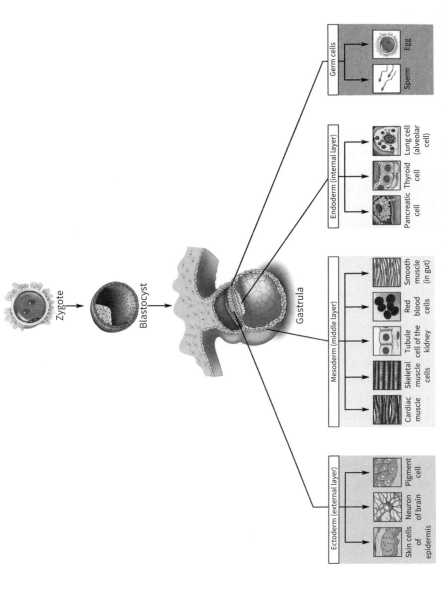

Figure 2.5 Cell differentiation. From the union of an egg and a sperm, nature creates hundreds of distinct cell types. A *zygote* is a fertilized egg. Zygote is derived from the ancient Greek meaning *yoked*. In time, the zygote divides to form a hollow cellular mass called a *blastocyst*. Blastocyst is from the Greek words for *sprout* and *cavity*. *Gastrula* denotes an early stage of animal embryo development in which three *germ layers* have formed. Gastrula is from the Greek for *little stomach*. The term germ layer refers to the three cellular layers—ectoderm, mesoderm, and endoderm—that will form all tissues of the embryo.
Source: Reproduced by permission from *Differentiation of Human Tissue*, Terese Winslow, LLC Medical Illustration.

The Human Cell Atlas

In 2016 a consortium of scientists from around the world convened in London to discuss how to build a Human Cell Atlas, a collection of maps and specifications to describe and define the cellular origins of both human health and human disease.[43] There are likely to be many benefits and surprises from this work. As one scientist said,

> So if you think of cancer—even before tumours heavily metastasise they start shedding off little cells that float around in the blood. You would not have any symptoms but you could start detecting these things. And then we can start working toward treatments that can block those cells, kill them, make them do something else. All these things will be possible. It would allow clinicians and pathologists to not only tell you, "you have a disease" but also say "right, you've got cells doing the following things wrong so we'll give you this particular drug."[44]

The initial draft will focus on tissues rather than whole organs. Researchers have posted the first large data file culled from over 530,000 cells from the immune system. The information comes from human umbilical cord blood and adult bone marrow cells.[45] The five-year goal—targeting 2022—is the analysis of 10 billion cells; the initial aim is a first draft detailing 30 million to 100 million cells. An objective of the Human Cell Atlas will be to discover which genes are active within each cell. That will be of great value for bioprinting and for everyone trying to understand how the body works at the cellular level.

Cells Live in Three Dimensions—Not Two Dimensions

Even with our inadequate knowledge of the rainbow of cells, there's so much we can do to get more out of what we *do* know about cells. A compelling feature of bioprinting is that it's implemented in three dimensions. Researchers have done lots of work with cells in two-dimensional confinement. Petri dishes are an example. Regardless, cells in the body live and thrive in three dimensions, not two dimensions. For many years there has been a growing consensus that to achieve relevant results—for example, new drugs that actually work in humans with minimal side effects—research on cells and tissues should be done in three dimensions. That's why scientists have increasingly turned to animal models. Mice, rats, fruit flies, worms, and fish are all three dimensional and give a more accurate picture of how cells work.

Biomedical engineer Mark Skylar-Scott is a research associate in Jennifer Lewis' lab at Harvard's Wyss Institute for Biologically Inspired Engineering. He spoke with passion about studying cells in three dimensions rather than two dimensions:

That's where all these pathways [molecules that work together to control cell functions] were found, where a lot of drugs were tested, and starting in the early 2000s, around the same time as the Genome Project, a few key organisms [the *Caenorhabditis elegans* worm, the fruit fly, the zebrafish, the mouse, and the chimpanzee] were selected for close study so that we could use them as three-dimensional models to gain an understanding of biology. So essentially you have these half-dozen studied-to-death organisms because biologists realized that 2-D cell culture was just not telling them how things really work. We don't exist as a single cell type on a flat surface. Cell-to-cell signaling isn't just a byproduct of biology; it *is* biology. And it doesn't exist in a dish.[46]

Cells Talk

Working with cells in three dimensions is an inherent feature of three-dimensional bioprinting. This allows scientists to leverage the cell–cell communication that *is* biology. Cells talk to one another. Cells chatter constantly. They speak to other cells far and near.

Far: To communicate with a remote zip code, certain cells send out hormones. These travel through the bloodstream to connect with cells in remote locations. Distant cell talk uses membrane proteins. The signaling molecules sent out into the bloodstream bind to receptor molecules resident on the proteins embedded in cell surfaces. After they bind there, they initiate a display of biochemical adroitness that opens a pore so the signal can be picked up in the inside of the cell. An example of this is growth hormone released by your pituitary gland. The hormone travels all over your body to talk to your skeleton and cartilage and stimulates growth by binding to receptors on the surface of bone and cartilage cells. Binding induces these cells to multiply by dividing—to make more copies of themselves.[47]

Near: We've mentioned the importance of scaling up bioprinted tissues, for getting large structures and high cell densities for transplantation. To make this happen, maybe the most important kind of cell talk is what almost all cells do: Cells speak to their next-door neighbor cells using channels that directly connect the inside of one cell with the inside of the next. The collections of such channels are known as gap junctions. They help to synchronize the beating of muscle cells in the heart. They help the contraction of the uterus during childbirth. They allow the pupil of the eye to adjust to varying levels of light.

Gap junctions are made up of proteins that combine to form a pore that directly connects the cytoplasm of one cell with the cytoplasm of another cell. This type of cell talk allows huge groups of cells to react to a signal that was initially transmitted to only one cell. A single gap junction is an enormous conglomeration of proteins, maybe 120,000 proteins in a single junction. [48]

"When Is a Kidney a Kidney?"

Our interest in cells is vested in bioprinting. We want to know how scientists bioprint cells and have them survive and function as they're supposed to in tissue constructs and, ultimately, in substitutes for all or part of an organ. So it's natural to ask if there's a minimum number of cells that we need to get to where we're going, be that a kidney or—if it's to survive—a vascularized kidney.

In a lengthy interview with Lydialyle Gibson, Jennifer Lewis and her team members, David Kolesky, Mark Skylar-Scott, and Kimberly Homan, responded to the question of how much you need to print to make a substitute for a kidney.[49] Kolesky observed,

> The smallest blood vessels that we print are usually about 400 microns in diameter [although they have gone down to around 75 microns]. But the smallest blood vessels in the body are 5 to 10 microns. If we have to build a tissue capillary by capillary, it's going to take forever. . . . Can we print the larger-scale blood vessels and have the smaller ones assemble on their own?[50]

The body produces new blood vessels from existing vasculature by the process called angiogenesis. To leverage angiogenesis, Lewis used a metaphor: "We lay down the highways and then let the cells create the smaller roads and driveways."[51] Skylar-Scott elaborated on this by adding,

> If you take your arteries and arterioles and veins and stretch them all out, they probably wouldn't reach the end of this room. But you exponentially grow in length when you get down to the scale of capillaries. That's when you're talking about hundreds of thousands of miles of those tiniest blood vessels throughout your body.[52]

It's for this reason that groups such as Lewis' hope to create vasculature by encouraging biology to do as much of the work as possible.

They intend that same division of labor for entire organs. As Lewis said,

> We want to elucidate what part of the organ we have to print, and what part we can leave to biology. The stem cells themselves are going to interact with the matrix around them and lay down their own matrix. They're going to grow and proliferate, and differentiate into different cell lineages. . . . We really don't want to construct an entire organ, cell by cell, capillary by capillary. That's a daunting challenge.
>
> But what we don't know yet, what we haven't yet teased out, is how far down the path we have to go before biology will do the rest, before you can implant something in a patient that the body will accept. What are the right cell types to print and

with what spatial precision and what density? How close to that billion cells per milliliter do we need to get to recapitulate function?[53,‡‡]

This is a central problem of scaling up a bioprinted construct. Scientists need to figure out how much of a printed kidney they have to build in order for it to succeed as a replacement for the body's failing kidney. How much of a kidney must researchers bioprint before the body will accept the implant? As interviewer Lydialyle Gibson deftly phrased the question, "When is a kidney a kidney?"[54]

Gibson's nimble sentence is very timely and echoes what we heard earlier from Dietmar Hutmacher, "We still need to learn a lot more about how we can direct the biology."[55]

Moore's Law

In this chapter we've tried to look at cells from many angles and to connect our reconnoitering to bioprinting. Despite the amount of cell biology of which scientists are in ignorance, they have to start someplace. In this vein, Lewis referred to Moore's law, although she didn't call this out explicitly in the interview.[56]

Physical chemist Gordon Moore became a research director in industry and then an executive at Intel Corporation, the silicon chip manufacturing company that he co-founded.[57] In 1965, Moore did some reckoning, the gist of which later became known as Moore's law.[58] He initially predicted that silicon computer chip complexity would regularly double each year for at least the following ten years. Moore later revised his estimate to doubling every two years.[59] Moore's law has been remarkably accurate for fifty years.

Moore's law is not a law in the sense that Newton's laws of motion describe physical reality. Still, many researchers have used it as the starting point to quantitatively examine whether the rate of past discoveries can be used to predict future ones and, more generally, to question the predictability of technological progress.[60,61] Keep in mind that the rate of scientific progress is unpredictable. Science is essentially a nonlinear endeavor. If you made a plot with time being the horizontal axis and accomplishments being the vertical axis, the fulfillment of many scientific quests would not appear as straight lines. They would appear as exponentially increasing lines, graphs with a line sloping sharply upward.

Moore's law describing computer chip intricacy is the quintessential example of such nonlinear progress. When Jennifer Lewis was interviewed and she instanced Moore's law, she had recently created the first bioprinted proximal convoluted tubule—a key component of a nephron, the basic structural and functional unit of

---

‡‡ This is a rough back-of-the-envelope calculation assuming an adult kidney contains about 150 billion cells and has a volume of approximately 150 cubic centimeters (= 150 milliliters). That works out to a cell density of about 1 billion cells per milliliter.

the kidney. Lewis compared a single nephron of a kidney to a single transistor, the dynamo of a computer chip. Lewis said,

> Right now we're trying to construct a single nephron.[§§] But once you have the nephron, then it gets interesting. Once you can say, "We can construct a nephron and it's fully functional"—and to be clear, we're not there yet—then it's like saying, "We have a transistor."[62]

Lewis brought up an image on her computer of a graph with a line that sloped dramatically upward. This rapid upward trajectory was a snapshot of Moore's law. "Here's the first transistor, in 1947," she said, pointing.[***] "It took 25 years just to get to this, a single solid-state transistor," the basis of all computer chips. She continued,

> And from there, it took very little time for this rapid upward trajectory, so that now you have computer chips made of *billions* of transistors. . . . Right now, we're still here at the bottom. We have a part of one nephron. But we're getting closer to having a complete one—and once we've got one, the key question is, will it follow this same trajectory? I don't know. But this is illustrative. It tells you what might be possible."[63]

Earlier, we borrowed from Christian de Duve by miniaturizing ourselves and swimming around and into a cell. In an alternative fancy, we can imagine those scientists conducting bioprinting as being spelunkers exploring a black cave very deep underground. They're scouring an unknown environment and eager to learn all they can. If they find previously unknown forms of life within the cave, that's great. All the same, in the end they want to get out of the cave and share their discoveries. In our context, scientists want to leave behind the proof-of-concept experiments they perform in their labs and have bioprinting see the light of day. They want to bioprint tissues and organs for human transplantation.

There's a tissue that causes a lot of humanity a lot of grief. Years ago, it seemed that particular tissue was a terrific candidate for being the first to be bioprinted and successfully transplanted into humans. As we're about to learn, bioprinting that tissue has proven elusive.

[§§] There are approximately one million nephrons in a human kidney.
[***] The solid-state transistor was invented at Bell Labs in 1947 and replaced the vacuum-tube transistor.

3
Bioprinted Cartilage
The Dream and the Devilish Details

The year is 1741 and we are lucky to be present in the anatomical theater of the University of Padua, Italy, about 22 miles west of Venice, at the autopsy of a woman who suffered from hip pain during her life. The wooden theater rises through two stories of the Bo Palace, the central building of the university. It's an almost funnel-like space and very compressed—only 33 feet at its widest point. At the bottom of the funnel rests the oval-shaped dissection table, surrounded by six steeply banked and narrow tiers with railings. The tiers are so narrow that there are no benches, only space for standing. These tiers are concentric and slightly elliptical, matching the shape of the dissection table. The Padua theater has a flat ceiling with no windows and so virtually no natural light. Our only illumination comes from two chandeliers of four candles each and eight or so other candles that students hold nearby the table.[1]

We're in with 200–300 spectators, all admitted according to rank. Of course, the first level is reserved for professors of anatomy, the rectors of the city and of the university, the councilors and members of the medical college, and representatives of the Venetian nobility.[2] Because of our academic status, we're fortunate to be standing among the medical students in the second and third rows. Since we arrived early, we were able to watch as a lift from a small chamber beneath the first tier raised up a woman's body on the dissection table. That was several days ago. Now, on the third day of dissection, following a musical entertainment during a welcome break, we're once again riveted on the work of Giovanni Battista Morgagni, Professor of Anatomy at the University of Padua, as we look down the precipitous sides of the theater to the narrow floor and its dissection table.[3,*]

We know that during her life the woman had awful sciatic pains in her right hip joint. The autopsy shows osteoarthritic cartilage that explains her misery all too well. The head of the right femur, the top of the thighbone that forms a joint at the hip, is not rounded but is depressed. And cartilage of a pale ash color covers it, not the appropriate smooth, white articular cartilage. In fact, this cartilage is totally absent in the posterior part of the head of the right femur. The bone is naked in that part, utterly lacking articular cartilage, and it forms many roundish and overhanging particles—bone spurs. In the superior portion of the acetabulum—the upper part of the socket

---

[*] Morgagni has been called the father of pathology and the father of pathological anatomy. Pathological anatomy tries to correlate the findings of a postmortem examination with a disease process in life.

of the hip bone where the femur fits—two bony structures are actually buried within the acetabulum itself. It's little wonder that she suffered so much during her lifetime.[4]

Articular Cartilage—A Heavily Challenged Tissue

Many times we don't really appreciate something (or someone) until it's gone. So it is with articular cartilage, cartilage that's present where bone meets bone.[†]

People who have degenerated articular cartilage, owing to aging or intensive sports, often have the best understanding of the importance of this tissue. It has an ostensible simplicity, composed of only a single cell type called chondrocytes that have a low density within their extracellular matrix. However, articular cartilage has very little capacity to repair itself; it lacks a blood supply, so nutrients and other healing substances can't get to the damaged region. Over time, injured cartilage is lost cartilage.

The sparse population of cartilage cells is arranged in distinct zones, and the cells are arrayed differently with respect to one another depending on the zone you consider. Moreover, the cartilage cells in each layer have diverse structures and express (i.e., produce) varying proteins. The main component of articular cartilage's extracellular matrix, the structural protein collagen, is also organized divergently in these zones. Articular cartilage, generally no thicker than a dime, helps our joints to remain strong in the presence of forces that can reach three times our own body weight with each step we take.[5] A professor of orthopedic surgery at Stanford put it this way: "Mother Nature did a brilliant job of engineering, to the point that it is difficult to re-create. This is one of the body's most complex tissues."[6] And he's talking about cartilage, for heaven's sake—not the heart or the kidney. It's no wonder that bioprinting is so challenging.

Jos Malda is a Professor of Orthopedics at the University of Utrecht in the Netherlands. His research specialty is articular cartilage. Malda described articular cartilage with a wry euphemism. He called it a heavily challenged tissue.[7] Its job is to absorb shock, transmit forces, and enable low-friction movement of joints. To accomplish all this requires both great resilience and high resistance to both compression and shear. The words resilience and compression are familiar and self-explanatory. But we need to look at shear resistance more closely. We skirted this topic in Chapter 1 in speaking about shear thinning. Shear will be a recurring concept as we encounter more bioprinting investigations.

When you use a pair of scissors or hedge trimmers or pruning shears, you're producing shear forces on some solid object. In bioprinting, we're interested in shear

---

[†] There are three types of cartilage: hyaline, fibro, and elastic. Articular cartilage is hyaline cartilage that's found on the articular surface of bones—the surfaces at joints. So articular cartilage merits its own name.

forces on flowing materials, such as the forces on living cells when the nozzle of a bio-printer unceremoniously ejects them. Unlike a solid, a fluid continuously deforms or flows when a force is applied. One way to think of what's happening to the fluid is to first visualize two solid blocks, one on top of the other, sliding against each other. Suppose that a force pushes toward the left on the top block and a force pushes to the right on the bottom block. The opposing forces on the two blocks cause the sliding motion.

In this thought experiment, imagine that instead of having two rigid blocks you have only one block—of Jell-O. If we think about applying the same forces as before to the block of Jell-O, the deformation is similar, only now we imagine that the Jell-O is sliding internally. The Jell-O will give or deform a little. Then, depending on the size of the force, the Jell-O will either break or stand its ground and resist the force, but it won't deform anymore. The sliding forces are exerting shear stress (internally) on the Jell-O.[8]

Now go a step further in your mind and replace the Jell-O with a fluid. It will have these internal sliding forces, this shear stress, and, being a fluid, will continuously deform. It will flow. Viscosity is how you measure the resistance of fluids to internal sliding forces. Fluids that are less viscous will deform more easily because of the sliding forces than fluids that are more viscous. Water is less viscous than honey, so water will deform or flow more easily than honey with the same applied force.[9]

Jos Malda teamed with Dietmar Hutmacher of the Queensland University of Technology in Australia and additional colleagues to perform bioprinting experiments that strikingly reveal the importance of viscosity in formulating bioinks for articular cartilage engineering.[10] They used a hydrogel made of gelatin, a water-soluble protein that they derived from collagen. After adding some chemical groups, they came up with a hydrogel with the name gelatin–methacrylamide. For simplicity we'll use its shortened name, gelMA. But low viscosity hampered the bioprinting of gelMA. Material ejected from the print head formed droplets that spread out over the printing platform. To see its performance, look at panels (a) and (c) in Figure 3.1. To make the hydrogel suitable for bioprinting, they increased its viscosity by adding just a small amount of something called hyaluronic acid.

As recently as the 1970s scientists thought hyaluronic acid was an unprepossessing part of the extracellular matrix of connective tissue. Then studies began to emerge that showed it wasn't "just a goo."[11] Now it's known to be an important component throughout the body, for example, in repair of skin tissue. And there is some evidence that injections of hyaluronic acid can ease the pain of knee osteoarthritis.

Hutmacher and Malda added a small amount of hyaluronic acid to their gelMA hydrogel and, like a chrysalis transformed into a butterfly, their hydrogel with dormant promise completed its metamorphosis to a bioprintable bioink, shown in Figure 3.1, panels (b) and (d).

I spoke with Hutmacher, who's had a peripatetic academic career that has taken him from his native Germany to the United States, Singapore, and now Australia.[12] At the University of Utah, he worked on a master's degree in the lab of a renowned Dutch

Figure 3.1 Effect of an additive on the viscosity of a hydrogel. (a) and (c) Without the additive, the hydrogel formed droplets at the nozzle and deposited in flat lines that spread out on the build platform. (b) and (d) With the additive, the hydrogel formed strands that could be deposited from the nozzle and the team formed four layers with these strands. The scale bars in (a)–(c) are 5 mm; the scale bar in (d) is 2 mm.

Source: Reproduced by permission from Wouter Schuurman et al., "Gelatin–Methacrylamide Hydrogels as Potential Biomaterials for Fabrication of Tissue-Engineered Cartilage Constructs," *Macromolecular Bioscience* 13, no. 5 (February 18, 2013): 551–61, https://doi.org/10.1002/mabi.201200471.

physician and inventor, Willem Johan Kolff, widely regarded as the father of artificial organs. Kolff's artificial kidney evolved into dialysis machines.

Bioprinted Hydrogels with Articular Cartilage Progenitor Cells

Jos Malda and colleagues carried out a bioprinting investigation in which they compared stem cells, cartilage cells, and cartilage progenitor cells for bioprinted articular cartilage.[13] In gelMA hydrogels, Malda's team found the articular cartilage progenitor cells outperformed the cartilage cell-laden hydrogels and also the stem cells. The outperformance was reflected in the amount and distribution of extracellular matrix produced. And there's another consideration.

As bioprinting researchers work to scale up the size of their constructs, they've found that the cells they use often lose their distinguishing characteristics after multiple cycles of reproduction. In the trade, this is called loss of phenotype. Scientists define phenotype as "the conglomerate of multiple cellular processes involving gene and protein expression that result in the elaboration of a cell's particular morphology and function."[14] Morphology refers to the size, shape, and structure of a cell.

The concern for bioprinting is that the specialized state, the phenotype, of a given tissue cell may not survive multiple divisions outside the body. For example, the shape of the cell may alter or the set of molecular markers on the cell's surface may not be the same after repeated cell divisions. Whether such a loss of phenotype occurs seems to depend very much on what form of a given cell you're using. In the case of articular cartilage that might be adult cartilage cells, cartilage stem cells, induced pluripotent stem cells that become cartilage cells, cartilage progenitor cells, and so on.

This loss of phenotype business is a serious matter in bioprinting. A fundamental question in biology is to unravel what makes individuals, populations, and species so unalike.[15] The word *phenotype* was coined over 100 years ago.[16] The literal meaning of the word, derived from ancient Greek, is *visible type*.

Phenotype describes your physical characteristics—as a novelist might begin to depict you. For example, she was

> a tall girl with wheat-gold hair and eyes as brightly blue as a November sky when the sun is shining on a frosty world. There was in them a little of November's cold glitter, too, for Joan had been through much in the last few years. . . . Her eyes were eyes that looked straight and challenged. They could thaw to the satin blue of the Mediterranean Sea, where it purrs about the little villages of Southern France; but they did not thaw for everybody. She looked what she was—a girl of action; a girl whom life had made both reckless and wary—wary of friendly advances, reckless when there was a venture afoot.[17]

Your phenotype is the way(s) in which you present yourself to the world: partly a reflection of your genotype—your inherited genetic identity—and also a result of your life's experience.[18]

Those engaged in bioprinting are constantly working with cells taken outside their body of origin. They try to make their tissue creations content after printing by placing them in bioreactors under the conditions that best mimic their natural home. In spite of these efforts, scientists have found that if you bioprint with a hydrogel seeded with stem cells, for example, you may find that a succession of reproductive stages can produce changes in the phenotype of the cell—so the phenotype isn't faithfully maintained. They've also seen this happen if they build a cartilage tissue using a hydrogel and cartilage cells, as another example: After a few population doublings they steadily lose their phenotypic identity.

Our discussion of shear comes into play here, specifically the forces that impact cells during their passage through the print head. Bioprinting studies have shown that above a certain shear stress, the cells' capacity to maintain their identity can be harmed. This is true even in the absence of harmful effects on cell viability, especially with viscous hydrogels used in the extrusion method of bioprinting.[19]

More generally, cells are subjected to a Coney Island rollercoaster ride during bioprinting with a range of shear stresses that depend on the size and shape of the extrusion nozzle, the particulars of the extrusion mechanism (e.g., mechanical or pressure-driven), and the bioink rheological properties. A promising result in Malda's experiments was that the gelMA-based bioink his team used seemed impervious to all these lurking threats.

The upshot of much of what we've said about these separate cartilage cell sources is that it's necessary to have a cell with the capability to preserve a stable phenotype. Malda's team compared hydrogels laden with native articular cartilage cells, bone marrow-derived stem cells, and also progenitor cells harvested from articular cartilage. They found it was the progenitor cells that did a superior job in maintaining phenotype, specifically the phenotype of the superficial/tangential zone of articular cartilage. That's important because the superficial zone is the happening place; the potential wear and tear in your knees and hips starts in the superficial zone.

Malda's group also combined progenitor cells and bone marrow-derived stem cells and found that they got a bioprinted model of articular cartilage that had well-defined superficial and deep regions (Figure 3.2), each with a distinct cellular and extracellular matrix composition.

In the cross-sectional diagrams of articular cartilage, panel (a) highlights the zonal distribution of the cartilage cells. In the superficial zone near the articular surface, the cartilage cells are flattened and oriented sideways. In the middle, the cartilage cells aren't so flattened and they're distributed pretty evenly, not preferring to be oriented sideways. In the deep zone, the cells are larger and round and laid out in an up–down array. These are the three zones of articular cartilage.

Panel (b) spotlights the collagen fibers of the extracellular matrix. The fibers are oriented sideways near the articular surface, more randomly in the middle, and vertically deeper down. Remember that collagen acts as a natural scaffold in your articular cartilage, and this zonal architecture of the collagen is what encourages the cartilage cells to array themselves likewise. In your mind's eye, superimpose panels (a) and (b).

If you're familiar with the construction of a wood frame house, carpenters put up vertical 2 × 4-inch lengths of lumber as a frame. Once those two-by-fours are up, they guide the position of the external covering plywood, called sheathing, and also the internal drywall. The collagen acts like the two-by-fours, and the cells act like the plywood sheathing or the drywall, take your pick.

As Malda reminds us,

While biofabrication holds the promise of introducing a new generation of regenerative therapies for cartilage, simply recapitulating the native structural and

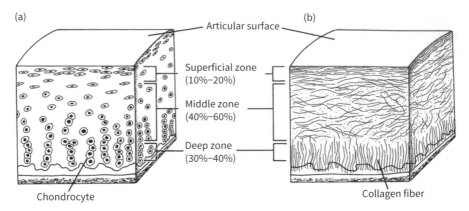

Figure 3.2 Cross-sectional diagrams of articular cartilage. (a) The cellular organization in the zones of articular cartilage. Chondrocyte is the technical term for a cartilage cell. There are three zones. The zone closest to the outer surface of the articular cartilage is called the superficial or tangential zone. The zone below that is called the middle zone. The zone furthest from the cartilage surface and closest to the bone that the cartilage is protecting is called the deep zone. (b) The collagen fiber portion of the extracellular matrix. It too has the same distinct architecture in different zones.

Source: Reproduced by permission from Alice J. Sophia Fox, Asheesh Bedi, and Scott A. Rodeo, "The Basic Science of Articular Cartilage: Structure, Composition, and Function," *Sports Health: A Multidisciplinary Approach* 1, no. 6 (November 2009): 461–68, https://doi.org/10.1177/1941738109350438.

cellular composition during the printing process is only the first step of many. Regeneration is driven by the encapsulated cells (and, upon implantation, by the interplay with the host biomechanical and biochemical environment) and occurs over time, after the initial bioprinting step. Thus, in the design of hydrogels, and consequently bioinks, harnessing the potential of regeneration-competent cells is paramount."[20]

And this ties in with our earlier emphasis: to be "regeneration-competent," a cell must be able to maintain its phenotype.

One of the principal reasons we're beginning our survey of bioprinted tissues with cartilage rather than skin or bone, for example, is that cartilage bushwhacked bioprinting scientists. When Malda and I chatted, he mentioned a paper by biological chemist Julie Glowacki of Harvard University.[21] Glowacki has made many important contributions in the area of musculoskeletal tissue. In 2000, Glowacki published a paper on engineering cartilage outside the body.[22] This was a time when tissue engineering was in its first youth and bioprinting of tissues and organs was just a gleam in the eyes of visionaries such as Gabor Forgacs, Vladimir Mironov, and Thomas Boland. Glowacki wrote, "Articular cartilage is a relatively simple tissue because of its cellular homogeneity and avascularity."[23] And that's true enough. Malda said he used to cite the Glowacki paper a lot in his own publications. He and everyone else have since found the bioprinting of articular cartilage to be a formidable undertaking.

It's difficult for a tissue to combine resilience and also resistance to compression and shear. Articular cartilage is able to do this because of the nature of its extracellular matrix. To expand on Glowacki's observations, there are numerous reasons researchers thought it might be a great candidate for bioprinting when it needs to be replaced: It's composed of a single cell type—chondrocytes; it's aneural—no nerves; it's avascular—no blood vessels; and it has low cellularity—a sparse population of cells in a vast expanse, relatively speaking, of extracellular matrix. Nonetheless, scientists believe that it's this very makeup of articular cartilage that's responsible for its one big downside—it has very limited capacity to regenerate itself. So when we wear out our knees, hips, hands, and shoulders we suffer from osteoarthritis.

Then along comes bioprinting and provides us with a substitute for our own articular cartilage? Not quite. Part of the reason we're not there yet is that natural articular cartilage is so subtle. As we've seen, its sparse population of cells belies the marvel of its zonal internal architecture. And the varied cell structures and extracellular collagen organization are testimony to nature's capacity for functional design.

Bioprinting of Collagen Hydrogels for Cartilage Substitutes

Collagen is commonly used as a scaffold material. It's also an exceptional biological molecule. People who prefer to look younger receive injections of collagen, which makes their skin appear smoother. Regardless of personal tastes, nature celebrates collagen as a natural scaffold material so much that the collagen family of proteins is the largest family of proteins in mammals.[24] Collagen self-assembles in a hierarchical or graded fashion from the molecular scale up to large fibrous bands that exhibit exceptional physical properties.[25]

Naturally occurring collagen molecules packed side-by-side form fibrils—small, slender, thread-like formations. Fibrils packed side-by-side form fibers. Gram for gram, so-called type 1 collagen has the tensile strength (resistance to breaking under tension) of steel.[26] Collagen is highly versatile, taking on different roles in contrasting parts of the body. In structural tissues such as bones and ligaments, collagen assumes the form of rope-like fibers that provide resistance to stretching and tearing forces. But in cartilage, which is mostly loaded in compression,

> collagen has more of a "holding" function, with the fibers arranged rather like a basket ... which forms the gliding surface that permits joint movement ... and yet its stiffness and shock-absorbing capability make it comparable to solid rubber. Moreover, the cartilage-on-cartilage sliding interface has lower friction than ice sliding on ice.[27]

Throughout the body, there are over 20 individual types of collagen, and it's not even known precisely what functions they all fulfill.

Materials scientist Larry Bonassar of Cornell University and colleagues have bio-printed collagen hydrogel constructs with distinct compositions and mechanical properties. The spotlight is on collagen as a scaffold in Bonassar's bioprinting. His team looked at a range of collagen concentrations to come up with the best bioink for the purpose. They evaluated their constructs by checking the viability of printed cells with a live/dead assay (explained later) and by the geometric accuracy of the printed structures. They also evaluated the structures' mechanical properties.[28]

The subject of Bonassar's analysis was the meniscus in the knees of sheep. The meniscus is not articular cartilage. It's a separate type of cartilage called fibrocartilage. Like articular cartilage, it's found in joints and it too has load-bearing responsibilities. It also has a zonal architecture but only consisting of an outer zone and an inner zone. The anatomy of the knee is such that if you injure your meniscus and it breaks down, then your articular cartilage is exposed to much higher impact and it too starts to break down, leading to osteoarthritis. We'll take an up-close look at Bonassar's bio-printing experiment and explain some of the language and procedures that also appear in other experiments.

Unlike many groups that use collagen protein, Bonassar's team developed high-density formulations of collagen gels that showed superior mechanical properties. This was critical for their success. A limitation of collagen hydrogels for bioprinting has been their inability to retain the desired shape after printing and to bear loads after transplantation in animal models—the mechanical properties aren't up to snuff. This problem seems to come from the collagen concentration in the hydrogels being too low. Bonassar's team tackled this by using higher concentrations.

They tested the geometrical accuracy of their bioconstructs using a convenient technique called micro-computed tomography or micro-CT. This is X-ray imaging in three dimensions just like you find in a hospital CAT scan but on a smaller scale and with much higher spatial resolution. Spatial resolution simply means how close two features can be in an image and still be seen as distinct. Micro-CT doesn't require special sample preparation (no need to cut super thin slices), and it's a non-destructive method so you get your sample back at the end.[29]

In the Bonassar research we're looking at, the excellent geometric/shape fidelity was likely due to hydrogels with greater collagen densities having greater viscosity, so the bioprinted layers were better able to retain their shape and adhere together. Bonassar's team found an optimal window of collagen concentration by careful observations. They found that the highest density of collagen they tested did not print well: The gel got gummed up in the print nozzle and only the liquid from the gel made it through. At the other end of the concentration spectrum, low densities of collagen weren't any good either: The deposited material simply spread out over the printing surface before it had a chance to firm up.

To test for compressive loading, the research team used a classic technique in which they essentially put a slice of the sample in a vice (formally, a dynamic mechanical

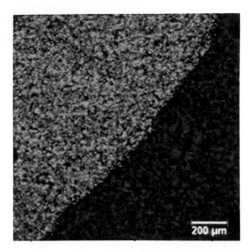

Figure 3.3 Microscope image of the interface between areas of varying collagen concentrations visualized with fluorescent microbeads after bioprinting.

Source: Reproduced by permission from Stephanie Rhee et al., "3D Bioprinting of Spatially Heterogeneous Collagen Constructs for Cartilage Tissue Engineering," *ACS Biomaterials Science & Engineering* 2, no. 10 (August 4, 2016): 1800–1805, https://doi.org/10.1021/acsbiomaterials.6b00288.

analyzer) and squeezed. This slightly deformed the sample, and then they released the pressure. They did this repeatedly, and the device measured how well their bioprinted sample responded to this squeezing that represented a compressive load. That is, the device gave them a quantitative readout of how easy it is to squash the sample. Within the optimal concentration window we mentioned, they found that as they increased the collagen concentration they increased the sample's ability to withstand a compressive load.

The principal objective of Bonassar's study was to evaluate a range of collagen concentrations to formulate bioinks for possible orthopedic applications. A secondary goal was to deposit distinct, adjacent domains of bioprinted collagen hydrogels laden with meniscal fibrochondrocyte cells. Fibrochondrocytes are the main cell type found in the fibrocartilage that makes up the meniscus of the knee. The bioprinting of adjacent domains with distinct composition and mechanical properties is a step toward mimicking the zonal architecture present in fibrocartilage.[30]

The team was successful in this secondary goal. The constructs with multiple domains showed sharply defined collagen density interfaces at boundaries, so the domains retained their compositional identity. Figure 3.3 shows the sharp change in collagen concentration as evidenced by the clearly defined interface between red and green fluorescing microbeads.[‡] The team also performed mechanical tests and confirmed that the mechanical stiffness for each domain was consistent with that of

---

‡ The group printed green and red fluorescing microbeads into neighboring domains for visualization.

single-domain constructs made of the respective collagen density material. However, the print accuracy of their system is on the order of millimeters. Ultimately, to create cartilage tissue for transplantation they need to improve the print accuracy so it's on the cellular scale—they must improve by a factor of 100 to get from millimeters to tens of microns.

Figure 3.3 is one of countless examples of science as art. It looks like a painting from the minimalist school collection in the Guggenheim Museum in New York City.

Bioprinted Ears from an Integrated Tissue–Organ Printer

Surgeon–scientist Anthony Atala's group at the Wake Forest Institute for Regenerative Medicine has bioprinted bone, cartilage, and skeletal muscle with their one-of-a-kind integrated tissue–organ printer, developed over the past decade.[31] Here, we present a portion of their results on bioprinted ear cartilage. The ears' cartilage is a tissue type called elastic cartilage, rather than hyaline cartilage or fibrocartilage.

Atala's team started with a CT scan of a human external ear and fed the information into a computer-aided design program. Then it was off to the races using a hydrogel laden with cartilage cells and a popular biodegradable support material—they chose a polymer that can be cross-linked, as we discussed in Chapter 1. The process is shown in Figure 3.4.

Atala's hydrogel had a potpourri of several fluids, such as gelatin and glycerol—a viscous liquid. Instead of collagen protein, he used a soluble protein that's bonded with sugar molecules and called fibrinogen. To say a substance is soluble means it can

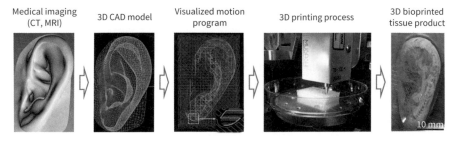

| Medical imaging (CT, MRI) | 3D CAD model | Visualized motion program | 3D printing process | 3D bioprinted tissue product |

Figure 3.4 Bioprinted ear construct from an integrated tissue–organ printer. Beginning with medical image data, a computer-aided design (CAD) program generates a so-called visualized motion program. This latter program includes instructions for 3D print head and build platform movements to create a bioprinted ear construct. CT, computed tomography; MRI, magnetic resonance imaging.

Source: Reproduced by permission from Hyun-Wook Kang et al., "A 3D Bioprinting System to Produce Human-Scale Tissue Constructs with Structural Integrity," *Nature Biotechnology* 34, no. 3 (February 15, 2016): 312–19, https://doi.org/10.1038/nbt.3413.

be dissolved in a fluid, whereas an insoluble substance can hang on to its identity and not disappear in a fluid.

Fibrinogen is circulating in your bloodstream as you read this. Should you cut yourself peeling a potato for dinner, a protein enzyme converts fibrinogen to an insoluble protein called fibrin that helps to form a blood clot.[§] In this bioprinting study, fibrinogen provided stability for the gel and a microenvironment conducive to cell adhesion and proliferation. Atala's team faced the same dilemma as did Bonassar's team in finding the right concentration of collagen. But Atala's group had to test various concentrations of each component in their composite hydrogel. This was vital to get proper printing resolution, dispensing uniformity, mechanical properties, and cell viability.

Everyone doing bioprinting uses an assay to check cell viability, and Atala's team was no exception. To check if cells are dead or alive is a simple staining procedure. The stain solution is a mixture of two fluorescent dyes. The dyes are fluorescent in that they light up in colors that are easily seen under a microscope. If the cell is alive—that is, viable—biochemical processes in the cell cooperate by causing one of the dyes to emit a green glow. If the cell is dead—its demise caused by the bioprinting process—the other dye will make its way through the ruptured cell membrane, bind to the cell's DNA, and the cell will emit a red glow. The actual counting of green cells versus red cells is typically automated. Atala's team achieved high cell viability of more than 90%, showing that the great majority of the cells within the hydrogel weren't significantly damaged during the printing process. This was true immediately after printing, and viability remained stable over the course of the experiments.

Atala's group also performed mechanical analysis. They found that after the cartilage tissue matured in mice, the retrieved tissue had strengthened. They did repeated bending and relaxation tests and the tissue could hold up under more of a load in bending than before it was transplanted into mice and was also much more resilient. As was true of mechanical testing in the Bonassar work, there's a special test fixture that does a so-called 3-point bend test and Atala used this for his mechanical assay of the ear.

Atala was drawn to medicine at an early age, inspired by his family's doctor. During his residency after medical school, he did a rotation in pediatric urologic surgery and fell in love with it.[32] While he was applying for fellowship programs, he got a call from Dr. Alan B. Retik, head of pediatric urology at Boston Children's Hospital, Harvard Medical School. Retik offered him a chance to do a research year or go straight through the clinical track. Atala chose to just do the clinical program.

But Retik saw Atala's potential as a researcher and persisted with follow-up calls. Finally, Retik asked to speak to Atala's wife. When his wife got off the phone, she encouraged Atala to do the research year and trust Retik's experience and what he thought would be best for his career. Ultimately, Atala agreed and has had a

---

[§] Enzymes are proteins that accelerate the rate of a chemical reaction.

spectacular research career while continuing to perform surgeries. As Atala said, "I was a reluctant researcher."[33]

Decellularized Extracellular Matrix Bioink

Decellularized extracellular matrix is simply the natural environment of a cell, sans cell. Remove the cells from a tissue and what you have left is the extracellular matrix— a labyrinth of protein, collagen, long sugar polymers, and other materials. It's the cells' haven, down to the family photos on the occasional table and the kids' drawings on the refrigerator door. The use of decellularized matrix in bioprinting has gained momentum. As the argument goes, no natural or man-made substitute material can recapitulate all the features of the natural extracellular matrix. Bear in mind, though, that the matrix of each tissue—cartilage, heart, fat—is unique with respect to composition and the arrangement of its parts.

A team led by mechanical engineer Dong-Woo Cho of Pohang University of Science and Technology (POSTECH) in South Korea has been at the vanguard in seizing decellularized matrix as a medium for bioprinting.[34] They applied their technique to hyaline cartilage (as well as fat and heart tissue) and in all cases found improvement in the spatial organization of matrix ingredients. Recalling our discussion of gene expression, Cho's group also demonstrated improved tissue-specific gene expression. Their goal in these experiments was to maximize the removal of cells while minimizing extracellular matrix loss and damage. To do this, they used a combination of physical and chemical means.

After removing the cells, the team had to use chemical treatments to break down the remaining structure of the extracellular matrix into a liquid form that could be dissolved in a solution. This preparative step is called solubilization. To improve the mechanical integrity of the printed constructs, Cho selectively combined cell-laden hydrogels with structurally stiffer synthetic polymers. For both cartilage and fat tissue, he used a supportive polymer; however, he used a thicker polymeric line width (200 microns) for cartilage than for fat tissue (100 microns). The rationale was more load-bearing support for cartilage and reduced stiffness for fat. For heart tissue, he printed only the decellularized extracellular matrix without any polymeric supporting framework.

In fabricating hybrid cell-laden hydrogel and polymeric structures using multiple print heads, the group first deposited a layer of polymer framework followed by a layer of cell-laden decellularized extracellular matrix positioned in the gaps between alternating polymer lines. They employed this technique to ensure an adequately porous structure, looking ahead to supplying a flow of nutrients and oxygen for the cells in future applications (Figure 3.5). The group laid down the cell-laden hydrogels as a viscous solution—that is, in a pre-gelation state. They repeated the process layer-by-layer with the plane of successive layers deposited at right angles to one another in a Lincoln Log design. For the heart cell-laden hydrogel they used the same design (but

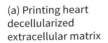

(a) Printing heart
decellularized
extracellular matrix

(b) Printing cartilage
decellularized
extracellular matrix

(c) Printing fat
decellularized
extracellular matrix

Figure 3.5 Printing heart, cartilage, and fat decellularized extracellular matrix.
(a) Printing heart decellularized extracellular matrix with no polymer framework.
(b) Printing cartilage decellularized extracellular matrix with a polymer framework
(white) having a 200-micron line width. (c) Printing fat decellularized extracellular
matrix with a polymer framework (blue) having a 100-micron line width.

Source: Reproduced with permission from Falguni Pati et al., "Printing Three-Dimensional Tissue
Analogues with Decellularized Extracellular Matrix Bioink," *Nature Communications* 5, no. 1 (June 2,
2014), https://doi.org/10.1038/ncomms4935.

without the polymer). After forming the final structure, they placed the constructs in
an incubator at body temperature to cause the hydrogels to form a gelled state. Figure
3.6 shows a schematic of their approach.

The decellularization process left the heart tissue extracellular matrix unchanged. For fat tissue, there was some minimal change in the matrix components.
Unfortunately, for cartilage tissue they had a big loss of both the structural protein
collagen and also the sugar polymers that are ubiquitous in cartilage's matrix. The
reason is that the decellularization procedure is harsher for cartilage than for the
other tissue types they worked with because cartilage has a denser extracellular matrix structure than heart or fat. So once again, the apparent simplicity of cartilage
belies its ease of use in bioprinting.

An issue central to Cho's approach was the extrusion of decellularized matrix bioink through the print head nozzle of his bioprinter. Cho's team needed to constitute
the bioink so that it could be extruded in filamentary form, then gel at physiological
temperature (37°C), and retain its faithful three-dimensional structure. They had to
tinker with the pH—a number that represents the concentration of hydrogen ions in
the substance—to get it right; it was initially too acidic.

As in the work of Hutmacher and Malda, Cho and friends did a careful job of evaluating rheological (flow) behavior of the pH-adjusted bioink. By measuring cell viability in printed constructs of greater than 95%, they showed that the shear stresses on
the cells were minimal.

Perhaps the greatest significance of Cho's work is in helping bring decellularization
to the negotiating table when devising bioinks.

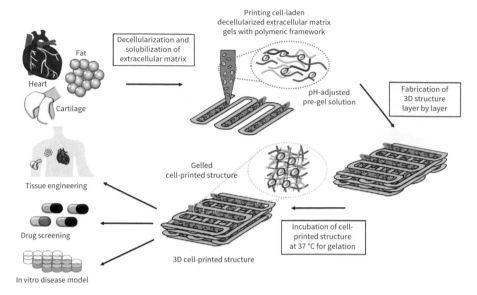

Figure 3.6 Bioprinting using decellularized extracellular matrix bioink. The researchers removed the cells from cartilage, fat, and heart tissues and used chemical treatment to break down the structure of the remaining extracellular matrix to dissolve it in a solution (solubilization). Adding the cells to each cell-specific matrix, they formed a supply of multiple cell-laden bioinks. First, they printed a polymer framework (for cartilage and fat only) and deposited the cell-laden hydrogels in alternating gaps to make the constructs porous for future nutrient and oxygen delivery. They printed successive layers at right angles in a Lincoln Log design. During printing, the hydrogels were in a viscous, pre-gelation state, but incubation of the final construct at body temperature resulted in a gelled structure.

Source: Reproduced by permission from Falguni Pati et al., "Printing Three-Dimensional Tissue Analogues with Decellularized Extracellular Matrix Bioink," *Nature Communications* 5, no. 1 (June 2, 2014), https://doi.org/10.1038/ncomms4935.

I found several surprises in my interview with Cho.[35] Despite his many scientific accomplishments, science didn't exert a strong draw on him in his youth. A colleague of his graduate advisor at Seoul National University recognized his ability and encouraged him to pursue a Ph.D. in mechanical engineering at the University of Wisconsin–Madison.** This he did. Then Cho landed a position as an assistant professor at POSTECH in Seoul. Although he was productive, he had not yet found his passion. In the mid-2000s, he became interested in 3D printing—and bioprinting really captured his imagination.

** Cho received a master's degree as well as a bachelor's degree at Seoul National University.

Scientists like Cho and others trying to realize the potential of bioprinting have benefited from research whose focus has been on decellularization but outside the immediate purview of bioprinting. Decellularized extracellular matrix has found favor with many bioprinting groups working with a variety of tissue types. We will try to shed a little light on decellularization studies in tissue engineering and regenerative medicine, aspects of which have now found expression in bioprinting. In doing so, we aren't taking a big detour since bioprinting is at the intersection of regenerative medicine, tissue engineering, and 3D printing.

So we segue from bioprinting using decellularized extracellular matrix to look at decellularization qua decellularization. And we'll meet two of the scientists whose names come up repeatedly when bioprinting researchers cite their sources.

Decellularization: Re-creating a Native Extracellular Matrix with Optimal Fidelity

Around 1850, it became generally accepted that cells made the intercellular substances that surrounds them—that is, the extracellular matrix.[36] Through the early to mid-twentieth century, scientists thought the matrix was present to perform pedestrian but necessary functions as a scaffold to fill the spaces between cells. Not so any longer. Some scientists have devoted their careers to mining the extracellular matrix for its potential in tissue engineering and regenerative medicine.[††]

What would you call someone who routinely harvests organs such as hearts and lungs from the newly dead and tries to bring the organs back to life? If you were nineteenth-century novelist Mary Wollstonecraft Shelley, you might call that person Dr. Frankenstein. Doris Taylor placidly accepts the nickname since she too harvests organs and strips them of their cells to work with what's left: the decellularized extracellular matrix.[37]

Taylor earned her Ph.D. in pharmacology, although her career has veered in many directions. She's now Director of the Center for Cell and Organ Biotechnology at the Texas Heart Institute and Texas A&M University. For years, Taylor has worked to engineer new organs that won't be rejected by the recipient's immune system. In the late 2000s, Taylor and Dr. Harald C. Ott, a thoracic surgeon, were working in a lab at the University of Minnesota. Their aim was to find a way to seed the heart's matrix so that stem cells would attach and grow into a working organ. They needed to find a new, improved way to decellularize cadaver or animal hearts. The existing methods degraded the extracellular matrix by removing the signaling molecules, proteins, and growth factors that enable communication with living cells. This meant that stem cells could not locate their proper places in the matrix or develop in coordination with other cells.

[††] Note that decellularization experiments have been reported since the 1970s, although the term *decellularization* was not necessarily employed.

Through intelligent trial and error, they found a way to remove a rat heart's original cells while keeping the matrix's signaling molecules intact. The remaining matrix, devoid of cells, was translucent, which led to the term "ghost heart." Taylor and colleagues seeded the decellularized heart with several types of cells, including neonatal cardiomyocytes, endothelial cells, and smooth muscle cells.[38,‡‡] They cultured these recellularized hearts in a bioreactor that simulated physiological conditions for organ maturation. Over a period of days, the recellularized ghost hearts formed heart tissue that beat and showed nascent pump function.[39]

Taylor and team noted that they've also applied their decellularization/recellularization technique to a variety of mammalian organs, including lung, liver, kidney, and muscle.[40]

If further rationale is needed to pivot to a discussion of decellularization in this chapter on cartilage, we note that in Cho's study he cites the work of Taylor and Ott: "Decellularization was conducted by following the protocol published elsewhere with slight modification."[41,42]

Taylor's hope to make new hearts for transplantation may well be a moonshot. Even so, it could lead to a better understanding of cell organization in the heart and perhaps new ideas of how to fix an ailing heart. A magazine features editor interviewing Taylor in her lab noted jars of ghost organs eerily suspended in solution and was reminded of novelist Shelley's gothic novel. But Taylor told him that these are the resources for her work. She added, "There are some days that I go, 'Oh my god, what have I gotten into?' On the other hand, all it takes is a kid calling you, saying, 'Can you help my mother?' and it makes it all worthwhile."[43]

Taylor's colleague, Harald Ott, is now at the Massachusetts General Hospital and the Harvard Medical School. Since establishing his own lab, Ott has started with decellularized lung and kidney to bioengineer and transplant versions of these organs in rats.[44,45] To regenerate the decellularized rat kidneys, Ott's team used human cells to seed the blood vessel portions of the organ, and they used kidney cells from newborn rats for other parts of the bioengineered kidney. The team placed the kidneys in an incubation chamber for up to five days to allow the tissues to grow before transplantation into rats. When transplanted into a rat, the bioengineered kidney produced urine through the ureteral duct.

As to Ott's lung work, his bioengineered rat lungs can perform the necessary gas-exchange functions when transplanted into living rats. The defining functional property of lung tissue is the capacity to allow gas exchange between circulating blood and inhaled air: the removal of carbon dioxide from the bloodstream and the delivery of oxygen. Now Ott and his team have scaled up to humans.

A 47-year-old New York man died of cardiac arrest. The Ott lab stripped his lungs of their cells and genetic material, leaving two cone-shaped objects made up of collagen and other proteins. These human lungs are now decellularized extracellular

---

‡‡ Neonatal means that the cells were taken from the hearts of newborn rats.

matrices. Ott's team is using these lungs as starting points to try to grow a set of bioengineered lungs in the lab. The team will seed the matrices with stem cells and hopes to direct the cells to differentiate into the various cell types found in the lung. The goal is to rebuild functioning lung tissue. [46,47]

Our transition from bioprinted cartilage to a brief interlude on decellularized extracellular matrix is to share some of the origins of a technique that has made its way into the hydrogel bioink toolkit. Elsewhere, Atala's group has used this approach to better address the biochemical mimicking of the in vivo (in a living organism) extracellular matrix in bioprinted liver constructs.[48,49] In fact, Atala's interest extends back to the days before bioprinting when he used an "acellular collagen matrix" as a biomaterial for urethral repair.[50] This was also before the days when decellularized extracellular matrix was the commonly used description and *acellular matrix* was the designation of choice.

Thanks to the work of many, the problems of the bioprinting of cartilage are in sharper focus and work proceeds apace on many fronts. Articular cartilage tissue receives its nourishment from the joint fluid that it's bathed in by means of a process called *diffusion*. This means that oxygen and nutrients spontaneously move from regions of higher concentration to regions of lower concentration. Such passive transport is limited to about 100–200 microns, which is the known diffusion limit of oxygen.[51] This is obviously a miniscule distance. It's about the size of a dust mite, those very, very little critters that binge on human skin flakes in mattresses, sofas, pillows, and blankets.

Most tissues in the body rely on the active transport of blood to supply cells with the oxygen and nutrients they require.[§§] Almost every cell in our body is close to a blood vessel, however small that blood vessel may be—a capillary, for example. Generally, a tissue requires new blood vessel formation to grow beyond 100–200 microns in size. With the exception of avascular tissues like articular cartilage, bioprinting of tissues requires vascularization, a supply of blood vessels. This is a significant job for scientists to re-create, though developmental biology has mastered the technique.

Occasionally, however, even biology's mastery of vascularization gets derailed. Frank Sinatra starred in a 1959 movie called *A Hole in the Head*. But when vascularization becomes overzealous, doctors may have to create a hole in your brain. That's what befell a young Australian doctor. Vascularization is an absolute necessity for our tissues but, as we're about to learn, a superfluity can be bad for your health.

[§§] Cells use oxygen to derive energy from the nutrients by the process known as cellular respiration.

4

Vascularization

Getting Blood from Here to There, Everywhere

Katy had a cavernoma, a symptomatic brain cavernoma. She named it Clive. Clive the Cavernoma. Naming her vascular anomaly was a smart coping mechanism, making it easier to accept what she couldn't change. A few years ago, Katy, a doctor in Australia, was finishing off a ward round—visiting each of her patients in the hospital unit—when she felt a wave of vertigo and lost all feeling in her right arm. Then the stroke call went out alerting staff that someone needed urgent attention for a potential stroke. An MRI scan showed bleeding on the left side of her brain caused by a cavernoma, a mass of abnormal blood vessels. It's not a cancerous mass. But a symptomatic cavernoma in the brain is serious stuff. Katy said, "Clive was not a happy Cavernoma."[1]

Most fortunately, the location and positioning of Clive allowed for surgical removal; sometimes surgery isn't possible. Looking at the MRI of Katy's brain after the surgery is startling. There's an obvious hole where Clive the Cavernoma once lived. "I have a hole in my brain. ... Some other wonderful neurons in my brain are now doing some weight lifting and have taken on jobs that the injured part can no longer do."[2] This speaks to the wonders of neuroplasticity: The brain can change and adapt. Nearby neurons helpfully take over a task previously performed by their brethren—for example, when there is a hole where there used to be a bit of brain.[3]

The Vasculature of the Human Body

Cavernomas are a type of vascular malformation, an unusual cluster of vessels with small bubbles, or caverns, filled with blood. Cavernomas look like raspberries. They're not as rare as you might think, although most often they're asymptomatic and cause no harm. They're estimated to occur in about 0.2% of the population.[4] If you tot up the numbers, in the United States, it comes to about 652,000 people. These vascular bloopers can occur anywhere; most are in the head and neck.

Nature isn't perfect. But when you consider how extensive our vascular system is, the vasculature of the human body does its job spectacularly well. If stretched end-to-end, the arteries, veins, and capillaries in an average adult's circulatory system would measure about 100,000 miles.[5] That would girdle the globe four times at the equator.

In bioprinting substitute parts for potential transplantation into the body, vasculature is vital. However, prior to taxiing our plane to the runway for our next bioprinting jaunt, we first need to get our boarding pass: We need to have some idea of what the body's vasculature is and what it isn't. Our focus is bioprinting. But it is only when we understand natural human vasculature that we can fathom its crucial importance in bioprinting and why vascularization of bioconstructs has been such a colossal concern.

Consider the blood's trajectory within the human body. For a historical perspective we begin with Galen, a Greek physician born in 129 AD. He inherited the conceptual systems of the ancient Greeks and codified, systematized, and built on that knowledge while also adding considerably to it.[6] Following Galen's monumental efforts, for roughly the next 1,500 years people believed that the blood's circulatory system—wasn't. For Galen, the words *circulatory system* would be oxymoronic. That's because he believed that the cardiovascular system was made up of two distinct networks: arteries and veins. His was an open-ended system in which blood and air dissipated at the ends of veins and arteries according to the needs of the local tissue. Blood assimilated into tissue was ultimately lost through invisible emanation. The tissues and organs received fresh supplies as required—from the liver, not the heart, in Galen's view. Blood didn't circulate. It ebbed and flowed much like the tide.[7]

The ancient Greeks took a big step toward modernity when they sought non-divine causes for natural phenomena. But Galen's belief that blood ebbs and flows within the body was incorrect, though it persisted until the 1600s. Then in 1628 an English physician named William Harvey showed that arteries and veins are functionally connected in the lungs and in the peripheral tissues and that blood circulates. Harvey's theory of blood circulation is widely regarded as the foundation for modern medicine.[8]

Figure 4.1 is a simplified schematic of our vasculature. It's common to refer to two interconnected circuits of blood flow that, taken together, form a closed, not an open-ended, system. The two circuits are usually called the pulmonary circuit and the systemic circuit. (Pulmonary means related to the lungs.) The systemic circuit is the collection of blood vessels that supply blood from the heart—blood that contains oxygen—to the tissues of the body and also returns that blood, now devoid of oxygen, back to the heart. Generally, arteries carry oxygen-rich blood away from the heart. An easy mnemonic is *a* for *artery* and *away* (from the heart). The exceptions to this general rule are the pulmonary vessels. The pulmonary veins transport oxygenated blood back to the heart from the lungs, while the pulmonary arteries move deoxygenated blood from the heart to the lungs. So the pulmonary arteries still move blood away from the heart; however, it's not the typical oxygenated blood but, rather, oxygen-poor blood.[9]

The heart is mostly made of specialized muscle tissue and acts as a sophisticated pump so it has quite a bit of punch to it. Keep in mind that the right side of the heart appears on the left when you look at Figure 4.1 since we're pretending that you're looking at the circulatory system of someone standing in front of you. The right side

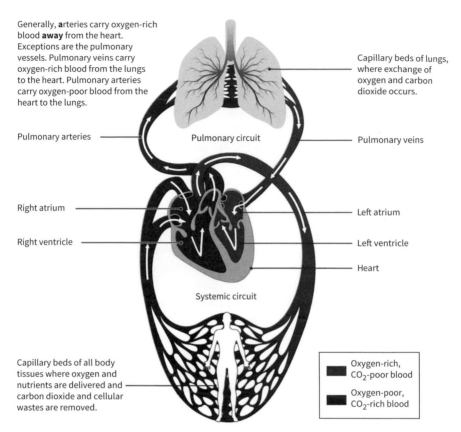

Generally, **a**rteries carry oxygen-rich blood **away** from the heart. Exceptions are the pulmonary vessels. Pulmonary veins carry oxygen-rich blood from the lungs to the heart. Pulmonary arteries carry oxygen-poor blood from the heart to the lungs.

Capillary beds of lungs, where exchange of oxygen and carbon dioxide occurs.

Pulmonary arteries

Pulmonary circuit

Pulmonary veins

Right atrium

Left atrium

Right ventricle

Left ventricle

Heart

Systemic circuit

Capillary beds of all body tissues where oxygen and nutrients are delivered and carbon dioxide and cellular wastes are removed.

Oxygen-rich, CO_2-poor blood

Oxygen-poor, CO_2-rich blood

Figure 4.1 The human vascular system showing both the pulmonary circuit and the systemic circuit including the heart, major arterial and venous connections, the lungs and their capillary beds, and the capillary beds of all body tissues.
Source: 123rf.com, Image ID 16255884; and Shutterstock.com, Image ID 449931577.

of the heart pushes red blood cells to your lungs. When the blood enters the lungs it goes through a network of extremely small blood vessels called capillaries and known collectively as pulmonary capillary beds. Capillary beds form connections between very small arteries called arterioles and very small veins called venules, as shown in Figure 4.2.* Moreover, the lungs' capillaries are in intimate contact with little air sacs called alveoli within the lungs, and the air sacs deliver oxygen to your red blood cells. *Alveoli* is the plural and the singular is *alveolus* from a Latin word meaning "small cavity."

The freshly oxygenated red blood cells leave the lungs through the pulmonary veins and return to the heart where they're quickly pushed out to deliver their

---

* Some capillaries have such a small diameter that they only allow one red blood cell at a time to get through.

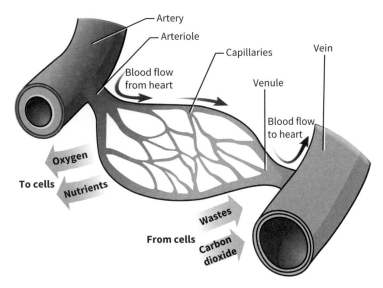

Figure 4.2 Capillary beds form connections between very small arteries called arterioles and very small veins called venules.

Source: Reproduced with permission from U.S. National Library of Medicine, "Capillaries," Nih.gov, 2015, https://ghr.nlm.nih.gov/art/large/capillaries.jpeg.

precious cargo to the body's tissues and organs. As oxygenated blood makes its way through your arteries—that progressively diminish in size to arterioles—an exchange takes place. The cells barter their carbon dioxide and metabolic waste products such as urea and creatinine for oxygen and nutrients such as glucose that they receive from the arteries. The superposition of an outline of a human at the bottom of Figure 4.1 is meant to convey that these systemic capillary beds are ubiquitous in your head and body.

That's a quick journey through your vasculature. We touched on the enormity of the burden to provide vasculature in Chapter 2: How much do you actually need to bioprint to be successful? Now that we've seen the hierarchical sophistication of native vasculature, clear down to arterioles and venules, we can sympathize with bioprinting researchers who have reservations about printing everything. Their hope is to encourage biology to build as much as possible of their tissue or organ of choice once they've provided the elusive starting place to initiate vascularization.

Blood Vessels Aren't Just Pipes

There have been a number of successful efforts to print constructs with channels that can be used to establish vasculature. It's not sufficient to simply have conduits running through a construct, though that's a start. For blood vessels to perform as they do in native tissues and organs, it's necessary to line these conduits with a specialized cell that nature has designated for the purpose of lining blood vessels. It's known as an endothelial cell, and the lining of blood vessels is composed of a single layer of endothelial cells.[†] Although only a monolayer thick, the endothelial cells have indispensable functionality: They control the passage of materials into and out of the bloodstream. The endothelial cells are the fundamental elements of the vascular system.[10] The exceedingly thin sheet of endothelial cells that line every blood vessel in our body is referred to as the endothelium.

Despite the fact that there are an estimated 620 billion endothelial cells in our body, there are few if any so-called molecular markers—that is, proteins or genes—that are both specifically and uniformly expressed in the endothelium.[11,12] By *uniformly* we mean the endothelium present in your lungs or liver or heart or kidney, for example, may not have common markers that identify it as the endothelium. It's as if 620 billion children were at Camp Hiawatha but, to their counselors' consternation, they weren't all wearing the regulation Camp Hiawatha T-shirt and cap.

The endothelium isn't all that's needed to create mature blood vessels. The vessels need to be surrounded by smooth muscle cells to give them elasticity. Smooth muscle is a type of muscle that contracts slowly—50 times slower than fast skeletal muscle—and automatically.[13] Since it's not dependent on nerve impulses to contract, it's sometimes referred to as involuntary muscle. Also, cells called fibroblasts are necessary to help with the blood vessels' mechanical integrity. Fibroblasts are the most common cells in connective tissue, the tissue that's spread all over our body and that surrounds other tissues and organs providing cohesion and support. From a blood vessel's perspective, the vessel needs to snake through spaces in the connective tissue. Connective tissue cells maintain conditions in these spaces that favor the exchange of nutrients and wastes between the tissues/organs and the blood vessels. So blood vessels and connective tissue cooperate with one another.

Bioprinting researchers have to factor fibroblasts into their constructs if they want to provide functional vasculature to the tissues they create. We said that fibroblasts are all over the place in connective tissue. Despite being so prevalent, fibroblasts are funny little cells to get a handle on. In 2008 a respected scientist interested in blood vessel development wrote that fibroblasts were originally described over 100 years ago and are still largely defined by their location "and what they are not—non-smooth muscle cells, non-endothelial cells, non-epithelial cells."[14] And others agree.[15] Yet fibroblasts are present in most tissues of the body. They produce components of the extracellular matrix, providing a physical support for other cells to become specialized and thus to perform their biological functions.

[†] Endothelial cells also line vessels of the lymphatic system, a collection of tissues and organs—for example, bone marrow and spleen—that carry white blood cells and fight infection.

To bioprint a vascular network that best mimics the blood vessels in our body we need endothelial cells, smooth muscle cells, and fibroblasts. Though as Gabor Forgacs has pointed out, even if smooth muscle cells and fibroblasts aren't in the bioprinted construct, they may be recruited from the surrounding tissue after implantation.[16]

Microchannel Networks Created with Fugitive Ink

Incorporating vasculature in bioprinted constructs was a huge hurdle for scientists for many years following the launch of the field. Finally a materials scientist made headway. She was working with non-biological matter and had not yet entered the bioprinting Sturm und Drang. In 2003, materials scientist Jennifer Lewis and her group, then at the University of Illinois at Urbana–Champaign, used 3D printing to make a three-dimensional network of microchannels. The group used an ink that would melt when it was heated. This substance is in the category of inks called fugitive inks because of their ephemeral nature.[17] Think of fugitive inks as placeholders. They are used to make sacrificial parts of a 3D printed (or bioprinted) construct that can be easily removed when their job is done. Their function is to occupy regions of the construct on a temporary basis, after which they're ousted. For example, in the Lewis work, fugitive inks are printed as cylindrical channels with diameters in the range of 10–300 microns. When the fugitive ink is removed, what you're left with are three-dimensional open channels. Lewis was well aware of how her fugitive ink approach might be applied to bioprinting. She wrote, "A 3D microvascular network provides an analog to the human circulatory system."[18,‡]

However, Lewis' initial microchannels were not within the confines of a hydrogel, the most common medium for bioprinted cell encapsulation. She addressed that in a subsequent study in which she and her colleagues printed biomimetic microvascular networks embedded within a hydrogel.[19]

So Lewis' team had an approach to potentially incorporate vasculature in bioconstructs in their back pocket. When they dived headfirst into bioprinting in 2013, her group tackled another related issue. In Figure 1.1, we displayed several examples of bioprinting techniques with a single print head or nozzle. We used single nozzles for ease of illustration. Even in the early years, c. 2000, groups (such as Forgacs' group) used four nozzle printers.[20] But if we contemplate bioprinting vasculature comparable to the hierarchy of the human circulatory system, a very large number of nozzles might be required.

---

‡ What Lewis refers to as *microvascular* we've called *microchannel*.

Multi-nozzle Print Heads—It's About Time

The term *high-throughput* crops up in all sorts of places yet the basic idea is the same: to speed things up, to get more samples tested or more genes sequenced in a shorter time. In 2013, Lewis' group, now relocated at the Wyss Institute for Biologically Inspired Engineering at Harvard University, created a print head with multi-nozzle arrays for high-throughput printing of single and multiple materials over large areas.[21] In this case a large area means about one square meter or around 11 square feet. This is roughly the area covered by an open large rain umbrella. Contrast this with the current bioprinting standards of approximately one square centimeter, about the area of a dime. Although her result was 3D printed rather than bioprinted, she explicitly calls out materials with embedded microvascularization as a potential application.

In fact, Lewis' team designed their multi-nozzle print heads to mimic fluid transport within biological vasculature that consists of hierarchically branching channels. In the case of Lewis' design, each print head distributes ink from a single microchannel into repeatedly bifurcated branches. This deposits ink simultaneously from multiple nozzles during printing.

Figure 4.3 shows the construction steps. In panel (a) Lewis' team machined the hierarchical network of microchannels into a block of acrylic, also known by its trade names such as Plexiglas or Lucite. In panel (b) they chemically bonded the patterned material to a flat acrylic substrate. As shown in panel (c), this forms a single block (with a threaded reservoir inlet). Panel (d) is an optical micrograph of a 64-nozzle array composed of square nozzles (with dimensions of 200 microns × 200 microns) and with a 400-micron center-to-center spacing. Figure 4.3 (e) is a schematic showing a side view of the six-generation 64-nozzle array. In this drawing, the bottommost 64 nozzles are a representation of what you see sticking out of the acrylic block on the far right of panel (c).

Before moving to another topic, we note that if you aspire to bioprint large tissues for transplantation you have to work fast because the viability of the cells you're printing is at risk. This is a compelling reason to strive for high-throughput.

Creating Vasculature in Bioprinted Tissues

In 2014 Lewis' group used a one-of-a-kind bioprinter with four print heads to create vascularized cell-laden constructs.[22] They designed both fugitive inks and cell-laden hydrogel inks. The particular fugitive ink they created for this work was removed by modest cooling rather than heating.

Note how unusual that is: The ink actually liquefies when the temperature is lowered, contrary to most of our everyday experience. Water solidifies when the temperature is lowered, and when we want to liquefy butter we heat it.

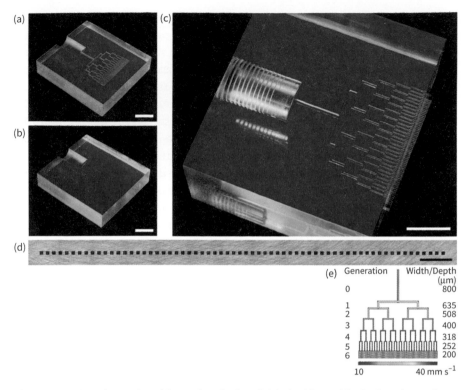

Figure 4.3 Microchannel multi-nozzle print head. (a) The hierarchical microchannel network milled into an acrylic block. (b) The patterned material is chemically bonded to a flat acrylic substrate. (c) An embedded microchannel structure within a monolithic block (with a threaded reservoir inlet). Scale bars = 10 mm. (d). Optical micrograph of 64-nozzle array composed of square nozzles (200 microns × 200 microns) with a 400-micron center-to-center spacing. Scale bar = 2 mm. (e) Side view of the six–generation 64-nozzle array.

Source: Reproduced by permission from Christopher J. Hansen et al., "High-Throughput Printing via Microvascular Multinozzle Arrays," *Advanced Materials* 25, no. 1 (October 26, 2012): 96–102, https://doi.org/10.1002/adma.201203321.

Biomedical engineer Kimberly Homan is a team member in Lewis' lab. Homan said, "The reason this ink was so critical is because other fugitive materials require high temperatures or other crazy treatments for removal that would kill the cells nearby." The Lewis ink requires only 10 minutes in the fridge.[23]

The group also formulated cell-laden gelatin–methacrylamide (gelMA) inks (the same substance used in the work of Hutmacher and Malda in Chapter 3) containing fluorescently labeled protein. The technique of using fluorescently labeled protein is widely used in bioprinting. Let's encapsulate the fluorescent protein story. In the 1960s a scientist analyzing bioluminescence—the ability of living creatures to generate light—discovered that a jellyfish that glowed green did so because of the existence of a protein called green fluorescent protein.[24] About 25 years passed before another scientist realized that such a protein could be valuable as a tracer molecule for other proteins—or viruses and other items of interest to biologists. He suggested attaching a gene for the fluorescent protein into the section of DNA that would code for the protein of interest, so now the cell would produce a protein with the fluorescing protein attached to it. That would make the protein of interest easy to detect. Subsequently, a third scientist succeeded in inserting the green fluorescent protein into living organisms (both bacteria and small worms) and successfully saw the green glow. In the worms, for example, gene expression of the green fluorescent protein caused certain neurons to light up as expected. Although green fluorescent proteins were the first to be deployed, biochemists soon developed a large color palette of fluorescent proteins.

It's also a good time to give you more familiarity with gelMA, building on what we already know. GelMA forms from the protein collagen after it's heated and/or chemically altered so the collagen no longer has the intricate folded structure it naturally takes on, although the ordering of its molecular subunits isn't changed. This is said to *denature* the collagen. In the final step to obtain gelMA, you add some chemical groups to the denatured collagen that allow the collagen strands in the gelMA to be cross-linked. Shining ultraviolet light on the gelMA construct after it's bioprinted produces the cross-linking and gives the material more stability. Living cells can readily adhere, remodel—that is, optimize their structure—and migrate through the gelMA.

Resuming the Lewis group study, they cross-linked the gelMA to provide greater stability to their construct. To remove the sacrificial fugitive ink they lowered the temperature and placed empty syringe tips into the inlet and outlet microchannels and aspirated the now-liquefied fugitive ink from the network of channels under a light vacuum. This left a system of channels devoid of the fugitive ink that enabled creation of those channels. Now it was necessary to promote endothelialization—that is, to line the channels with the single layer of endothelial cells that every blood vessel must have to be a certifiable blood vessel.

To create the endothelium, the team employed human umbilical vein endothelial cells that expressed red fluorescent protein—the better to see where the endothelial cells were localized after printing. The team also made use of green fluorescent protein to illuminate human neonatal dermal fibroblasts. This was the other cell-laden ink they created to provide fibroblasts in their construct for the mechanical integrity they supplied.

The Lewis group highlighted an important point concerning cell viability. We've spoken before about concerns that shear forces on cells exiting the print head nozzle can harm those cells, and great care is taken in nozzle design to address this. But Lewis called attention to the so-called total build time to fabricate the tissue construct, suggesting that there is a maximum time over which cell-laden inks can be stored in the ink reservoir of the bioprinter before being damaged. Their intention going forward is to examine this issue carefully. And in this context, they pointed out that using the high-throughput 64-multi-nozzle array they've already designed would vastly reduce the build time.

In sum, Lewis created a new bioprinting approach for creating vascularized tissue constructs containing multiple cell types, and it's an approach that can readily be scaled up in size. However, their inability to directly supply blood to these tissues limited the tissue thickness to only one to two millimeters and limited the times the tissues were kept alive in the lab to less than 14 days.

They wished to do better. They were on the hunt to vascularize thick tissue.

The Thickness Dilemma of Vascularized Tissue

Bioprinting overlaps with the larger field known as tissue engineering that seeks to develop biological substitutes to restore, maintain, or improve tissue function. For decades, those in tissue engineering have battled the challenge of bioengineering thick tissues—the vascularization problem. In fact, this has been called "the thick tissue dilemma." [25] The thickness dilemma is: Without perfused vascular channels, a three-dimensional engineered tissue with a dense population of cells will quickly develop a necrotic core. Recall from Chapter 1 that a necrotic core means large numbers of dead cells in the interior of the printed construct. Cells that die by necrosis swell and burst. And this releases their contents and damages nearby cells. [26]

Cells in the body or in bioprinted tissue must be in the region of 100–200 microns of a blood vessel; diffusion of oxygen and nutrients from blood vessels to cells is limited to this distance. [§] Comparatively large organisms such as humans have transport systems that minimize the distance that essential gases and liquids have to diffuse to reach their cells: Our vasculature performs this function.

Responding to the Thickness Dilemma in Bioprinted Tissue

In further studies, Lewis' group provided one response to the thickness dilemma of bioprinted vascularized tissue. [27] They reported a method for bioprinting cell-laden, vascularized tissues in excess of 1 centimeter thick that can be kept alive for more than six weeks. That's significant because 1 centimeter is roughly 50 times the nominal diffusion limit. So their vascularization design can bring oxygen and nutrients to cells in thick tissues, mimetic of what our natural vasculature accomplishes. Lewis team member David Kolesky said in an interview, "That's about 10 times thicker than

---

[§] This diffusion limit is only a rule of thumb and doesn't apply equally well to all tissue types. The diffusion limit is particularly important for tissues such as solid organs. But, for example, articular knee cartilage can be much thicker than the diffusion limit would dictate because the density of cells in cartilage is low, as is their metabolic activity.

what people have been doing—and we're not sure what the limit is, or if there is a limit."[28] Their study built on the work we looked at earlier on creating vasculature in bioprinted tissues and took it to a decidedly new level.

The bioprinting study we'll be looking at is a major step in human tissue generation. The reason it was so successful is the method used to cross-link the constituents of the extracellular matrix. Cross-linking the matrix is essential for providing mechanical and thermal stability for long-term perfusion. But the oft-used ultraviolet light method for cross-linking won't work for thick tissues because ultraviolet light has a low penetration depth in tissue. Instead, Lewis used two enzymes to achieve cross-linking.

What they did was three-dimensional bioprinting coupled with casting.** The material they cast was an extracellular matrix based on a gelatin and fibrinogen blend. Lewis' team cross-linked the gelatin/fibrinogen matrix by a dual-enzymatic strategy that allowed them to fabricate arbitrarily thick tissues. The group designed biological, fugitive, and elastomeric inks for multimaterial bioprinting. The elastomeric ink is silicone, a polymer in the class known as elastomers—a portmanteau made from the words *elastic* and *polymer*. It's a rubbery material capable of recovering its original shape after being stretched a good deal. Lewis used the silicone ink to 3D print a so-called perfusion chip on a glass substrate, and it's within the confines of this perfusion chip that both bioprinting and casting take place.

They turned to fugitive ink to create channels for perfusion—that is, for circulating fluid—and lined these channels with endothelial cells to give them the potential to provide oxygen and other nutrients to the tissues surrounding the channels. Their extracellular matrix contained fibroblasts and adult human stem cells. Fibroblasts serve as model cells that surround the stem cells and vascular network. These model cells could be replaced with either support cells, such as smooth muscle cells, or tissue-specific cells—liver cells, for example—in future experiments. They chose to induce the stem cells to differentiate into bone tissue as a demonstration that they could direct stem cell-laden, vascularized tissues down a specific lineage to produce a desired human tissue (in this case, bone).

Since this experiment was a big breakthrough, we're going to dissect it further with the aid of images showing the steps in the experiment.

The Splendid Scheme

Figure 4.4 shows the first step. Lewis' group used their custom-designed multimaterial bioprinter equipped with four independently controlled print heads to good advantage. They sequentially printed the multiple inks necessary to build the silicone perfusion chip housing and two sets of channels. The group printed layers of vascular

---

** The casting process consists of molten material poured or forced into a mold and allowed to harden.

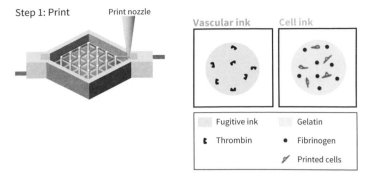

Figure 4.4 Creating thick vascularized tissue. Step 1: Print. In the left column, within a printed silicone perfusion layer housing, Lewis' group printed layers of vascular channels (blue) filled with a fugitive ink and layers of a second set of channels (yellow) containing living stem cells. In the middle column, looking down a vascular ink channel, we see fugitive ink and the enzyme thrombin. In the right column, looking down a cell-laden ink channel, we see gelatin, the enzyme fibrinogen, and printed cells.

Source: Reproduced with permission from David B. Kolesky et al., "Three-Dimensional Bioprinting of Thick Vascularized Tissues," *Proceedings of the National Academy of Sciences of the USA* 113, no. 12 (March 7, 2016): 3179–84, https://doi.org/10.1073/pnas.1521342113.

channels filled with fugitive ink and layers of a second set of channels containing living stem cells. The layered channels are in an interdigitated network like locked fingers.[29] What we see in the leftmost column is the perfusion chip housing. The blue print nozzle indicates the building of the vascular channels; the nozzle that built the layers of cell channels isn't shown.

In the second column of Figure 4.4 we're looking down a vascular channel-to-be that still contains fugitive ink. Lewis labels this column as *vascular ink* because that's where they will construct the vascular channels, including the incorporation of an endothelial lining. In addition to its sacrificial component—that is, the fugitive ink itself—this vascular ink contains the enzyme thrombin, as you can see by the component labeling beneath the column.

In the third column we're peering down a cell-laden ink channel containing gelatin, fibrinogen, and cells. Fibrinogen is the protein that thrombin will enzymatically convert to fibrin in just a moment. Note the cellular component is labeled below the cartoons as *printed cells*.

Up next: the fireworks. A quick glance at Figure 4.5 is like glimpsing a person and finding their appearance strange. You manage a furtive peek—like Seinfeld would do in his sitcom. Then you recognize the absence of any wrinkles on the person's forehead; it's really smooth. Another stealthy Seinfeldesque squint and you notice that the hands are veiny. Botox! We will have a similar moment of epiphany with Figure 4.5.

First, notice the leftmost column in Figure 4.5, telling us that they've cast an extracellular matrix. Observe that the inner walls are now a beige color, indicating

Step 2: Cast

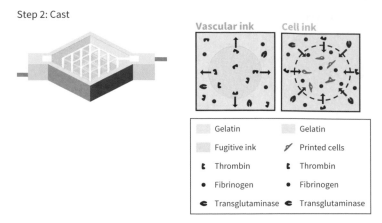

Gelatin	Gelatin
Fugitive ink	Printed cells
Thrombin	Thrombin
Fibrinogen	Fibrinogen
Transglutaminase	Transglutaminase

Figure 4.5 Creating thick vascularized tissue. Step 2: Cast. In the left column, Lewis' group cast an extracellular matrix encompassing everything within the perfusion chip housing but outside of the dual network of channels. The extracellular matrix is made of gelatin, fibrinogen, cells, the enzyme thrombin, and the enzyme transglutaminase. In the middle column, thrombin diffuses from the vascular ink into the extracellular matrix, where it will convert the fibrinogen to fibrin. In the right column, the enzyme thrombin will induce fibrinogen to polymerize into fibrin in both the extracellular matrix and, via diffusion, the printed cell ink. The enzyme transglutaminase diffuses from the matrix and cross-links the gelatin and fibrin. All enzymes diffuse from regions of high concentration to regions of low concentration. Enzyme conversions are implied here, and their outcome is made explicit in Figure 4.6.

Source: Reproduced with permission from David B. Kolesky et al., "Three-Dimensional Bioprinting of Thick Vascularized Tissues," *Proceedings of the National Academy of Sciences of the USA* 113, no. 12 (March 7, 2016): 3179–84, https://doi.org/10.1073/pnas.1521342113.

that the extracellular matrix (that contains gelatin, fibrinogen, cells, the enzyme thrombin, and the enzyme transglutaminase) encompasses everything outside of the dual network of channels. But what's up elsewhere with the cornucopia of odd ornamentation—the dashed ring of lines, and arrows pointing inward and pointing outward like a motorist's traffic circle nightmare?

As shown in the middle column, after casting, thrombin diffuses from the vascular ink into the extracellular matrix (moving from a region of high concentration to a region of low concentration). Once in the extracellular matrix, the thrombin converts the fibrinogen to fibrin. (The enzymatic conversions are implied in this figure and their outcome is shown explicitly in the figure to follow.) In the right column, transglutaminase diffuses from the molten casting matrix into the printed cell ink and slowly cross-links the gelatin and fibrin. (This cross-linking occurs within the cast matrix, too.) Also in the right column, thrombin induces fibrinogen cleavage and rapid polymerization into fibrin in both the cast extracellular matrix and, through diffusion, in the printed cell ink. Let's expand on all the enzymatic intrigue.

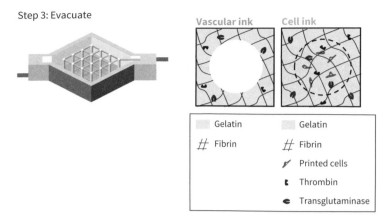

Figure 4.6 Creating thick vascularized tissue. Step 3: Evacuate. In this figure we see that they've cooled the whole construct causing the fugitive ink to liquefy. Then they evacuated the vascular channels with syringes under a slight vacuum—hence the white circle in the middle column.

Source: Reproduced with permission from David B. Kolesky et al., "Three-Dimensional Bioprinting of Thick Vascularized Tissues," *Proceedings of the National Academy of Sciences of the USA* 113, no. 12 (March 7, 2016): 3179–84, https://doi.org/10.1073/pnas.1521342113.

Thrombin is the enzyme in blood plasma that's at the center of blood clotting; it catalyzes the conversion of fibrinogen to fibrin, a fact we didn't call out explicitly when we first introduced fibrinogen. The second enzyme at play is transglutaminase. The name is likely unfamiliar. Even so, if you stand in front of the meat cases in your local supermarket you'll see transglutaminase at work. It's known generically as meat glue, and it's used to connect pieces of meat together to make more uniform, attractive servings. In 2005 a research team in Taiwan reported that transglutaminase was an enzyme capable of cross-linking collagen-based matrices.[30] Lewis was able to leverage their result.

So the moment of epiphany has arrived: The outward pointing and inward pointing arrows in the second and third columns are meant to convey movement of thrombin and transglutaminase from regions of high concentration to regions of low concentration. The dashed ring of lines in the third column reminds us that the cell-laden ink is contained in filament channels; movement between the cell-laden filament and the surrounding extracellular matrix occurs by diffusion across the boundary demarcated by the dashed lines. You can appreciate how well this static diagram conveys the several dynamic processes that are occurring simultaneously!

In Figure 4.6 they've cooled the whole construct, causing the fugitive ink to liquefy. At that point they evacuated the vascular channels with syringes under a slight vacuum—hence the white circle in the middle column.

The first column of Figure 4.7 shows the chip as an external pump supplies perfusion. In the second column, they line the vascular network with human umbilical

Step 4: Perfuse

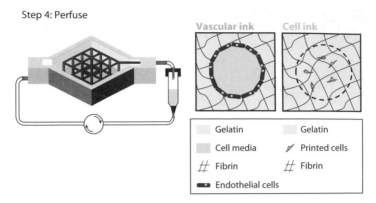

Figure 4.7 Creating thick vascularized tissue. Step 4: Perfuse. The first column shows the chip as an external pump supplies perfusion. In the second column, Lewis' group lined the vascular ink channel with endothelial cells by injecting the cells via a pipette. Afterward, they incubated the chip at body temperature to enable endothelial cell adhesion to the inner walls of the open channels.

Source: Reproduced with permission from David B. Kolesky et al., "Three-Dimensional Bioprinting of Thick Vascularized Tissues," *Proceedings of the National Academy of Sciences of the USA* 113, no. 12 (March 7, 2016): 3179–84, https://doi.org/10.1073/pnas.1521342113.

vein endothelial cells by injecting the cells via a pipette. Afterward they incubate the chip at body temperature to enable endothelial cell adhesion to the inner walls of the open channels.

Perfused Vasculature Brings the Gift of Life

Lewis and colleagues wanted to demonstrate the stability of the vasculature that's established using their scheme during long-term perfusion. Figure 4.8 cartoons a single vascular channel lined with human umbilical vein endothelial cells in a fibroblast-laden extracellular matrix. They housed this construct in a simple perfusion chip and monitored it microscopically for 45 days.

Green fluorescent protein and red fluorescent protein bring a Christmassy look to the microscope images up next. Figure 4.9 (a) shows the single vascular channel lined with human umbilical vein endothelial cells (label HUVEC) expressing red fluorescent protein. The human neonatal dermal fibroblasts (label HNDFs) are expressing green fluorescent protein. These microscope images are taken top-down and, to the far right, in cross-section. The label *lumen* shows the central cavity of the vascular tube; *lumen* is from a Latin word meaning an *opening*.

Figure 4.9 (b) shows an important piece of information. The graph labels the fluorescence intensity in arbitrary units (the *a.u.* on the vertical axis) of the green fluorescent-tagged fibroblasts. What this is telling us is that the fibroblasts are alive

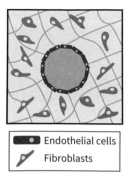

Endothelial cells

Fibroblasts

Figure 4.8 Cross-sectional view of a single vascular channel lined with endothelial cells and surrounded by a fibroblast-laden extracellular matrix. The construct is housed within a perfusable chip.

Source: Reproduced with permission from David B. Kolesky et al., "Three-Dimensional Bioprinting of Thick Vascularized Tissues," *Proceedings of the National Academy of Sciences of the USA* 113, no. 12 (March 7, 2016): 3179–84, https://doi.org/10.1073/pnas.1521342113.

within about one millimeter to either side of the vascular channel (within the yellow bars). Further away from the channel the fluorescence drops off considerably, showing that the fibroblasts are no longer viable. This speaks to the very raison d'être of the experiment: "Perfusable vasculature is critical to support living tissues thicker than 1 millimeter over long time periods."[31]

To reinforce the content of Figure 4.9 (b), look at Figure 4.9 (c) taken from the on-line Supporting Information. Lewis performed a similar experiment to construct four vascular channels. Again, fibroblasts are expressing green fluorescent protein (label GFP HNDF), and four perfusion channels lined with endothelial cells are expressing red fluorescent protein (label RFP HUVECs). The cell density as measured by green fluorescent protein levels is highest near the channels and decreases in the central region far from the channels. After 14 days, viability measurements showed that cells within one millimeter of vascular channels retained high viability and formed a dense, interconnected network. But cells beyond this region showed marked loss of viability and change in morphology.

Response to the Thickness Dilemma, Take Two

Anthony Atala's group at the Wake Forest Institute for Regenerative Medicine also incorporated vasculature in thick tissue. They used the same fugitive ink as Lewis' group, although they positioned it in another way.

The group placed the fugitive ink on the outside as an external sacrificial scaffold. That is, they printed the fugitive ink as an outer casing to help the inner tissue construct retain its shape throughout the printing process. Once the printing was complete, they immediately washed out the shell of fugitive ink.

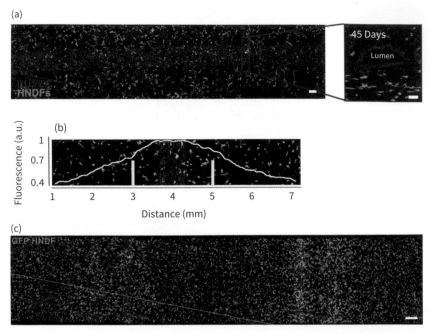

Figure 4.9 (a) Endothelial cell-lined vascular network (red) supporting fibroblast-laden extracellular matrix (green) shown by top-down (left) and cross-sectional (right) microscopy after 45 days. Scale bars = 100 microns. The label "Lumen" denotes the opening of the vascular construct. (b) Distribution of human neonatal dermal fibroblasts within the extracellular matrix shown by fluorescent intensity in arbitrary units (a.u.) as a function of distance from the vasculature. (c) Four perfused channels within fibroblast-laden gelatin–fibrin hydrogel showing cell density (as measured by fluorescence levels) is highest near the channels and decreases in the central region far from the channels. Scale bar = 500 microns.

Source: Reproduced by permission from David B. Kolesky et al., "Three-Dimensional Bioprinting of Thick Vascularized Tissues," *Proceedings of the National Academy of Sciences of the USA* 113, no. 12 (March 7, 2016): 3179–84, https://doi.org/10.1073/pnas.1521342113.

In the previous chapter we described how Atala's team used their integrated tissue and organ printer to create bioprinted ears. We'll now elaborate on that to show how they included vasculature.[32]

Atala and colleagues made two sets of bioconstructs. In one they designed a lattice of microchannels made from their polymer support material. This was a simplified mimic of your blood vessels but with the same goal in mind: to provide the bioprinted tissues with nutrients and oxygen. In the control set of printed constructs there were no microchannels. Both sets were placed in a medium conducive to the growth and proliferation of elastic cartilage cells for five weeks before further evaluation.

Atala's team reported the constructs with microchannels showed enhanced tissue formation; they produced new viable cartilaginous extracellular matrix throughout the entire ear constructs. The constructs without microchannels showed tissue formation limited to their periphery, likely owing to the lack of nutrients and oxygen reaching the cells deeper in the constructs. The cells in the newly formed tissues demonstrated similar structural characteristics to those in native ear cartilage.

To determine whether their bioprinted ear constructs would mature inside the body, Atala implanted them under the skin of genetically manipulated laboratory mice that have an inhibited immune system so they will not reject foreign tissues such as the bioprinted human ear constructs.

The bioprinted tissues were retrieved after one-month and two-month intervals. The shape of the ears was well maintained with lots of new cartilage formation. Microscopic inspection revealed that the tissues had created substantial new extracellular matrix proteins. There was also a suggestion of vascularization in the outer region of the tissues after implantation. The group inferred this from the proteins they identified—proteins that are expressed by the endothelial cells that line blood vessels.

To enlarge on this, remember our discussion of gene expression/protein creation in Chapter 2: DNA plus RNA in the cells that line blood vessels produce proteins that are characteristic of these cells. The proteins Atala found are a tip-off that blood vessel formation occurred. (The inner regions of Atala's constructs were avascular as in native cartilage. However, the outer portion of our external ears is covered with skin and supports vasculature.)

Atala's group used slightly different language to describe the benefits of the microchannels in bringing oxygen and nutrients to distant cells: They spoke of extending the diffusion limit.[33] This is an alternative way of saying that their microchannels provided sufficient vasculature in proximity to the cells to provide them with nourishment. They also pointed out that the ear cartilage constructs showed no sign of necrosis after five weeks in the lab prior to implantation in the mice. Necrosis is death of an area of tissue, but it's a form of premature tissue death rather than natural death or wearing out of the tissue. Necrosis is distinct from so-called programmed cell death that is a natural occurrence in cells.

Finally, recall that when Atala rolled out his own one-of-a-kind bioprinter he demonstrated its versatility by also bioprinting bone and skeletal muscle. The team reported that the bioprinted bone and muscle constructs with microchannels that were implanted in mice also showed evidence of vascularization without necrosis.

We've seen that bioprinting has made progress in incorporating a vascular system into printed tissue. Nerve fibers as well as blood vessels are distributed throughout the entirety of the human body, and innervation is essential for normal tissue function. As an example, human bone is both vascularized and innervated, and blood vessels closely interact with nerve fibers supporting bone development and fracture healing.[34]

At present, bioprinting's notable contributions with respect to nerve tissue have involved techniques for bridging gaps in severed nerves. This capability will be invaluable for any future progress in transplanting bioprinted, innervated and vascularized tissue into humans. Neuronal ingrowth and integration into a recipient's nervous system is essential to gain ultimate functionality of any future bioprinted construct. By *neuronal ingrowth* we mean that the construct's nerve pathways must connect up to the host's nerve pathways following transplantation. We might look at this as the ultimate instance of ruptured nerve repair: Nerves don't get any more detached than that. And the developing ability to repair severed nerves with bioprinting will prove invaluable.

Bioprinting has been used in animal models to repair nerve damage. Such nerve injuries often occur in regions of bifurcating nerve corridors—nerve corridors dividing into two branches. This results in tricky nerve gap geometries, and conventional methods to regenerate these damaged nerves are challenging. Bioprinting holds promise for personalized nerve regeneration in humans that is designed to match the native nerve tissue.[35]

There are intrinsic advantages to bioprinting in repairing severed nerves: highly automated fabrication, precision deposition of separate cell types and biomaterials at predefined positions. Supplementing these advantages are treasures from other endeavors. One such treasure we'll describe is the birth of a radically new concept in biology: growth factors—proteins that stimulate the growth of specific tissues. It all began with nerve growth factor.

That tale is an example of how a skilled observer can create a concept out of apparent chaos. And how that indomitable scientist triumphed over fascism and war, hiding underground and clutching her precious microscope while the bombs dropped overhead.

5
Innervation

The Body's Internet

Rita Levi-Montalcini was busy setting up a minuscule biological laboratory in the bedroom of her home in Turin, Italy. It was more like an artist's garret than a space to study the function and structure of the nervous system of chick embryos.[1] The year was 1940 and she was tingling with the desire she'd had since earliest childhood to "undertake a voyage of adventure to unknown lands."[2] Levi-Montalcini had graduated from medical school in 1936 (with a summa cum laude degree). She wasn't sure if she wanted to fully devote herself to medicine or pursue basic research in neurology at the same time. So she enrolled in a three-year specialization in neurology and psychiatry at the University of Turin.[3] Then in 1938, Benito Mussolini, fascist dictator of Italy, issued his infamous manifesto that abolished Italian citizenship for all Italian Jews. The anti-Semitic laws that followed forbade any Italian Jews from holding governmental and professional positions.[4] Levi-Montalcini couldn't take part in university activities so in December of 1939 she decided to practice medicine—secretively, since this was also forbidden to her. Even so, the obstacle of having to turn to Aryan doctors to have her prescriptions signed forced Levi-Montalcini to abandon her practice.[5]

So there she was in her austere bedroom laboratory tackling a research problem in the embryonic development of the nervous system. But by the end of 1942 life in Turin had become insupportable. British planes were systematically bombing the cities of northern Italy—Turin in particular because of its automotive industries. Almost every night the whine of sirens warning of the planes overhead forced her to carry her precious Zeiss microscope and equally precious embryonic samples to her basement vigil, despite the very real risk of being buried under the ruins of bombed buildings. She remained for hours while the air pulsated with the drone of warplanes and the detonating bombs further shattered the stillness of the night.[6]

It was too much. Like the majority of people in Turin, she and her family moved out of town. In a country cottage in the hills, an hour away from Turin, she relocated her laboratory on a small table in the corner of a room that was also the family dining area and sitting room and continued her experiments. One year after relocating to the cottage, the Nazi army invaded Italy. This new danger forced her family to flee to Florence, where they lived undercover until the end of the war.

Nerve Growth Factor

Bioprinting is chock full of experiments that use what are called growth factors, and our knowledge of their very existence all began with Rita Levi-Montalcini, who discovered nerve growth factor, the granddaddy of all growth factors. Growth factors are signaling molecules that control cellular responses through binding with transmembrane receptors on target cells.[7] Biologists knew that organs could communicate via substances that went into the bloodstream. Insulin is a prime example. Nerve growth factor showed there was another way for cells to communicate at short range.

Levi-Montalcini decided early on that her research challenge was to find out how nerves emerging from an embryo's developing spinal cord find their way to the budding limbs they will eventually innervate.[8] In 1940 she found accounts of work done years earlier by embryologist Viktor Hamburger, then at the University of Chicago, that captured her attention.[9] Hamburger had also been examining how embryonic nerve tissue differentiates into specialized types. He believed that the process of specialization of nerve cells depends on the destination of their main branching fiber.[10],* In one experiment he cut off an embryo's developing wing bud before elongated nerve fibers reached it in order to see how this would affect the growth and differentiation of the nerve cells meant to extend to that destination. What he found was that the long fibers withered and died, as did the cell bodies. He concluded that there was probably some *inductive* factor in the severed limb that could no longer transmit signals to the nerve cells.[†] As a result, the cells couldn't mature and differentiate nor could they extend fibers into the region where the wing bud had been.

Levi-Montalcini repeated Hamburger's experiments and came to an alternative conclusion. She believed that some unknown chemical made in the wing bud attracted the nerve cell fibers and encouraged their growth rather than making them differentiate. Without this unknown chemical, the nerve fibers and their cell bodies could not survive.[11]

The difference in their conclusions was highly significant. Levi-Montalcini surmised that nerve cell differentiation *did* occur despite the removal of the wing bud and that the cells soon died because they no longer received a sustaining trophic factor.[‡] Unlike Hamburger, she did not believe that the limb contributed to differentiation by providing an inductive factor. Rather, it produced a chemical that nourished nerve cells that were already specialized.[12]

Hamburger learned of her results and in 1947 invited her to visit his lab, now at Washington University in St. Louis, and to repeat and extend her experiments. This she did and became further convinced of her hypothesis of an unknown trophic

---

* Nerve cells have a cell body that includes the nucleus but, unlike most types of cells, nerve cells also have a major branching fiber that can extend far away from the main cell body.

† Induction is the ability of a cell or tissue to influence the fate of nearby cells or tissues by chemical signals during development.

‡ A trophic factor is a biomolecule that nourishes cells and supports their survival.

factor; Hamburger later came to agree with her interpretation. In subsequent exper-iments, Levi-Montalcini transplanted limb bud-sized pieces of certain cancerous tumors onto chick embryos to re-create the work of one of Hamburger's former stu-dents. She observed wild nerve growth—webs of sprouting nerve fibers so thick that she thought she might be hallucinating.[13] Additional experiments confirmed her be-lief that the unknown trophic factor was a *nerve-growth promoting agent* as she had begun to call it.[14] This factor not only made nerves grow, it also attracted them to itself. Later, she and biochemist co-worker Stanley Cohen bestowed the name *nerve growth factor* on the still-mysterious substance.[15]

Next came the Herculean task of figuring out what the growth-promoting agent was. She tried and tried to reproduce the nerve growth by injecting extracts from the tumors into whole chick embryos. Her every attempt met with failure. So she changed course and decided to grow the embryo nerve tissue along with pieces of the tumor tissue in vitro (outside a living organism). She'd learned how to grow embryonic nerve tissue in vitro during her medical school studies. But the Hamburger lab in St. Louis had no experience with in vitro culturing of cells. So Levi-Montalcini wrote to a co-worker from her days in Turin who had since moved to Rio de Janeiro in Brazil and set up an in vitro culture unit there.[16] In 1952 she famously traveled to her friend's lab at the University of Rio de Janeiro, in the company of two specially prepped labo-ratory white mice in a small cardboard box with holes that she nestled in her overcoat pocket.[17]

Her stay of four months in Brazil had its share of successes. Revealing the identity of the growth-promoting agent was not one of them. On returning to the University of Washington she met the recently hired Stanley Cohen, and the two formed a pro-ductive working partnership to pursue the identity and chemical nature of the growth factor. They worked together on the problem for the following six years and were able to isolate nerve growth factor and determine its chemical structure.§

Eventually, despite the relentless skepticism of the scientific community, Levi-Montalcini and Cohen convinced their critics and established the phenomenon of growth factors. In her pursuit of nerve growth factor, Levi-Montalcini had followed a chimera. With her unshakable resolve she turned a fabrication of her mind into a re-ality. "And as imagination bodies forth/The forms of things unknown, the poet's pen/ Turns them to shapes and gives to airy nothing/A local habitation and a name," quoth Shakespeare.[18]

Since the discovery of nerve growth factor, scientists have identified hundreds of other growth factors that supply signaling pathways for all manner of cells. Examples abound—growth factors for skin cells, for blood vessel development, for liver cells, for bone cells, to name a few. Levi-Montalcini won the Nobel Prize for Physiology or Medicine in 1986, awarded jointly to Stanley Cohen "for their discoveries of 'growth factors.'"[19]

---

§ Cohen subsequently moved to Vanderbilt University, where he found a second growth factor known as epidermal growth factor that he also purified and chemically analyzed.

In this chapter we'll see examples of growth factors in bioprinting, including nerve growth factor. One dramatic result is the use of two separate growth factors that express differently in sensory and motor nerves. Intelligent deployment of these growth factors enabled regeneration of a demanding peripheral nerve injury that contained bifurcating sensory and motor nerve pathways. As was true of the vascular system, we need to have a background of the basics of the human nervous system, so we now define terms such as peripheral, sensory, and motor nerves.

The Nervous System of the Human Body

Think of the nervous system as our body's internet. It is our means of communicating with the outside world.[**] We have two parts to our nervous system: The central nervous system consists of all the nerves in our brain and spinal cord, and the peripheral nervous system consists of all the nerves in diverse locations in our bodies. As was true of the circulatory system in the previous chapter, these divisions are no more than a convention, a convenient way for us to view the whole. There are three types of peripheral nerves in our bodies. With bioprinting studies in mind, we'll only mention two of these. They are motor nerves and sensory nerves.

Motor nerves send impulses from the brain and spinal cord to all muscles of the body allowing us to play at hopscotch, catch a Frisbee, or bend over to pick a four-leaf clover. Sensory nerves send messages in the other direction: from our sense receptors back to the spinal cord and the brain. Sensory nerves allow us to hear the vibrato of a cello, to feel a light summer breeze, or to know if our last step off a ladder has put us safely on the ground.[††,20] Nerve cells, also called neurons, are the cells at play.

Neurons are among the most unique cells in our bodies. Like some fantastic imaginary animal in a bestiary from the Middle Ages, a nerve cell defies expectations of what any self-respecting cell should look like. The nerve cell's body is ordinary but it's spiked with antennae, called dendrites, to receive signals, and it has a long projection called an axon for transmission. In humans, axons can be three feet long. Dendrites branch out into tree-like structures forming multiple connections (called synapses) with the axons of surrounding neurons. It's a marvel that axons and dendrites find their right partners and form synapses to become a functional network. When we rupture neurons—in a traffic accident or on a battlefield—a gap of over 3 centimeters (just over an inch) leaves the nerve unable to regenerate.

[**] The nervous system also controls the inner working of much of our body.
[††] We have sense receptors for vision, taste, smell, hearing, and balance that are in our eyes, mouth, nose, and ears (for both sound and balance). We have sense receptors in our skin and distributed throughout our body for senses like touch, temperature, and pain.

Bioprinting of a Cellular Nerve Guide

We first met Gabor Forgacs in the Preface. He was one of the people holding a spade and wearing a hard hat at the groundbreaking ceremonies to mark the start of bioprinting near the turn of the millennium. In 2009, Forgacs and his colleagues showed how to create vessels of tubular shape and hierarchical trees that combine tubes of distinct diameters.[21] The research built on his own efforts and those of two other bioprinting groundbreakers, Vladimir Mironov and Thomas Boland, as well as additional collaborators.[22,23,24,25]

They began by forming cell aggregates—clusters of cells containing thousands of individual cells—and placed them in precise positions with respect to one another.[26] Initially they placed them in a circle and waited until the cell aggregates fused and formed rings. After forming additional rings, the team placed the rings vertically close together and the rings fused to form tubes.

As their techniques evolved, they found other ways to bioprint the tubular structures they needed, and we'll show how they did this to form a nerve graft. First though, we provide a backdrop for the stage before the performance begins. When a peripheral nerve is damaged or severed, the axons furthest from the damaged site degenerate. In the best situation, the body's natural repair mechanism can re-innervate the damaged nerve. If that does not occur, surgery is the next option.

Surgically suturing the two ends together is sometimes possible though not always. When it's not possible, the quintessence for nerve repair is to take a nerve segment from another part of the injured patient and transplant it to the damaged site. But this presents some issues: matching the diameter and mechanical properties of the nerve, risking damage to the donor site, and the overall risk of multiple surgeries.[27] Also, grafts taken from another part of the injured patient (known as autografts) are typically derived from sensory nerves, with the result that when motor nerves are the ones that are injured their regeneration is less than optimal.

Apart from the actual surgery, there is the question of how to encourage the two ends of the severed nerve to functionally rejoin. The question is answered by a nerve guide to bridge the gap between the end of the axon that's still attached to the nerve cell body and the part of the axon that is not connected to the nerve cell body. This enables proper regrowth of axons and return of motor and sensory function.[28] The most successful nerve guides have been made of collagen.

A nerve guide should also be able to provide biomolecules such as nerve growth factors that can improve regeneration. Adding them to the nerve guide directly can accomplish this, just as you might put an antibiotic ointment over a cut on your finger before wrapping a gauze bandage around it. An alternative is to seed the nerve guide with the major cell type that supports nerve cells in the peripheral nervous system, cells that are called Schwann cells.

Figure 5.1 is a diagram of a typical peripheral nerve cell. Schwann cells wrap themselves around axons in the peripheral nervous system and in so doing form a myelin

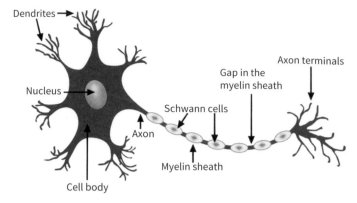

Figure 5.1 Structure of a typical neuron in the peripheral nervous system. In addition to a cell body containing a nucleus, a neuron has specialized cell parts called axons and dendrites. Axons (typically one axon per neuron) carry electrical signals away from the cell body and dendrites bring electrical signals to the cell body. Many neurons have an insulating sheath of a fatty substance called myelin wrapped around most of the length of their axon. In the peripheral nervous system, so-called Schwann cells form the myelin. (Neurons in the central nervous system derive their myelin from another source.)
Source: Reproduced by permission from the National Cancer Institute SEER Program.

sheath with gaps punctuating the entirety of the sheath.[29] Myelin is an insulating layer that enables the transmission of nerve signals at high velocity. We'll have more to say about myelination later.

Bioprinting has much to offer a damaged nervous system. Forgacs' group used a bioprinter to fabricate a cellular nerve guide and compared their construct to the two benchmarks: autograft and collagen tube. Figure 5.2 shows their strategy.[30] They used two cellular bioinks. One bioink contained mouse bone marrow stem cells, shown in red. They also employed a combination of bone marrow stem cells and Schwann cells in the ratio of 9:1, shown in green. Schwann cells are known to favor nerve regeneration. They employed the hydrogel agarose as a temporary support structure.

To make the cellular cylinders, Forgacs' team grew the required cells in cell culture media and spun this down in a centrifuge.[‡‡] They recovered the cell pellet and aspirated it (drew it in by suction) into 0.5-millimeter inner diameter capillary tubes. After incubation in cell media, the contents of the capillary tubes formed the cellular cylinders of the two bioinks shown in Figure 5.2.

---

‡‡ In a centrifuge, substances are spun rapidly (around the center of the machine) in tubes that are placed in angled positions with the bottom of the tubes furthest to the outside. During centrifugation, materials that have higher density move to the bottom of the tubes and materials that have lower density are displaced and move to the top of the tubes. Gravity can accomplish this but using so-called centrifugal force is quicker. (A perforated rotating drum in a washing machine that removes excess water from clothes uses centrifugal force.)

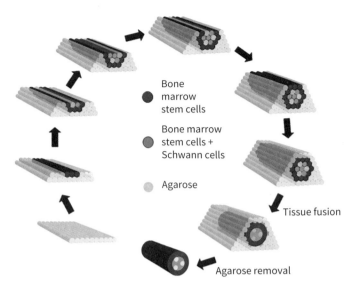

Figure 5.2 Building of a nerve graft by bioprinting. Scheme for layer-by-layer deposition of cellular (red and green) and agarose (beige) cylinders to build a three-channel nerve graft tube. Removal of the agarose rods provides three hollow channels to serve as conduits for nerve fiber bundles.

Source: Reproduced by permission from Françoise Marga et al., "Toward Engineering Functional Organ Modules by Additive Manufacturing," *Biofabrication* 4, no. 2 (March 12, 2012): 022001, https://doi.org/10.1088/1758-5082/4/2/022001.

Following the steps clockwise in the figure, you can see how the additive manufacturing of each layer in the bioprinting process formed the structure in the 3:00 position. Fusion of the bioink cylinders completed the structure—the 5:00 position—and finally the agarose rods were removed manually.[31] When they removed the agarose rods from within the fused construct, they were left with three hollow channels designed to provide conduits for nerve fiber bundles. After a maturation period of seven days, the multichannel construct had sufficient mechanical integrity to be implanted into rats with sciatic nerve lesions.

Implanted Nerve Guides

Forgacs performed tests on rats with a one-centimeter section of their sciatic nerve removed. He chose the sciatic nerve because it's what is known as a mixed-function nerve: It's made up of the axons of both sensory and motor nerves. The researchers compared three groups of animals. One had the simplest possible autograft: The nerve section that was removed was rotated 180 degrees and sutured back in. A second group had a commercial collagen nerve guide to promote nerve regeneration, while the third used the bioprinted nerve guide for regeneration.

For all three nerve guide groups, Forgacs and colleagues used electrical studies of muscle function to test both motor and sensory nerve response. To examine motor nerve function, they stimulated a large number of individual nerve fibers of a motor nerve simultaneously and recorded the aggregate electrical activity. To test sensory responses, they evaluated the increase in blood pressure after electrical stimulation of the same nerve—a cardiovascular response to activation of muscle sensory nerves. This may seem roundabout, but it's actually a standard procedure; I looked it up.[32] The details of their electrophysiological experiment would lead us too far astray. Suffice it to say that with respect to motor nerve function, they found that the autograft faired best and the bioprinted nerve guide seemed to score better than the collagen nerve guide. As to the sensory nerve function, they didn't find significant differences among the three groups.

This was pioneering work. We'll soon see a more recent bioprinting experiment to form a nerve guide. But Forgacs' group led the way. They stressed that their objective was to develop a proof-of-concept for a new type of nerve graft and that many adjustments would be needed to optimize the performance of such a graft. An improvement they suggested was the possibility of adding growth factors, and this is exactly what was done in the next nerve guide experiment we present.

A Bioprinted Nerve Guide Using Growth Factors as Biochemical Cues

If you mention the Brothers Grimm fairy tale of Hansel and Gretel, most people will think first of the breadcrumbs that Hansel scattered along their path as they went deep into the forest. The breadcrumbs failed to lead them back out of the woods because a great many birds had devoured them. The children's idea of marking their way was a good one, but breadcrumbs failed to get the job done. Bioprinting mavens know better. They use biochemical breadcrumbs—growth factors. Since Levi-Montalcini's initial discovery of the first nerve growth factor, researchers have found other neuronal growth factors. In the context of the nervous system, growth factors are often referred to as neurotrophic factors.

Michael McAlpine, trained as a chemist, is in the Department of Mechanical Engineering at the University of Minnesota. McAlpine was one of the leaders in a large collaborative effort involving a team assembled from five universities to design and test a nerve guide.[33] Xiaofeng Jia of the Department of Neurosurgery at the University of Maryland School of Medicine was a co-leader of the work. Like Forgacs' group, McAlpine's team addressed bifurcating mixed nerves: nerve bundles containing both sensory and motor nerves that form a fork or division in their geometry. But McAlpine's guides used both physical cues in the form of microgrooves and also path-specific biochemical cues in the form of spatiotemporal growth factor gradients—brainy biochemical breadcrumbs.

The biochemical breadcrumbs are neurotrophic factors that the team distributed in their nerve guide along the paths they wanted nerve regeneration to occur. Let's explain the word *spatiotemporal*. Spatiotemporal refers to both space and time. We'll look at time first. The researchers knew that the typical regeneration period of a peripheral nerve injury is at least three to four weeks. So they used standard mathematical drug release models to determine the rate of release of the neurotrophic proteins they distributed along the nerve guide. In doing this, they established a temporal gradient. That is, the rate of release of the proteins along the nerve guide wasn't constant in time. The release rate of the neurotrophic proteins was fastest at the left of Figure 5.3 and slowest at the right of the figure. This contributed a temporal component to the neurotrophic protein concentration gradient in Figure 5.3.

Now considering the space component of spatiotemporal, they distributed the quantity of neurotrophic factors in the form of a gradient; that is, they didn't simply put a constant amount of protein at every point along the intended nerve regeneration paths. Rather, they deposited a continuously enriching attractant. This established a spatial gradient. You may wonder why they used a gradient.

Granted, in these bioprinted nerve guides there is not a lot of competition by enticing signals to lure axons hither and yon. Still, this experiment is designed for broader applicability in the human body where there may indeed be such competition in the wooing of directional axon growth. It's as if you were taking a day hike in the backcountry just for the pleasure of the out-of-doors. Suppose you were on fairly flat terrain, no significant differences in steepness whichever way you went. But you did notice that changing your direction slightly to the right gave you better views of the horizon, or more plentiful wild flowers, or fewer gnats. If this desirable trend increased as you gently angled to the right, you would likely follow that more attractive route. So it is with our axons.

One more circumstance makes neurotrophic gradients a particularly felicitous choice in the McAlpine experiments. In the late 1970s, scientists confirmed a hunch that Levi-Montalcini long harbored: In embryonic tissues such as limbs that are destined to receive nerves, nerve growth factor is released by the target tissue and flows back to the spinal cord to guide nerve fibers from the cord to their ultimate destination. That is, the fibers grow from areas of low nerve growth factor concentration near the spinal cord toward areas where the concentration of growth factor is higher, in the limbs and other tissues on the body's periphery. So in the developing embryo, nerve growth factor directs growing nerve fibers along a concentration gradient. If we acknowledge science as a form of art, McAlpine's experiments are an example of art imitating life.[34,35]

McAlpine's team selected nerve growth factor as the supporting sensory path cue and glial cell line-derived neurotrophic factor (explained in a moment) as the motor path cue. They bioprinted the hydrogels laden with these two nerve growth enhancers along the bottoms of the bifurcated channels—the upper channel for the sensory nerves and the lower one for the motor nerves (Figure 5.3). The result was a hollow printed silicone conduit containing gradient patterns of hydrogel droplets

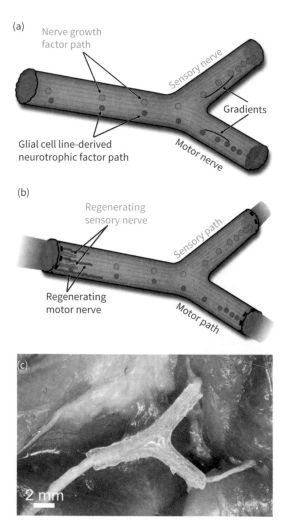

Figure 5.3 Functionalization of nerve pathways with path-specific biochemical gradients. (a) Schematic of the path-specific incorporation of gradient distributions of supporting biochemical cues—nerve growth factor and glial cell line-derived neurotrophic factor—in the sensory and motor paths, respectively. (b) Schematic of implanted nerve guide showing incipient nerve regeneration of sensory and motor nerve paths. (c) Photograph of an implanted 3D printed nerve guide prior to suturing.

Source: Reproduced by permission from Blake N. Johnson et al., "3D Printed Anatomical Nerve Regeneration Pathways," *Advanced Functional Materials* 25, no. 39 (September 18, 2015): 6205–17, https://doi.org/10.1002/adfm.201501760.

printed along the inner silicone wall, which is adjacent to regenerating nerve on the inside of the conduit.[§§] McAlpine's group chose silicone as the nerve guide material because earlier studies had shown it was a good choice for peripheral nerve guide applications.[36]

Glial cell line-derived neurotrophic factor is another protein that promotes survival of various types of neurons. For our historical enlightenment, a few words on this protein are in order. For decades, the protein discovered by Levi-Montalcini and which she named nerve growth factor reigned as the only agent known to spur neuronal development and repair. Years after her seminal work, a stream of discoveries appeared in scientific journals revealing many additional nerve-nurturing proteins.[37] One of the most heralded of these was glial cell line-derived neurotrophic factor.[38]

Glial cells are cells in the brain but they are not neurons. The word *glia* comes from the ancient Greek word for *glue* and for a long time it was supposed that glia were only supportive cells for the neurons.[***] This probably sounds like a story you've heard before—remember what we said about the extracellular matrix and how scientists long believed that it was nothing more than Styrofoam packing for cells. Glial cells have a variety of functions that influence neurotransmission and are approximately as numerous as neurons in the human brain, although for decades they were thought to vastly outnumber neurons.[39,40]

One concluding remark on the meaning of the word *supportive*, a semantic yet substantive distinction: In the central nervous system, oligodendrocytes are the glial subtype responsible for myelin production (we'll unpack the word oligodendrocyte in just a bit).[41] Oligodendrocytes are said to provide both trophic support and mechanical support to the wrapped neurons. When McAlpine chose nerve growth factor as the supporting sensory path cue, for example, it was to leverage the growth factor's trophic support—further evidence that glial cells have come a long way from being regarded as simply a mechanical prop. They're like a friend whose emotional sustenance you can count on as well as someone whose shoulder you can literally lean on.

The team selected gelMA hydrogel as the medium in which to encapsulate the two nerve path protein cues for bioprinting. This was an apt choice because gelMA has a good track record for controlled drug release and a high degree of neurocompatibility. Figure 5.3 (a) shows a schematic of the path-specific incorporation of spatial gradient distributions of the biochemical cues. As you can see, proceeding from left to right in the figure, the spacing between the green droplets becomes smaller and the same is true for the red droplets. Each droplet is a small amount of bioprinted hydrogel containing the relevant growth factor. So the concentration—the number of

---

[§§] In Figure 5.3, the red and green circles denote locations on the inside of the hollow silicone conduit.
[***] *Glial cells* and *glia* are used synonymously in the literature.

droplets per unit length—increases as you proceed to the right. This is what we meant earlier by a continuously enriching attractant.

What about the microgroove physical cues we alluded to? First, it's notable that the physical cues provided by McAlpine's team qualitatively resemble naturally occurring physical cues that surgeons have found in degraded nerve pathways. These have the exotic name the bands of Büngner. They consist of aligned Schwann cell strands that guide selectively regrowing axons.[42] Microgrooves have been shown to affect the orientations of the two main components of regenerating peripheral nerves—axons and Schwann cells.[43,44]

The team did control studies with and without microgrooves and verified that the physical cues were indeed vital. There was a high degree of alignment between nerve projections and the grooves. But in the absence of grooves the nerve projections simply formed a randomly distributed network—or, as Voltaire wrote, "wandered about at random, like dogs that have lost the scent."[45]

McAlpine's team used rats as their animal model. They examined the regeneration of bifurcated nerve injuries across a one-centimeter nerve gap and showed that the nerve guides produced a greater return of function of the regenerated nerve than in the control studies. They tested functionality by a detailed gait analysis. These were quantitative measurements of a recognized standard called the gait duty cycle. A *gait* refers to a particular sequence of lifting and placing the feet during legged locomotion—walking or running, for example. Each repetition of the sequence is called a *gait cycle*.

The gait duty cycle is what you get when you divide the stand time by the total amount of time the limb is either standing or swinging. Stand time is the duration of time that the rat's paw is in contact with the glass, and swing time is the duration of time that the paw is not in contact with the glass.[46]

For a walking human the duty cycle will be greater than 0.5 because there's an overlap time when both feet are on the ground. For a running human the duty factor is less than 0.5 since there's a time when both feet are in the air.

For gait analysis, McAlpine and company trained the rats—prior to any nerve injury—to cross a walkway for seven days. The walkway consisted of a glass plate and two plexiglass walls. Each rat walked freely and traversed from one side to the other of the glass plate. Researchers placed two infrared beams to detect the arrival of the rat at the finish line and to control the start and end of the data acquisition. The team performed these recordings in a darkened room but the interior of the crosswalk was illuminated from within. Contact of the rats' paws with the glass plate caused light to be reflected downward, and the illuminated area was recorded with a camera below the glass plate and connected to a computer. The team performed this automated gait analysis 12 weeks after the nerve guide was surgically implanted and compared the rats' gait to what it had been before the nerve injury.

The outcome was encouraging: There was significant improvement in the functional return of the limbs treated with the nerve guides that used the biochemical gradients compared to the control group that lacked the biochemical gradients.

Bioprinting Takes Aim at the Central Nervous System—Spinal Cord Tissue

Up to now we've been looking at bioprinting's potential value for insults to the peripheral nervous system. Spinal cord injuries in the central nervous system are another matter. Several groups have sought to rebuild functional axon connections across areas of central nervous system tissue damage. For example, McAlpine formed another team including neurosurgeon and Ph.D. Ann M. Parr from the Department of Neurosurgery at the University of Minnesota.[47]

As co-leader, Parr said,

> There's a perception that people with spinal cord injuries will only be happy if they can walk again. In reality, most want simple things like bladder control or to be able to stop uncontrollable movement of their legs. These simple improvements in function could greatly improve their lives.[48]

Regardless of the type of injury—sports-related, falls, violence, automobile accidents—a physical insult disrupts the highly organized cell architecture of the spinal cord, damages numerous motor and sensory pathways, and causes cell death at the site of the injury.[49]

Teaming with physicians like Parr was one of the virtues of being at the University of Minnesota. When discussing his bioprinting work with McAlpine, I learned something I didn't know from the literature.[50] Life's vicissitudes had brought him to the University of Minnesota. His previous career had been almost pure Ivy League: undergraduate chemistry major at Brown University, chemistry graduate student at Harvard, and after a chemistry post-doc stint at California Institute of Technology it was back to the Ivies as a faculty member in mechanical and aerospace engineering at Princeton. But life doesn't have anything comparable to the microgrooves that help align axons and so we are wont to wander from the straight and narrow path. Instead of staying at Princeton, he found himself a member of the mechanical engineering faculty at the University of Minnesota. That turned out to be terrific. His lab in the Engineering Department is right across the street from the University of Minnesota Medical School. By contrast, Princeton doesn't have a medical school, let alone one right across the street. The medical school has been a wonderful opportunity for collaborative work and given him an insight into what scientists from that background want to see done with bioprinting.

Joining forces with Parr, McAlpine was well positioned to confront the taxing obstacle of spinal cord injuries. Unlike peripheral nerve injuries, when nerves are injured in the spinal column they do not have the inherent capacity to regenerate themselves. So McAlpine and Parr not only had to provide a structural framework to guide new nerve growth but also had to provide the cells that would grow to become these new nerves. In addition, to be successful they required precise positioning of the cells within the construct. The researchers attempted this by bioprinting the cells as part of their construct rather than subsequently seeding the cells onto the construct.

For spinal cord repair, most techniques to date have used direct injection of cells at the site of the injury. Neural stem cells or neural progenitor cells injected at the site of a spinal cord contusion in rodents have shown some functional recovery in loco-motion. But injecting cells doesn't provide for a support or scaffolding of the injected cells and doesn't place the injected cells in positions relative to one another in a way that mimics the intricate cell architecture of the spinal cord.[51] Bioprinting might overcome these problems.

In the present study, the team used three types of cells: fibroblasts, spinal neuronal progenitor cells, and oligodendrocyte progenitor cells. Fibroblasts were part of a recipe that Parr had worked out for making oligodendrocyte progenitor cells from induced pluripotent stem cells.[52] Fibroblasts were also used as a reference cell type to check that their printing system was compatible with cell viability. Both human spinal neuronal progenitor cells and mouse oligodendrocyte progenitor cells were derived from induced pluripotent stem cells. The group exhaustively considered an assort-ment of extracellular matrix hydrogels in which to print their cells as well as a great many scaffold materials.

Their long-term goal was to find the correct combination of hydrogels, scaffolds, and print conditions to enable the spinal neuronal progenitor cells to become mature neurons. The aim was to have these neurons sprout axons that would grow along the open channels of the scaffold and to have the oligodendrocyte progenitor cells be-come mature oligodendrocytes that would provide myelination for the axons to opti-mize the axons' signal transmission.

These are huge ambitions. To understand what occurred, we begin by defining some new terms. After that we consider the trials the group faced in finding the best hydrogel and the optimal scaffold material. Subsequently, we turn to the placement of the cell-laden hydrogels in the scaffold and look at images of the team's results. This work draws on our accumulated knowledge—for example, progenitor cells, induced pluripotent stem cells, and fluorescent markers—and at the same time provides new subjects such as oligodendrocytes, alginate-based scaffolds, and an elaboration of the process of myelination of axons.

Nervous System Nomenclature

Now (finally!) we unpack the word oligodendrocyte. The prefix *oligo-*, from the ancient Greek for *small*, means *having few*. *Dendro-* comes from the Greek word *dendron*, meaning *tree*. The suffix *-cyte* is from the Greek for *vessel* and denotes a *mature cell*. When we put it all together and confer the appropriate context, we have a mature cell that has a few tree-like protrusions that are usually called processes.[†††] Oligodendrocytes are one type of glial cell found in the central nervous system.

[†††] Historically, scientists generated such compound terms when laboring over optical microscopes and trying their level best to find a way to stain the tissue sample at hand in order to delineate its fine structure. They often used words that tried to convey a pictorial sense of what they were seeing.

Recall that glial cells were once thought of as mere glue holding the neurons in place. We now know better. Oligodendrocytes are an enormously important type of glial cell (justifying their byzantine name) because, within the central nervous system, they provide the axons of neurons with the wrapping material called myelin. Myelin is a fatty material that acts as an electrical insulator and increases the velocity of the electrical signal from one nerve cell body to another in the central nervous system.

A typical simile is wrapping tape around a leaky water hose. One neuroscience book goes further and asks with breezy insouciance, "What's up with myelin?"[53] To address this question, first have a look back at Figure 2.3 (a) in Chapter 2, the protein pump that controls the sodium–potassium passage into and out of our cells. The sodium–potassium pump establishes a concentration gradient across the cell membrane.[‡‡‡] Though Shakespeare warns us of the tedium of a twice-told tale, we need to reiterate what we said in Chapter 2.[54] What happens in our nerve cells is that the axons deplete themselves of sodium ions and then use special sodium channels (*not* shown in Figure 2.3) to allow the ions to rush back in during a nerve impulse. (A nerve impulse is an electrical signal transmitted along an axon.) The sodium–potassium protein pump keeps the axons ready for the next nerve signal by establishing a gradient of sodium ions across the nerve cell membranes—more sodium ions outside the cell than inside the cell.[§§§] And this gives us pause for thought.

We know that the entire nerve cell, inclusive of cell body, dendrites, and axon, is encased by a cell membrane that's traversed by sodium–potassium protein pumps, sodium channel proteins, and numerous other protein goodies that enable it to connect to its environment. But if the nerve's axon is additionally encased by a myelin sheath you might imagine that the myelin would block the flow of ions.

That would be true if the myelin sheath was continuous. It turns out that there are gaps in the myelin wrapping. And all the sodium–potassium protein pumps and sodium protein channels are jammed together at the roughly regularly occurring gaps.[55] As a result, the controlled flow of sodium and potassium ions across the cell membrane *does* sustain the generation and propagation of nerve signals. And in a novel fashion. The nerve impulse hops from one gap in the myelin to the next gap all along the axon. This conduction mode is called saltatory conduction from the Latin word *saltare*, meaning *to leap as in a dance*.

Yes, the nerve impulse leaps like a lamb in springtime from gap to gap, dancing its way along the axon.[56,57] This nerve conduction is actually considerably faster than in nerve fibers that don't have a myelin sheath. For example, in unmyelinated nerve fibers that conduct pain or temperature, the conduction velocity along the axon— no hopping—is 0.5–2.0 meters/second, which is about as fast as you walk or jog. By

---

[‡‡‡] In fact, it establishes both a concentration gradient and an electrical gradient across the cell membrane.
[§§§] The sodium–potassium pump is only one of a complement of ion channels involved in re-establishing gradients.

contrast, conduction velocity in the most thoroughly myelinated axons—many myelin layers—is 80–120 meters/second, which is race car speed.[58]

Tidying up what feels like a loose end, Figure 5.1 (a) shows Schwann cells and myelin sheaths and, as such, the picture is of a neuron in the peripheral nervous system, not the central nervous system. Schwann cells produce myelin in the peripheral nervous system while oligodendrocytes produce myelin in the central nervous system. But both Schwann cells and oligodendrocytes belong to the glial family of cells, so myelin production remains a family-run business.

Optimizing Bioink for Induced Pluripotent Stem Cell-Derived Neural Progenitor Cells

McAlpine and Parr's lofty goals were accompanied by a strong dose of adversity. They had to struggle to find a hydrogel bioink formulation to sustain the neural progenitor cells they wished to bioprint. Initially they tried both gelMA and gelatin mixed with fibrin, both of which we've seen used in other bioprinting experiments. However, the cell viability of their neural progenitor cells in these common hydrogels was not good. What's more, the printed cells did not proliferate and they observed no axon propagation.

To better understand what was going wrong, the team used human fibroblasts as a reference cell type using the same printing conditions, cell densities, and hydrogel. They found that the human fibroblasts maintained good cell viability and the cells proliferated. This established that the printing process was not inherently incompatible with cell viability. They concluded that the poor results with neural progenitor cells might be due to a greater sensitivity to the printing procedure or a lack of biological components to encourage neural cell proliferation.

At this point they evaluated a widely used gelatinous protein mixture called Matrigel that contains growth factors and other proteins that mimic the basement membrane. The basement membrane is a specialized form of extracellular matrix that comprises the supporting layer for epithelial and endothelial monolayers, thus separating these monolayers from the underlying connective tissue.

Trying this was a good move. Since Matrigel is a liquid at 4°C and polymerizes (transitions to a network of polymer chains) at higher temperatures, they designed a cooling system to maintain the printing temperature at 4°C and also lowered the printing pressure. With these precautions in place, they avoided clogging or gelation in the print nozzle during printing.

However, the law of unintended consequences reared its head and they found diverse cell structures sprung up for both the neural progenitor and oligodendrocyte progenitor cells when they used varying concentrations of Matrigel diluted with cell culture media. After testing a range of concentrations, they settled on one that gave a reasonable balance of cell viability and stable structure.

Building Spinal Cord Scaffolds

There was more heavy weather ahead for McAlpine and Parr. Their initial choice of scaffold material was biocompatible silicone. They worked to optimize the printing process time for scaffold construction because prolonged scaffold printing can cause drying of the cell suspension in the hydrogel bioink and this compromises cell viability. The team found that cell viability went way down with scaffold print times of more than 15 minutes. So they designed the printing process to be completed within 15 minutes.

But there was another concern. Although the neuronal progenitor cells survived and differentiated in the silicone scaffold, the scaffold material was not biodegradable. In order to rapidly prototype neurocompatible scaffolds for simultaneous scaffold construction with printed progenitor cells, they turned to a contrasting scaffold material: a biocompatible and biodegradable soft hydrogel called alginate. Nonetheless, alginate in pure form didn't have the appropriate stiffness and printing fidelity. They blended the alginate with another material and tested many blend compositions to optimize the print quality. Through diligent experimentation, they hit upon an optimized mix that gave them the right viscoelasticity for their purposes.**** They reported, "Indeed, for the first time, our approach allowed simultaneous 3D matrix and bioprinting of human induced pluripotent stem cell-derived neural progenitors during scaffold assembly to enable precise placement of cells."[59]

Cell Types Within Spinal Cord Scaffolds

McAlpine and Parr printed both spinal neural progenitor cells and oligodendrocyte progenitor cells directly in the scaffold channels, with each cell type contained in Matrigel as the carrier bioink. They used a type of optical microscopy to determine the morphology and alignment of cells in the scaffold channels that were about 150 microns in diameter. Panel (a) of Figure 5.4 shows a composite representation of the scaffold and channel arrangement. In some experiments they co-deposited both spinal neural progenitor cells and oligodendrocyte progenitor cells, and in others they deposited only one cell type. The growth factors in the illustration are part of the Matrigel extracellular matrix. Panel (b) shows the layer-by-layer scaffold built up from the basic unit shown in panel (a).

**** Recall that viscosity characterizes the response of a liquid-like material to a force that causes it to flow. Elasticity characterizes the response of a solid-like material to a force that deforms it. It's common to speak of viscoelastic properties. Surprisingly, viscoelasticity is not simply an abstruse term. It has subtle relevance in our daily lives. If you look at the ingredients in a can of Kraft Cheez Whiz, you won't find any actual cheese but you will find sodium alginate. Alginate gives the product a smooth texture while still having a solid-like shape after it's expelled from the can.

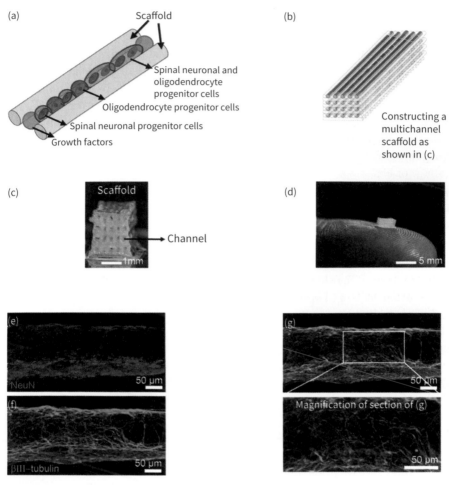

Figure 5.4 (a) Composite representation of the scaffold and channel arrangement. (b) Layer-by-layer scaffold built up from the basic unit in panel (a). (c) Scaffold photograph and (d) scaffold on a finger showing the scale of a scaffold. (e) Neuronal progenitor cell expressing the mature neuron marker NeuN. (f) Neuronal axon projections expressing βIII-tubulin. (g) Merged image of (e) and (f) and magnification of a section of (g) showing proximity of nerve cells and axonal projections.

Source: Reproduced by permission from Daeha Joung et al., "3D Printed Stem-Cell Derived Neural Progenitors Generate Spinal Cord Scaffolds," *Advanced Functional Materials* 28, no. 39 (August 9, 2018): 1801850, https://doi.org/10.1002/adfm.201801850.

After many experiments, the team finally found formulations and conditions to enable the bioprinted spinal neuronal progenitor cells to become neurons and extend axons throughout scaffold channels. The bioprinted oligodendrocyte progenitor cells were not as cooperative as the neuronal progenitor cells. At the conclusion of the experiments, the team had yet to find oligodendrocyte progenitor cells differentiating into myelinating oligodendrocytes. So while McAlpine and Parr have produced mature neurons, there is not yet evidence of the oligodendrocyte progenitor cells maturing into myelinating oligodendrocytes nor is there evidence that the neuronal axons are wrapped in a myelin sheath. The long-term goal remains: to myelinate the axons to achieve an effective relay across the spinal cord injury site.

Still, the sprouting of axons from growing spinal cord neurons is a major step. This work is the first demonstration of bioprinted neuronal progenitor cells that differentiated into axons propagating in a designed scaffold. We present their successful imaging of neurons and their emergent axons. After seven days of culture, in panels (e)–(g) of Figure 5.4 we see the contents of a scaffold channel in which the neural progenitor cells differentiated into neurons. They're identified as neurons because they express the mature neuron marker NeuN that shows up red in panel (e) when detected by the antibody to NeuN—a protein found in mature neurons.[††††] In (f) we see the axonal projections of the neurons detected by the antibody to the protein βIII-tubulin found in the axonal projections from neural cells. Panel (g) is a merged image of the previous two panels, and beneath it is a higher magnification of a section of panel (g) that better displays the intimate proximity of the nerve cells and axonal projections from those cells.

The use of NeuN and βIII-tubulin are applications of a technique that has the broad-shouldered name of immunocytochemistry. *Immuno-* means *related to the body's immune system.* Immunocytochemistry is a lab test that uses antibodies to check for certain antigens in a sample of cells. An antibody is a protein produced by the body's immune system in response to the presence of a foreign substance called an antigen. The antibodies are often linked to a fluorescent dye. When the antibodies bind to the antigens in the cell sample, the dye is activated and the antigens become visible under a microscope.[60] So the green in the microscope images doesn't come from green fluorescent protein that we learned about in the previous chapter but, rather, from this alternative technique. In this way, the cell surface antigens serve as markers to help identify and classify cells.

Even better than mature neurons sprouting axonal extensions, the team confirmed the functional activity of these neuronal networks by watching fluorescently tagged calcium ion flow, as we now discuss.

Calcium Imaging of Bioprinted Spinal Neurons

After all this investment in finding the best scaffold, finding the best bioink, and using multiple means to view the distribution of neurons and oligodendrocytes, McAlpine and Parr's team wanted to determine if the neural networks they built were functional. They weren't prepared at this point to use animal models, though that is one of their intentions in work to follow: transplanting the spinal cord scaffold into an

---

[††††] NeuN stands for neuronal nuclei, so it is highlighting the nucleus of the mature neurons.

animal with a contusion spinal cord injury to appraise the survival of the cells and their effect on the animal's functional recovery.[61]

Instead, to test the performance of the neural networks, the team watched fluorescently tagged calcium ion flow. This has become widely employed as a tool in molecular biology for the measurement of intracellular calcium ion levels and to visualize the physiological release of calcium. And it's this latter use that McAlpine and Parr jumped on for their experiments. It so happens that intracellular calcium—calcium within cells—controls key cellular functions in all types of neurons.[62]

The fact is all cells have a social life and neurons are no exception, so cell signaling is of great importance. As a signaling molecule, calcium is positively a social butterfly. Calcium ions (calcium missing two electrons) stimulate a great number of biological processes. For example, nerve cells cause a release of calcium ions that initiate muscle contraction. Calcium ions achieve this feat by binding to and influencing the activity of various calcium ion-responsive proteins.[63]

McAlpine and Parr incubated their bioprinted cells with a fluorescent calcium indicator and monitored calcium flow surges by changes in fluorescent intensity. They did this using the technique known as confocal microscopy, a standard means of imaging tissues labeled with fluorescent probes.[‡‡‡‡] The calcium intracellular influxes provided the evidence they needed of active neurons in the scaffold. The calcium also showed that the bioprinted neuronal progenitors not only survived but also differentiated into functionally mature neurons.

Summarizing this nascent effort to bioprint spinal cord tissue, the group concluded that by directing the patterning and combination of neuronal and glial cells they could rebuild operational axonal connections, while acknowledging that much remains to be accomplished with the glial component.

Let's take stock. We've considered the historical antecedents to bioprinting, notably 3D printing; a medley of cell types that give rise to the differentiated tissue and organ cells in our body; cartilage, a tissue type that scientists are aiming to bioprint; the gnarly question of vascularization; and the thorny task of innervation.

Now that we have come of age, it's time to embark on a grand tour inspired by the eighteenth- and nineteenth-century British custom, to polish our education by traveling to farther reaches and fulfill our quest to become learned persons on the subject of bioprinting. Dr. Anthony Atala espoused a view held by many that, in principle, the progress toward printing transplantable tissues with increasing sophistication may be said to begin with essentially two-dimensional tissues such as skin. Then "through to

---

‡‡‡‡ Confocal microscopy is a type of fluorescence microscopy in which a screen with a pinhole is used so that only a tiny portion of the sample is illuminated by laser light at a given time (the laser is scanned across the entire sample to get a full image). A confocal microscope gives improved resolution compared to a conventional microscope because the pinhole screen blocks out-of-focus fluorescent light.

hollow tubes such as blood vessels, to hollow non-tubular organs such as the bladder, and finally to solid organs such as the kidney."[64] In this pursuit, the problems will become increasingly difficult, "including tissue maturation and functionality, and appropriate vascularization and innervation."[65]

In accord with this vision, in the following pages we begin our grand tour. We appraise a handful of bioprinting studies of skin, bone, skeletal muscle, and neuromuscular junctions. However, the tour begins inauspiciously. We enter a Dantesque purgatory of the dead and the dying, buried in the acrid smell of charred flesh, and the poignancy of horribly burned children.

6
Skin and Bones—and Muscle Too

Ioannis Yannas was shaken. "I was shocked beyond belief in what I was seeing there."[1] He was seeing children in agony, clinging to life or dying before his eyes in a burn unit at the Shriners Burns Institute of Massachusetts General Hospital in December of 1969. Yannas, a materials scientist at the Massachusetts Institute of Technology, was shocked, but many children were in *physiological* shock. They suffered progressive circulatory failure from third-degree burns that had scorched the entire thickness of their skin. A full-thickness burn destroys blood vessels, sweat glands, oil-secreting glands, and hair follicles. What was skin is now a hard, dry crust, black or turning black, and when it falls off, a deep, putrid ulcer.[2] Yet the carnage goes much deeper than that. The entire body runs riot.

Blood pressure is too low to get oxygen everywhere it's needed because the heart isn't pumping enough or because there isn't enough blood to pump. Lack of oxygen to the cells causes the body temperature to drop; cells pour acidic compounds into the bloodstream damaging tissues and organs; hemorrhaging is rampant as blood clotting malfunctions and fails to stop the deluge. Emergency medicine has a name for hypothermia, metabolic acidosis, and coagulopathy: the trauma triad of death.[3] To achieve immediate wound closure of massive burns—50% or more of the body—you need to have prompt availability of skin or a skin substitute; there is not enough unburned skin on the patient to harvest and graft it onto the burned area.

Dr. John Burke was chief of surgery at the Shriners Burns Institute. In 1969 Burke's team established the world's first frozen skin bank. Cadavers were the source. But applying skin grafts from a cadaver is no different than transplanting a kidney from a donor to a patient in need. The recipient must be placed on immunosuppressant drugs or their body will reject the foreign tissue. And this cadaver skin is temporary. When the patient is well enough, more surgery must be performed to harvest skin from the patient's own body to replace the cadaver skin. A cascade of surgeries.

Desperate for a way to quickly close large burn wounds so his young patients might survive the onslaught, he called in an SOS to the Massachusetts Institute of Technology. Ioannis Yannas of the Mechanical Engineering Department answered the call.[4] Burke took Yannas on a tour of one of the wards. As Yannas recalled, "These pictures in my mind have stayed with me. They became the reason I started working on this project."[5] The materials scientist and the medical man teamed up.

Yannas and his students worked through the early 1970s synthesizing polymers from separate chemical families—plastics, rubber, textile fibers, and human tissues as well—and put these dressings in foils. They took the subway from Cambridge to

Boston, where a technician placed the dressings on animals with burn injuries in the Shriners animal laboratory facility. Dressings made from chemical family after chemical family failed; they had no effect on the speed of closure of the wounds. The project to date had been a total failure. Yannas decided that perhaps he should try a collagen matrix from human tissues that he'd worked on for his Ph.D. thesis. To his astonishment, instead of speeding up the rate of closure of the wound, the collagen dressing delayed the sealing so the lesion would hardly close at all. The team was terribly disappointed.

"We were basically going the wrong way—instead of making it close faster we were making it close slower." Yannas and Burke decided "to not just let go and go home and cry, but to start trying to understand what had happened."[6] It was then they saw that the animals treated with the collagen dressings were not making scar tissue as in every previous experiment. The collagen was actually causing new skin to be formed—skin not scar. They reached this understanding in the mid-1970s and referred to their work as artificial skin.[7] In essence, the team was using a burn victim's body as a bioreactor to grow new tissue.[8] This would give the children a chance to regrow the skin they had lost. It was way, way beyond what the team had expected. It was a transformative change in burn care. They were hesitant to publish their results because it would seem so inconceivable to the scientific community.

The team announced the first human application of their creation in 1981.[9] Their invention was soon recognized as the first example of organ regeneration in adults. It was a regeneration template for the dermis, the layer of skin beneath the epidermis, the skin's outermost layer. A damaged epidermis can grow back. A damaged dermis cannot regenerate itself on its own. Yannas and Burke's invention became commercially available as Integra and has saved many, many lives.

Give Me Some (Bioprinted) Skin

The regeneration template for the dermis protected burn victims from infection and dehydration and wasn't rejected by the patient's immune system, though as Yannas said, it wasn't perfect. For example, the new skin grows in without sweat glands so patients need to forgo vigorous exercise and sun exposure. Years later the search for improved replacement skin continues.

Materials scientist Brian Derby of the University of Manchester was one of the first to embrace bioprinting as a route to replacement skin. In 2005 Yannas hailed Derby's work as a significant achievement and went on to say "Dr. Derby's process promises to greatly simplify cell-seeding of scaffolds that are used to induce organ regeneration."[10] Derby's work was of particular value in carefully exploring the effects of the printing process on cell viability.[11,12]

When I spoke to Derby, he shared some grim humor from that time. In 2005 he received a lot of publicity for speaking and writing about the prospect of bioprinted skin and his preliminary work in that direction. He got a phone call from the university press office saying that the BBC wanted to visit his lab and do a TV story on his work. The film crew came and Derby showed them around the lab. As they were about to leave, Derby asked the crew chief, "What made you come and see us today?" The answer was,

> Oh, we had a crew out to do a murder trial [the major court for the city of Manchester is about a half mile from Derby's lab] but they're going to do a plea bargain—it's only manslaughter now, down from murder. So since we have the crew out, we asked the press office if there's anything interesting to do.

Derby's droll observation on this was, "So then I knew where our research stood: It was less interesting than murder but more interesting than manslaughter."[13]

Many others have joined the search for the best that bioprinted skin can offer. Those laboring in the trenches of bioprinting are well aware that a suitable surrogate for a tissue or organ doesn't necessarily have to be an indistinguishable replica of that which occurs naturally. Regardless, recapitulating function and a semblance of form remains a daunting assignment.

We consider a bioprinted skin investigation from Nanyang Technological University in Singapore, an increasing presence in the bioprinting world. Mechanical engineer Wai Yee Yeong of Nanyang and May Win Naing of the Singapore Institute of Manufacturing Technology led the team. They pursued a strategy for bioprinting pigmented human skin constructs with *uniform* skin pigmentation—an objective that has been elusive.[14] One difference between this work and many other bioprinted skin studies was that Yeong's team used melanocytes rather than only fibroblasts and keratinocytes.[15,16] Clearly, the time has come to give you the skinny on the language of skin.

Bioprinted Skin with Uniform Pigmentation

Figure 6.1 is a simplified cartoon of our skin in cross section.[17] The epidermis is the outermost region and contains melanocytes. These are cells that produce the pigment melanin and help protect us against the damaging effects of ultraviolet radiation. *Melanin* is derived from the ancient Greek word for *black*. On the basement membrane at the epidermal–dermal junction, melanocytes have long finger-like projections that push out into the gaps between keratinocytes. Keratinocytes are cells that make the protein keratin and in so doing provide a barrier that defends us against all manner of environmental threats, including bacteria and viruses. Keratinocytes are densely packed with varying degrees of differentiation (specialization) across the epidermal region. *Keratin* is derived from the Greek word for *horn of an animal.**

---

* One day near the middle of the nineteenth-century, German chemist Johann Franz Simon was boiling pulverized horn, nails, hair, and epidermis in a mixture of alcohol and ether. In so doing, he isolated and

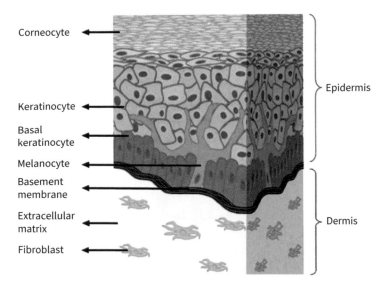

Corneocyte

Keratinocyte

Basal
keratinocyte

Melanocyte

Basement
membrane

Extracellular
matrix

Fibroblast

Epidermis

Dermis

Figure 6.1 Schematic representation of human skin showing an epidermal region with keratinocytes and melanocytes and a dermal region with predominantly extracellular matrix and relatively low fibroblast density.

Source: Reproduced by permission from Wei Long Ng et al., "Skin Bioprinting: Impending Reality or Fantasy?" *Trends in Biotechnology* 34, no. 9 (September 2016): 689–99, https://doi.org/10.1016/ j.tibtech.2016.04.006.

In the topmost layer of keratinocytes where the skin meets the ambient environment, the cells are terminally differentiated keratinocytes and they're given their own special name: corneocytes. The prefix *corneo-* comes from a Latin word meaning *made of horn*, so it's much of a muchness with keratin.[18],[†]

A sobriquet that's been applied to New York City is "the city that never sleeps." Our bodies merit a similar tag. Figure 6.1 is just a still photo of a dynamic process caught in the act. In the epidermis, for example, basal keratinocytes are continuously pushed toward the outermost layer of the epidermis where they become corneocytes in a period of about 30 days.[19] After joining the ranks of the corneocytes, in due time they're sloughed off—thither to slouch back to their elemental forms.

Below the epidermis is the skin compartment called the dermis and it's predominantly extracellular matrix with a relatively low fibroblast density. Dermal fibroblasts generate connective tissue and produce protein molecules that comprise the extracellular matrix.

named a protein—keratin—that is common to horn, nails, hair, and epidermis. After Simon published his results, the word *keratin* took its place in the scientific literature.

[†] With all these new words taking up residence in the epidermis, we add one collective designation to our vocabulary. The epidermis often uses the nom de guerre *epithelium* and skin cells of the epidermis are referred to as *epithelial cells*. We'll encounter this usage later in the book.

Yeong's team developed a bioprinting strategy to fabricate pigmented human skin constructs with uniform pigmentation. Secondarily, they fashioned pigmented skin by a contrasting technique known as manual casting and compared the results to the bioprinted skin. The bioprinted skin was superior by every measure they applied. Their approach is shown in Figure 6.2, where we see a condensed version of the group's two-step bioprinting process.

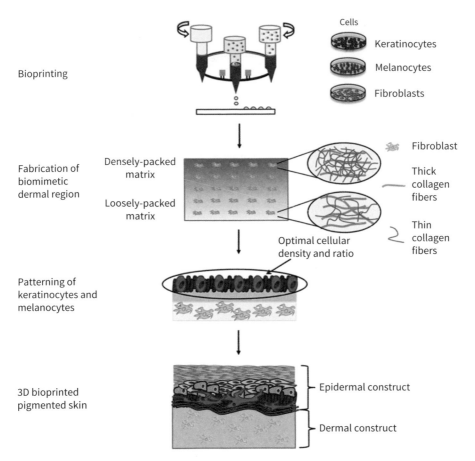

Figure 6.2 Bioprinting strategy for fabrication of uniformly pigmented human skin construct. First we see bioprinting of porous, collagen-based, extracellular matrix containing fibroblasts to mimic the skin's dermal layer. Then we see deposition of the epidermal cells, keratinocytes and melanocytes, in a patterned array. After 39 days in various culture media (not shown), the final construct mimics the epidermal and dermal regions of uniformly pigmented skin.

Source: Reproduced by permission from Wei Long Ng et al., "Proof-of-Concept: 3D Bioprinting of Pigmented Human Skin Constructs," *Biofabrication* 10, no. 2 (January 23, 2018): 025005, https://doi.org/10.1088/1758-5090/aa9e1e.

The team first bioprinted an extracellular matrix of collagen and fibroblasts and cultured this in fibroblast growth medium for four days. This formed the dermal region of the skin substitute. Their second step was to bioprint the epidermal region consisting of melanocytes and keratinocytes directly onto the surface of the dermal construct. They cultured this in melanocyte/keratinocyte co-culture medium for seven days, followed by a further culturing step that lasted for four weeks.

A task critical to achieving uniform pigmentation of the bioprinted skin was to ensure the transfer of melanin from the melanocytes to the surrounding keratinocytes. To accomplish this, they determined an optimal keratinocyte density for bioprinting, and to develop faithful skin mimics they printed an optimized ratio of melanocytes to basal keratinocytes. The authors took that ratio to be approximately 1:20 in human skin.[20]

The Singapore group also used immunohistochemical analysis of their constructs. *Histo-* is from the ancient Greek for *web* or *tissue*.[21] Immunohistochemistry is a way to identify the location of proteins in thin tissue sections. The difference between immunocytochemistry that we mentioned in Chapter 5 and immunohistochemistry is that the *cyto-* assay is done with intact cells but with their surrounding extracellular matrix removed. The *histo-* assay looks at cells within the context of their familiar extracellular matrix—that is, cells as they actually reside in tissues.[22]

Yeong's team found a particular type of collagen (known as collagen VII) in the basement membrane of their bioprinted constructs, a so-called anchoring protein at the epidermis–dermis junction. This was important because this anchoring form of collagen is critical for the attachment and positional orientation of the melanocytes at the junction.[23]

Their efforts successfully demonstrated a well-developed epidermal region and uniform distribution of melanin granules therein. So they achieved their proof-of-concept of uniformly pigmented skin.

A final comment on Yeong's bioprinted skin is appropriate. Features that were missing from skin substitutes decades ago are still missing in bioprinted skin. This is a reminder of the sophistication of every tissue type in the human body. The authors are clear that their results, as in all other bioprinted skin experiments, lack sweat glands. This is not a negligible omission: Sweating is our primary means of controlling our body temperature. In fact, heat intolerance occurs in survivors of large-scale, deep burns even when skin is restored because of the failure of complete sweat gland regeneration.[24]

The good news is that quite recently bioprinting experiments have begun to address this need. The bioprinting of innovatively constructed extracellular matrix has resulted in the functional restoration of sweat glands in the burned paws of mice.[25,26] In time it might be possible to incorporate sweat glands in bioprinted human skin.

Figure 6.3 Photographic series showing the compression and recovery of a 1-cm-diameter cylinder of the printed biomaterial hyperelastic bone over a single compression cycle.

Source: Reproduced by permission from Adam E. Jakus et al., "Hyperelastic 'Bone': A Highly Versatile, Growth Factor–Free, Osteoregenerative, Scalable, and Surgically Friendly Biomaterial," *Science Translational Medicine* 8, no. 358 (September 28, 2016): 358ra127, https://doi.org/10.1126/scitranslmed.aaf7704.

Throw Me a (Bioprinted) Bone

We'll look at two approaches to printing bone that propose alternatives to conventional bone grafting (transplanting of bone tissue) but come at the venture from disparate directions. A team from Northwestern University headed by biomaterials scientist Ramille N. Shah created a new synthetic osteoregenerative biomaterial—a material able to regenerate bone—that they call hyperelastic bone.[27] And we note that while hyperelastic bone is a purely 3D printed (not bioprinted) construct, the group also conducted studies with their synthetic bone seeded with stem cells. With our broad use of the term *bioprinting*, this supported inclusion of their work in our book about bioprinting.

Hyperelastic bone can endure a large compressive force and recover its original shape. Figure 6.3 is a photographic series showing the compression and recovery of a one-centimeter-diameter cylinder of hyperelastic bone over a single compression cycle lasting two seconds. Looking at the figure makes you wonder how on earth this could be used as an alternative to bone grafting. We'll try to provide perspective.

Shah's team printed hyperelastic bone from a blend of several substances, each of which is a safe and clinically used medical material. The predominant material (90%) is called hydroxyapatite, a ceramic material and the principal calcium mineral of our bones.

There was just one problem. Hydroxyapatite is quite brittle. But Shah's background was a good fit in coming to grips with this situation, as I became aware of when speaking with her. As an undergraduate at Northwestern University she started out in civil engineering, a far cry from bioengineering. Her first project was work on a new, highly flowable concrete and this engaged her with materials science. She enjoyed trying to understand how the microstructure of matter and the method of synthesis affects properties and performance, and she switched her major to materials science and engineering. At that point she had her first taste of biomaterials in course work and she never looked back.[28]

So when Shah's group was confronted with the brittleness of hydroxyapatite, they found they could add 10% of either of two possible biocompatible, biodegradable polymers to provide elasticity. Hyperelastic bone was born.

Hyperelastic bone has highly anisotropic properties—that is, its properties vary greatly depending on the direction you're considering. Suppose you were to poke your finger into your thigh from the side. Hyperelastic bone has squishiness in this direction. That means a surgeon can deform it by hand. This is good for fitting printed hyperelastic bone into a bone defect in craniofacial and other non-direct load bearing applications. Returning to your thigh, in the direction that supports the weight of your body, hyperelastic bone is mechanically strong and stiff.

The team made a 4-centimeter-long piece of an anatomically correct construct in the form of an adult mid-femur, a mid-upper leg bone (Figure 6.4). In the axial direction, the width of the printed piece was on the order of 2.8 centimeters. So the shape was that of a short cylinder with its width about three-quarters the length of its height. They carried out mechanical load tests and showed that the construct remained exceptionally strong and stiff along its length but was still compliant and elastic in the axial direction.[29]

Hyperelastic bone can be cut, rolled, folded, and pressed into a bone defect (a body area that's missing bone material). The bone substitute is also highly porous and absorbent and so encourages the growth of blood vessels into the surgery area.[30] In one

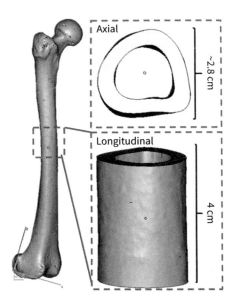

Figure 6.4 Anatomically correct hyperelastic bone construct in the form of an adult mid-femur.

Source: Reproduced by permission from Adam E. Jakus et al., "Hyperelastic 'Bone': A Highly Versatile, Growth Factor–Free, Osteoregenerative, Scalable, and Surgically Friendly Biomaterial," *Science Translational Medicine* 8, no. 358 (September 28, 2016): 358ra127, https://doi.org/10.1126/scitranslmed.aaf7704.

experiment, Shah's team printed hyperelastic bone as a scaffold and seeded it with stem cells. The cells grew well, filled the entire scaffold volume in four weeks, and produced bone mineral. The team reported that this growth and differentiation was performed without the addition of any osteogenic growth factors.[31] This is a demonstration of osteoinduction. In other words, the synthetic bone construct induced actual bone growth by causing the stem cells to differentiate into human bone cells.[32] Lead author Adam Jakus said, "It went from a synthetic scaffold to natural minerals being created by the cells themselves."[33,‡]

The team evaluated hyperelastic bone in a mouse implant, in a rat spinal fusion, and in a primate (a rhesus macaque) that had a cranial bone that was in need of repair.[34] In all cases, hyperelastic bone did not provoke a negative immune response, it was readily vascularized, and it rapidly supported new bone growth. In the vertebral spinal fusion experiment on rats, the hyperelastic bone fused the vertebrae more efficiently than the controls they used.[35] In the case of the macaque, the hyperelastic bone was printed during the surgery following removal of the damaged bone. This allowed the group to determine the actual size and shape of the hole, demonstrating how hyperelastic bone can print designer bone on demand.[36] Postdoctoral fellow Jakus said, "Within four weeks, the implant had fully integrated, fully vascularized with the monkey's own skull. And there is actually evidence of new bone formation."[37]

To overcome technical and surgical quandaries, Shah's team did extensive comparisons of a range of ink formulations and printing methods to get the recipe right. Just why the recipe is right is still a bit of a mystery. As Jakus said, "Cells might actually see it as maybe incomplete bone. So it spurs them even further to remodel it and make it into natural bone."[38]

When speaking to Shah I learned that she chose the Massachusetts Institute of Technology for graduate school because at the time it was one of the few universities that had a graduate program in biomaterials science. Her thesis advisor's lab was located at the Veterans Hospital. Being near patients, physicians, and surgeons was a reminder of why she was doing her research and she loved that environment: It really connected the translational aspect of the research.[39,§]

Morphing Cartilage to Bone: Emulating Embryonic Development

It's time to learn of a bone bioprinting effort where researchers employed a tactic so rooted in developmental biology that the approach is almost the antithesis of hyperelastic bone. A group in Ireland led by biomedical engineer Daniel J. Kelly of Trinity College, Dublin used a precursor to bone and transformed it into vascularized bone.[40]

‡ Note that osteoinduction is the initial step in the formation of bone; mineralization is a later step.
§ Shah has left Northwestern University and is now on the faculty of the Department of Bioengineering at the University of Illinois at Chicago.

The precursor was a cartilage template modified using stem cells. The team incorporated biodegradable polymer microfibers for mechanical reinforcement.

Conceptually, this style of bioprinting is poles apart from using synthetic materials or combinations of synthetic and natural materials to print organs. Kelly's team took a developmental engineering perspective: Bioprint the developmental precursor of a more complicated organ. Afterward, if they provide the bioprinted implant with sufficient mechanical support, it may act as a template for organogenesis. In tissue engineering (which encompasses bioprinting), organogenesis refers to the design and development of engineered tissue.[41]

As Kelly said,

> Our bones begin life as a simpler cartilage template, which develops into a more complex tissue as we grow. So we have instead used bioprinting technology to fabricate mechanically reinforced cartilage templates in the shape of an adult bone, and demonstrated that these tissues develop into functional bone organs following implantation into the body.[42]

Kelly continued, "Here the bioinks are designed to provide an environment that enables the conversion of this precursor tissue into a more complex organ."

When the human skeleton forms in the growing embryo, the long bones of the body form from cells called chondrocytes within the developing limb bud.** *Chondro-* is derived from the Greek word for *cartilage*. The chondrocytes multiply in number and increase in size. This provides a growing template for the formation of bone. There are canals within the template that serve as channels for vascularization, and this supply of nutrients enables the transformation of the template into bone during the formation of your skeleton.[43]

To summarize what we've said, chondrocytes become cartilage, and cartilage becomes bone (in Kelly's work).†† Scientists have successfully replicated this natural embryonic process outside the body by using human stem cells. As Kelly explained,

> It has been demonstrated that cartilaginous templates generated in vitro using . . . stem cells . . . are vascularized and form bone following implantation, suggesting that such engineered tissues could be used for the reconstruction of large bone defects.[44]

The toughie Kelly's team faced was that in both an embryo and in vitro, the tissues are soft and immature. They're fine when in the low load-bearing environment of the developing limb within the womb, but they need additional mechanical support to function in an adult. What Kelly's team did was to solve this problem with

---

** In an embryo, a limb bud is a small protuberance from which a limb develops.
†† This is meant to describe Kelly's work. This is not meant to show the only possible developmental sequence because cartilage is an end result in its own right.

bioprinting. They engineered mechanically reinforced cartilage templates that developed in the body of mice. Over time these templates changed into bone.

The group tried a range of hydrogel bioinks to find the best choice to assist the transformation of the stem cells into chondrocytes and to assist in the creation of bone tissue from cartilage. They also printed a reinforcing network of fibers that gave them a 350-fold increase in stiffness, and they created microchannels to permit vascularization of their constructs. In addition, the team printed a mechanically reinforced cartilage template that was bioinspired by the geometry of vertebrae. They implanted these vertebral-shaped constructs in mice and found that over time the cartilaginous template developed into vascularized bone tissue. The vascularization of the implants was the result of in vivo ingrowth.

The team has high ambitions. As Kelly said,

> We see this as a platform technology for treating a range of diseases and injuries to the musculoskeletal system. We are currently working on developing this technology to bioprint biological implants that could be used to regenerate diseased synovial joints. Such strategies may eventually be used as an alternative to metal and polymer joint replacement prostheses.[45]

Synovial joints are the most common joints in mammals and are characterized by the presence of an articular cavity filled with the lubricant synovial fluid and surrounded by a joint capsule.[46] Examples of synovial joints are our knees, hips, shoulders, wrists, and elbows.

Kelly also led a multinational team from Ireland, India, and the United States in a collaboration that used a gene activated bioink to improve the outcomes of bioprinted bone in regenerating large and challenging anatomical defects.[47] This is an example of gene therapy applied to bioprinted cells. We now give a précis of gene therapy.

A Succinct Sketch of Gene Therapy

Gene therapy does not refer to a strand of DNA emoting on a psychoanalyst's couch. Gene therapy is an experimental technique that introduces new DNA into a person to help treat or prevent disease. In some cases gene therapy adds a working copy of a gene to cells that have only dysfunctional copies. Other times, just adding a gene won't solve the problem and scientists try other routes such as preventing the cell from making the protein that the gene encodes, repairing the gene, or finding a way to block or eliminate the protein.[48] To say "a gene encodes a protein" means that a portion of DNA contains the genetic message for the assembly of the correct subunits of the protein.

A vector is the most common way to deliver a gene to a cell. In the language of molecular biology, a vector is a segment of DNA that carries genetic material into a cell. There are several types of such vectors and the one we're interested in is called a plasmid, a small circular DNA molecule found in bacteria. After you purchase or synthesize the gene of interest, you insert that gene into the plasmid vector and the vector delivers the gene to the cells. This enables the cells to produce the desired protein. Introducing a piece of foreign DNA via a vector is known as *transfection*, and the cells that receive the foreign DNA are *transfected*.

Gene Activated Bioinks

Kelly combined a sophisticated bioink with mesenchymal stem cells. (For the record, there is concern in the scientific community on the use of the term mesenchymal stem cell.)[49] These adult stem cells occur in bone marrow and other tissues as well. They are capable of making and repairing skeletal tissues such as cartilage and bone.[50]

The team used two bone growth factor proteins that they introduced into the bone marrow-derived mesenchymal stem cells via plasmids. This gene activation allowed the stem cells themselves to make the proteins. It's more traditional to introduce the growth factors by bioprinting these proteins along with the tissue construct though there are limitations to this method. One is that this may result in a burst release profile, which means that a large amount of the growth factors are released immediately.[51,52,53] This can lead to concentrations of the proteins far in excess of what is normally found in the body and is undesirable. Researchers in bioprinting are trying to establish procedures that not only enhance their constructs but also will continue to be valuable if/when the day comes that bioprinting can be performed directly on or in the patient. Kelly's method avoided the burst release profile problem, and he demonstrated sustained therapeutic protein delivery up to 14 days after bioprinting.

Kelly's group printed alginate as a carrier for their stem cells. They knew that alginate provides a cell-friendly environment yet its mechanical properties are wanting. So they co-printed it with a supporting mesh of a biodegradable polymer to provide superior mechanical stability. The team also incorporated a bioactive material that occurs naturally in bone and has been shown to promote the differentiation of stem cells into bone cells. After bioprinting, Kelly found osteogenic differentiation of the stem cells with increased levels of bone mineralization.‡‡

These successful experiments were done in vitro. At that point the team went one step further and implanted their gene activated, stem cell-laden constructs under the skin of mice—so they upped the ante by doing an in vivo experiment. Once again they were successful: In a living organism they found enhanced differentiation of stem cells into bone cells and formation of vascularized and mineralized bone-like tissue compared to their control experiments.

---

‡‡ Mineralization is the process of laying down minerals, chiefly calcium and phosphorus, in an ordered fashion within the framework of the bone and contributes to the strength and stiffness of the bone structure.

I had a revealing discussion with Kelly and he told me of his serendipitous journey to becoming a bioprinting researcher in academia. In his final year as an undergraduate at Trinity College, Dublin he did a project in biomechanics that he very much enjoyed. But the summer after graduating he wasn't thinking much about getting a graduate degree. In fact, he spent that summer in Toronto, Canada, working in a shop called Groovy Shoes, selling sneakers. At that point, the young professor with whom Kelly had done his senior "capstone" project received a research grant that fit in nicely with Kelly's capstone project. The professor invited him to return to Trinity to take a master's degree.

About a year into his master's work, his advisor garnered two big grants in tissue engineering. This turned out to be the framework for Kelly's doctoral research. In spite of that, he wasn't convinced he wanted to become an academic researcher, so he worked in industry for a year. And once again, he was invited back to Trinity College by the same professor. Kelly became a Lecturer and later Professor of Tissue Engineering and Director of the Trinity Centre for Bioengineering. Kelly characterized himself as an accidental academic—shades of Anthony Atala's self-characterization as a reluctant researcher.[54]

Bioprinted Skeletal Muscle

Skeletal muscles move bones. The smallest *functional* unit in skeletal muscle is called a myofibril, and bundles of myofibrils form myofibers, also called muscle fibers, and they're strongly oriented in one direction. It's these bundles that are considered the basic *structural* element of muscle tissue. Muscle is not only highly vascularized but also heavily innervated, and this innervation enables the cooperative contraction of myofibers to implement force and move our bones.[55]

So from bioprinted bones we now move on to bioprinted skeletal muscle. Chemical engineer Sang Jin Lee of the Wake Forest Institute for Regenerative Medicine led a team that bioprinted a muscle construct composed of human muscle progenitor cells.[56] As we've seen before with other tissue types, structural organization is paramount and muscle architecture is a primary determinant of muscle function.[57]

If the two proteins that perform muscle contraction within myofibers were to enter their micro-gymnastics routine in the micro-Olympic Games, there is no doubt that all the judges would award them a perfect score of 10—whence they would enjoy great success as motivational speakers and have their pictures on boxes of your favorite breakfast cereal.

The proteins go by the names myosin and actin and they're each present in the form of microfilaments—long chains of proteins. Their choreographed routine is known as the sliding filament theory: Actin filaments, while tethered at their outer ends, slide past myosin filaments. In so doing, the actin filaments are transiently shortened while the myosin remains constant in length. This actin shortening creates muscle shortening—that is, muscle contraction.[58,59]

To envision what happens, consider this analogy for sliding filaments producing muscle contraction.[60] Suppose you're standing between two large bookcases loaded with books. The bookcases are several yards apart and they ride on rails so they can be easily moved. You may have seen these space-saving solutions at libraries or elsewhere. Let's say you want to bring the bookcases together but you can only use your arms and two ropes—each rope tethered to one of the bookcases. You stand centered between the bookcases and with each arm bent at the elbow you pull the ropes toward you. As you pull on each rope you regrasp it and then pull again. As you progress through the length of rope the bookcases move together and approach you. In this analogy your arms play the part of the myosin filaments, the ropes are the actin filaments, and the actin is tethered to the bookcases. Similar to the way you would remain centered between the bookcases, the myosin filaments remain centered and constant in length during normal muscle contraction. Similar to the way the ropes shorten, the actin filaments shorten and, as a result, the muscle shortens as it contracts.

Returning to bioprinting, Lee's team printed human skeletal muscle constructs consisting of three components. One was a hydrogel bioink laden with human muscle progenitor cells along with fibrin.[§§] This cell-laden ink also contained gelatin, and a few other ingredients seasoned to taste. A second hydrogel was a sacrificial ink lacking any cells. It was composed principally of gelatin (with small amounts of additional constituents). The third ink contained a rigid polymer used to make pillars to encourage both the muscle cells and the sacrificial gelatin to form aligned longitudinal beams.

Lee's design concept appears in Figure 6.5. We see a cross section of the long rows of muscle progenitor cells in their hydrogel medium alternating with long rows of sacrificial gelatin (peach color). The polymer pillars (shown in the same color as the cell-laden beams) were essential for cellular structure. The cell-laden bioprinted beams were anchored to the pillars at both ends. The pillars placed tension on the cells in a longitudinal direction and induced cellular stretching. The pillars also provided mechanical cues to align the cells and kept the cell-laden beams taut during cell growth and differentiation.

During the printing process, the sacrificial gelatin hydrogel patterns provided structural stability for the cell-laden beam patterns. After cross-linking, the group dissolved the sacrificial ink by gentle heating (to normal body temperature) and the now-voided spaces became open microchannels in the construct. So the bioprinted muscle constructs assumed the form of cellular beams with microchannels of 300–400 microns between the patterns of myofiber bundles as shown in Figure 6.5. The cellular constructs with microchannels remained viable in vitro because the microchannel structure carried oxygen and nutrients from culture media to the cells via diffusion.

[§§] The group started with fibrinogen and after printing they used thrombin to cross-link the fibrinogen to convert it to fibrin. If you're either dubious or curious, we described this transition in Chapter 4.

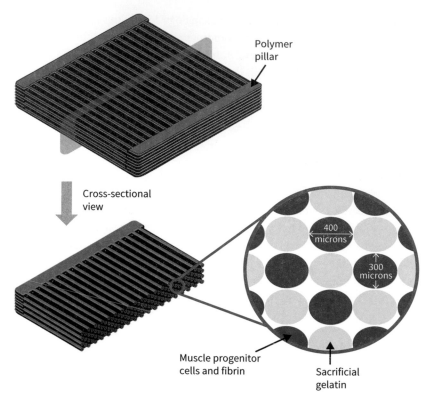

Polymer pillar

Cross-sectional view

400 microns

300 microns

Muscle progenitor cells and fibrin

Sacrificial gelatin

Figure 6.5 Bioprinting of skeletal muscle. The design consists of three components: a bioink laden with human muscle progenitor cells (dark purple beams), a sacrificial ink of gelatin hydrogel (peach color), and an ink containing a polymer to act as a frame, that Lee's team called a pillar (dark purple). In the magnified cross section, we are looking along the length of the cell-laden beams alternating with sacrificial gelatin beams. They later removed the gelatin beams to form open microchannels that provide nutrients to the cellular beams.

Source: Reproduced by permission from Ji Hyun Kim et al., "3D Bioprinted Human Skeletal Muscle Constructs for Muscle Function Restoration," *Scientific Reports* 8, no. 1 (August 17, 2018), https://doi.org/10.1038/s41598-018-29968-5.

Figure 6.6 shows immunofluorescently stained images of their results seven days after printing. The figure also shows control experiments (labeled *Non-printed*). In the controls, the human muscle progenitor cells were in a hydrogel of the same composition but were not bioprinted—that is, the cells were randomly distributed. The group detected myofibers in the bioprinted constructs at a greater than 11-fold increase compared to the control constructs after one week of cell differentiation. And only the myofibers in the bioprinted constructs were densely packed and aligned—as you can see in panel (c) of Figure 6.6. This is how myosin fibers are aligned in our muscles. Think sardines in a can.

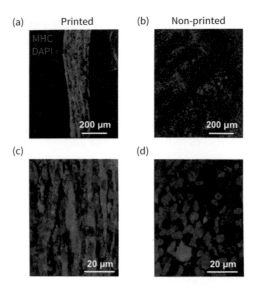

Figure 6.6 Immunofluorescently stained images of aligned and densely packed skeletal muscle myofibers (printed) and randomly distributed myofibers (non-printed). Red stain shows a particular protein in the muscle cells' cytoplasm. Blue stain shows the DNA within the cells' nuclei. DAPI, 4′,6-diamidino-2-phenylindole; MHC, myosin heavy chain.

Source: Reproduced by permission from Ji Hyun Kim et al., "3D Bioprinted Human Skeletal Muscle Constructs for Muscle Function Restoration," *Scientific Reports* 8, no. 1 (August 17, 2018), https://doi.org/10.1038/s41598-018-29968-5.

Lee's team performed evaluations of their bioprinted skeletal muscle constructs in rats. The rats had large muscle defects (30–40%) in the tibialis anterior muscle of their left hind limbs. Eight weeks after the bioprinted construct was implanted, the regenerated muscle was highly mature and organized with both vascular and neural integrity. Muscle force was recovered up to 85% of normal muscle force. Lee's team determined this by measuring sustained muscle contraction produced by nerve stimulation when they attached the anesthetized rat's left foot to a footplate that moved when the tested limb pushed against it.

The Twitch

Lee's group also weighed in on vascularization and innervation of their bioprinted constructs. One salient point was that eight weeks after implantation the bioprinted muscle constructs were well integrated with the rat's vascular and neural networks. The authors observed that for ultimate clinical success of bioprinted skeletal muscle constructs, the most important factor is functional integration of the bioengineered tissue by the host. Simply put, the researchers need to make sure the construct works in vivo; otherwise it's useless. In the vascular category they identified a large protein found in endothelial cells, the cells that line blood vessels. In the neural class they saw neurofilaments—protein polymers found in neurons—as well as acetylcholine receptors and neuromuscular junctions.

Acetylcholine may seem like a word you'd find on the label of a shampoo bottle and stare at with glazed eyes. However, it's a preeminent example of a neurotransmitter. It's a chemical synthesized by neurons that acts as a signal across a synapse—for example, from the axon of one nerve cell to the dendrites of another nerve cell. Alternatively, it transmits a signal from a nerve cell to a receptor on another type of cell, for instance, at a neuromuscular junction. A neuromuscular junction is, as the word suggests, the place where a motor neuron and a muscle fiber come together and, via neurotransmitter signaling, cause muscles to contract.

Lee and colleagues stressed that skeletal muscle is meant to twitch and to have success with bioprinted constructs you'd better do everything possible to make sure that it is capable of twitching. Specifically, they point to future bioprinting work in which neuronal components that facilitate muscle cell survival, differentiation, and innervation can be incorporated to improve muscle function. They call out the neurotrophins that we considered in Chapter 5 as well as neural cells and neurotransmitters as potentially helpful additions to accelerate functional recovery after implantation of their bioprinted skeletal muscle.

Apropos of neuromuscular junctions, our last bioprinting experiment in this chapter is an attempt to bioprint them.

Bioprinted Neuromuscular Junctions—Inks from the Silkworm *Bombyx mori*

Silk fibroin is a structural protein like collagen but with a novel feature: Living organisms, namely spiders and silkworms, produce it.[61] Silk fibroin has made its way into high-tech bioprinting via silkworm silk, not spider silk. The reason is that if spiders wore T-shirts they would not say, "Plays well with others." They would say, "Runs with scissors." There are no commercial options for spider silk largely, as it has been euphemistically phrased, "because of the more aggressive nature of spiders."[62] Spiders are territorial and cannibalistic. It's not uncommon for spiders to kill other spiders, so they can't be farmed like silkworms. It's all too clear: Most spiders that have read E. B. White's *Charlotte's Web* failed to learn kindness to fellow creatures.

In contrast to spiders, as long as you provide silkworms with mulberry leaves to munch, their behavior is exemplary. The larva or caterpillar of the silkmoth, *Bombyx mori*, encloses itself in a cocoon prior to emerging as a silk moth. The silk from these cocoons is used in bioprinting.

Biomedical engineer David L. Kaplan of Tufts University wanted to construct a three-dimensional model using silk fibroin to evaluate human neuromuscular junctions. *** 63 The neuromuscular junction is a synapse, a minute gap (about 20 nanometers) between a motor neuron and the cell membrane of a skeletal muscle cell.64 The motor neuron releases the chemical neurotransmitter acetylcholine and these molecules bind to receptors on the muscle cell membrane. This is the trigger that results in muscle fiber contraction. When large numbers of skeletal muscle cells are triggered, they collectively generate macroscopic movement.

Neuromuscular junction pathology contributes to a spectrum of human disorders, such as neuromuscular, motor neuron, and dystrophic (muscle wasting) disease. Kaplan was eager to make his work translational to the clinic in the future, so he decided to work with human cell lines and to work in three dimensions. His team used human skeletal myoblasts—embryonic precursor cells that differentiate into muscle cells. They also used motor neuron-like cells derived from human-induced neural stem cells. Their desire was to re-create the connection between motor neurons and skeletal muscle. The group hoped to observe functional acetylcholine receptors on myofibers with synapses similar to those found at neuromuscular junctions in our bodies.

Kaplan printed pairs of silk fibroin T-shaped objects in well plates containing cell culture growth media. Well plates are flat plastic plates with multiple shallow concavities called *wells*. A growth medium is a liquid or semisolid designed to support the growth of a particular type of cell.

To mimic cell alignment in our muscles, the team employed a deft means to three-dimensionally align myoblasts: They pipetted hydrogels containing myoblasts along the periphery of the wells around the pairs of T-shapes. Through a delicately elaborate interaction between cells and the surrounding hydrogel matrix, the gels compacted (more on this later). As the gels compacted, they formed loops of myofiber bundles that migrated up the long axis of the Ts (Figure 6.7).††† These suspended bundles anchored under the crossbar of the pairs of silk Ts (causing displacement of the previously upright Ts). Subsequently they added a differentiation medium—a medium that promotes cells to change to a more specialized cell type—and after three weeks under these conditions the compacted gels contained densely aligned myofibers.

Kaplan's images of unidirectionally aligned myofibers (not shown) look very similar to Figure 6.6 that we presented when discussing Lee's work on bioprinted skeletal muscle.

*** Materials other than silk fibroin could be used; silicone polymers are an example. But Kaplan points to the ease with which silk biopolymers can be printed in arbitrary geometries and the absence of undesirable chemical side effects when using silk.
††† Silk fibroin T-shapes were five millimeters high, two millimeters wide, and six millimeters across at their widest point. Pairs of T-shapes were separated by six millimeters.

Bioprinted
silk fibroin —
T-shape

Myoblast-laden gel

Figure 6.7 Cartoon showing pairs of silk fibroin T-shapes displaced by the compaction of a myoblast-laden hydrogel. Scale bar = 2 mm.

Source: Reproduced by permission from Thomas Anthony Dixon et al., "Bioinspired Three-Dimensional Human Neuromuscular Junction Development in Suspended Hydrogel Arrays," *Tissue Engineering Part C: Methods* 24, no. 6 (June 2018): 346–59, https://doi.org/10.1089/ten.tec.2018.0062.

Skeletal Muscle Alphabet Soup

We need to call for a brief time-out. The work of Kaplan's team on trying to bioprint neuromuscular junctions is neat but the descriptors in the formation of muscle tissue can be overwhelming. It's not only the myocytes, myoblasts, and myofibers we've already encountered. There are also myotubes and *mature* myotubes and *differentiated* myoblasts and *mononucleated* myoblasts and *multinucleated* muscle fibers and more. We want to enjoy learning about this innovative effort to bioprint neuromuscular junctions. It wouldn't be useful if the prose got tangled up in explanations about individual stages of muscle cell differentiation and their attributes.

So going forward, we'll use the language that Kaplan's team used without attempting to clarify the nuances involved in the detailed steps of cell differentiation. (For those interested, the overarching idea is that mononucleated myogenic precursor cells called myoblasts fuse to become incorporated into multinucleated myotubes that later mature into multinucleated myofibers.)

Cat's Cradle

These compacted loops of myofibers are one of the fascinating features of Kaplan's work and result from the crosstalk[‡‡‡] between the myoblast cells and their hydrogel environment. It's a cellular game of cat's cradle. The myoblast-laden hydrogels take the role of string and the silk fibroin flexible posts take the role of your fingers. There is a remarkable and subtle interplay between the cells and the biopolymer hydrogel matrix in which they're encapsulated.

[‡‡‡] Signals from the cells that modify the hydrogel and vice versa.

What happens is that the myoblasts impose forces on the hydrogel matrix causing it to compact. In so doing, both the muscle cells and the hydrogel they're in become highly directional along the axis defined by the hydrogel loop; the myoblast-seeded gels produce anisotropic alignment of myocytes. The epigram is: the compaction of gels by cells.[65] The issues at play are part of a wider field called mechanobiology, which studies how cells sense and use mechanical information provided by the extracellular matrix environment (whether their native matrix or otherwise) to make decisions about growth, movement, and differentiation.[66]

The Kaplan group had to do a great deal of work to find conditions under which they could co-culture both neuronal cells and muscle cells, but they were tenacious and succeeded. After that it was on to their objective of forming innervated, functionally active muscle fibers. They directly pipetted differentiated motor neuron-like cells on top of the myoblast-seeded constructs. After adding acetylcholine, they used immunofluorescent staining and identified acetylcholine receptors clustered on mature myotubes. They also found populations of neurites. Neurites are extensions from the cell body of a neuron, such as axons or dendrites. And significantly, they observed transient calcium fluxes arising from neuromuscular connections—the same calcium phenomenon we discussed in Chapter 5 when speaking of the detection of neuronal activity and functionality in studies of bioprinted spinal cord scaffolds.

The calcium fluxes occurred because of the binding of acetylcholine molecules to their receptors on myotubes and myofibers. The binding produced a nerve impulse and also enhanced fluorescence that the experimenters identified by video analysis.

However, they didn't quite reach their goal. The model they've constructed has so far not demonstrated force production of human muscle cells in culture. So as yet it cannot detect any potential contraction force *change* with the addition of co-cultured motor neurons. The calcium transients demonstrated myocyte activation (opening of calcium ion channels) but the team wasn't able to observe any resulting muscular contractions that produced a measurable force. What they look forward to in future work is force detection through measurement of displacement of the silk T-shapes with the addition of motor neurons.

They are actively pursuing several strategies to improve their model and achieve this end. One is to induce the development of more mature clusters of acetylcholine receptors. Other groups have observed that these mature clusters tend to densely aggregate. [67] This structural change into a so-called pretzel shape (of acetylcholine receptor clusters) may well improve the functionality of Kaplan's model by improving connectivity between motor neurons and myofibers at the synapse. Kaplan also hopes to integrate Schwann cells into his construct, cells that are known to be present in human neuromuscular junctions.

So this creative work has some acknowledged limitations at present. However, the model has demonstrated visualization of the cytoarchitecture of both human motor neuron-like cells and differentiated myoblasts as well as the ability to measure signal transduction through both cell types. And they have provided evidence of initial neuromuscular junction development.

Kaplan told me a story about the relationship between his unusual work with cocoons and his life outside the lab. He and his wife have a daughter, and when she was in elementary school (fourth or fifth grade) he brought cocoons and caterpillars to her class. He gave a cocoon to each child and told the children they would win a prize if they could pull apart a cocoon by simple means—wetting it, for example—but they were not allowed to cut it. And none of them could take the cocoons apart because they were so strong.

He also gave them each a large green caterpillar that his group happened to be studying at the time. Kaplan let the children take the caterpillars home with them; they were native to the area and could be safely released in the wild. For years afterward whenever he and his wife went to a parents' day event, invariably someone would come up to him and say, "So *you're* the one who let my kid bring home that caterpillar."[68]

In a casual game of golf among friends, a *gimme* is a putt that's close enough to the hole that it's assumed the player can make it; no effort is required. As we learned long ago when looking at cartilage, there are no gimmes in bioprinting. As is true of cartilage, so it goes with skin, bones, skeletal muscle, and neuromuscular junctions. And in quoting Dr. Anthony Atala in Chapter 5, we acknowledged that the degree of difficulty likely reaches its pinnacle with solid organs. The liver is one such solid organ. Nevertheless, bioprinting researchers are hardy souls. In the next pages our grand tour continues as we consider attempts to bioprint liver tissue.

We'll also learn how the tragedy that befalls one person can fire the compassion of another to a selfless act of courage and sacrifice.

7
The Liver

Regenerative Tissue That Can Almost Bioprint Itself

Heather Krueger was a spirited young woman from Tinley Park, Illinois, working as a nursing assistant. In March of 2014 she turned 25. Several days later she was told she had stage 4 liver cancer and a 50% chance of living a few more months. Then she went into liver failure and her body started to shut down. The cancer horrified her, as if she were in a locked room with all the walls moving in toward her. Krueger's cancer was the result of autoimmune hepatitis, a condition in which your body's immune system turns against your liver cells. The cause is unknown but it's thought to be a combination of genetic and environmental factors.[1,2]

Krueger's cousin, Jack Dwyer, worked at the water department for the nearby village of Frankfort, Illinois, and was distraught over her plight. She couldn't find a living donor and she was too far down on the state's transplant recipient list. She would likely die before a matching liver from a donor became available.[3] One day Dwyer told a co-worker the sad story while they sat in the break room. Sitting nearby was Chris Dempsey. He overheard the conversation.

Dempsey served for four years in the Marine Corps. After his honorable discharge he was hired to be the code enforcement officer for the village of Frankfort, Illinois.[4] It was pure chance that he happened to be in the Frankfort break room during lunch. He rarely took his lunch in the break room. And Krueger's cousin—a seasonal worker—just happened to be working that day.

Overhearing the story, Dempsey was moved. And moved to act. "When I heard about her situation, I just put myself in that situation, thinking if this was one of my family members or me, I would want somebody to help me out," he told a reporter.[5] He got the contact information for the state transplant coordinator and went in for testing.[6] He underwent blood work, a liver biopsy, an MRI, and both physical and psychological testing.[7] Dempsey not only needed to be healthy, he had to have the same blood type and a similarly sized liver as the intended recipient. By good luck he was a match. Dempsey called Krueger personally to share the news. As she recalled to an interviewer, "I got off the phone and ran down the hallway, and my mother and I were both crying our eyes out in disbelief."[8]

Overall the surgeries went well. The donor and recipient were in nearby rooms at the University of Illinois Hospital. Dempsey's surgery for partial removal of his liver took 8 hours. Krueger's surgery took 12 hours. Both were in the hospital for 12 days following surgery and both had complications and were briefly hospitalized again.[9]

Dempsey gave Krueger 55% of his liver. He was unable to work for two months and felt back to normal six to eight months following the surgery.

The story was picked up worldwide and reported in numerous magazines, newspapers, and television shows. The distinguishing feature of these versions of the story was not only Dempsey's remarkable act of kindness and generosity in giving part of himself to a complete stranger but also the fact that they fell in love and became husband and wife. However, for "We few, we happy few, we band of brothers [and sisters]" who aspire to join the bioprinting cognoscenti, the singular element is not the recitation of vows; it's the regeneration of vital organs.[10]

Pause and consider: The remaining 45% of Dempsey's liver regenerated to its former size in 30 days. Krueger's new liver regenerated to its appropriate size in 60 days.[11] It's as though our liver is able to bioprint itself.

The Hectic Life of the White Rabbit

The regenerative capacity of the liver is the stuff of legend. Literally. Many articles on the subject of regenerative medicine cite the ancient myth of the Greek Titan—or Elder God—Prometheus. He angered Zeus, the ruler of gods and men, by stealing the secret of fire from the gods and Zeus punished him. Prometheus was chained to a rock in the Caucasus Mountains where an eagle preyed on his liver—which renewed itself as fast as it was devoured. Most who cite this tale infer that the ancient Greeks knew of the regenerative capacity of the liver, though not everyone agrees with that inference.[12]

However, it's true enough that the liver can restore itself, although the word *regenerate* is something of a misnomer. In the early twentieth-century, experimental pathologists removed two-thirds of the liver of a rat; in about one week the remaining liver enlarged until the original liver mass was restored.[13] In this surgery, the segments of the liver that are removed don't grow back. Rather, the remaining functioning liver cells mature and replicate themselves; recruitment of liver stem cells or progenitor cells isn't required.[14] The liver is the main detoxifying organ of the body, and liver regrowth is presumed to have evolved to protect animals in the wild from the catastrophic results of liver loss that could result from ingesting food toxins.[15]

In addition to its exceptional regenerative power, the liver is like the White Rabbit in *Alice's Adventures in Wonderland*. It's not taking a watch out of its waistcoat-pocket while fretting about being late. But the liver's agenda of over 500 known functions does make for a very hectic life.[16] Liver specialist Dr. Anna Lok of the University of Michigan explained,

> We have mechanical ventilators to breathe for you if your lungs fail, dialysis
> machines if your kidneys fail, and the heart is mostly just a pump, so we have an

artificial heart. But if your liver fails, there's no machine to replace all its different functions, and the best you can hope for is a transplant.[17]

There is a great deal about this complicated organ that is yet unknown and unanticipated observations continue to surprise, as if we were looking at the rings of Saturn from a space probe. Even so, one fact presented early on. It's very hard to grow liver cells outside the body. Just ask Dr. Sangeeta Bhatia.

Cytoarchitecture

Bhatia is a biomedical engineer at the Massachusetts Institute of Technology (MIT) as well as a physician, a member of the 3D Organ Engineering Initiative at the Wyss Institute for Biologically Inspired Engineering at Harvard University, and an Associate Wyss faculty member.

When she was a graduate student at Harvard–MIT Division of Health Sciences and Technology, her thesis advisor was looking for a way to make a device similar to a dialysis machine but for patients with acute liver failure.[18] In the late 1980s and early 1990s, others had led foundational work on growing liver cells outside the body and preserving their functionality for many weeks. They had developed an in vitro system that maintained liver function by sandwiching hepatocytes (the main type of liver cell) between two layers of a collagen gel.[19,20,21] Bhatia's thesis advisor became an integral member of the team in their hepatocyte studies. He also had the wisdom to recognize Bhatia's talent, though she was still only a graduate student. He tasked Bhatia with sustaining liver cells within a mechanical cartridge operating outside the body to offer liver support: a multi-plate device with each plate supporting rows of liver cells and channels in between to provide for nutrient flow.[22]

Bhatia tried for two years to impose order on large numbers of liver cells in vitro. She failed. More, she had hit a dead end. Then someone suggested she try micro-patterning techniques commonly used in the semiconductor industry. It worked.[23] Using computer-chip manufacturing equipment, she patterned her surfaces with straight collagen lines etched onto glass slides as a physical structure to organize the cells. Her initial success was with rat liver cells. She moved on to human liver cells and succeeded again.[24]

A recurrent theme in this book is that cytoarchitecture—the details of how cells are arranged in a particular tissue or organ—is vitally important to the function of that tissue, whether in the body or outside it. The zonal architecture of articular cartilage is an example we've already encountered. The liver takes this idea and runs with it. If you're not especially careful in how you organize its various cells, liver tissue will refuse to survive, let alone perform, outside the body. So Bhatia's breakthroughs were vital to the bioprinting studies we will be considering.

When it comes to bioprinting, the liver presents at least one additional obstacle: its unique vasculature. Before we present our first appraisal of printed liver-like cells from induced pluripotent stem cells, we should know a bit more about the liver.

Liver Cells Are Swimming in Blood

The liver is supplied with blood from two distinct sources: The hepatic artery brings oxygen-rich blood from the heart and the portal vein brings oxygen and nutrient-rich blood from the gastrointestinal tract.[25] Structurally, the liver consists of four sections called lobes with connective tissue in the form of a ligament demarcating the division on its anterior surface (Figure 7.1). There are also two lobes on the posterior/visceral surface and located between the right and left lobes (not visible in Figure 7.1).[26] The lobes are further organized into what are called lobules (small lobes) that are in the shape of hexagonal prisms. A hexagonal prism is a three-dimensional structure where the hexagonally shaped top and bottom have six edges and the six sides are rectangles.

The major portion of a liver lobule is taken up by hepatocytes, the principal type of liver cell. The central vein you see in the figure of a single lobule is a branch of the hepatic vein (not shown) that carries blood out of the liver. Although hepatocytes are

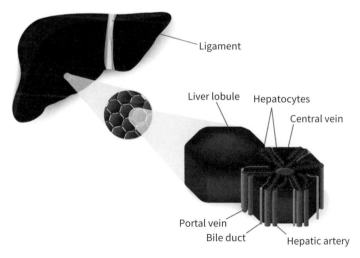

Figure 7.1 The liver's architecture is predominantly hexagonal with its large-scale lobes further organized into lobules—small lobes—in the shape of hexagonal prisms. The main portion of a liver lobule is made up of cells called hepatocytes. There is also an intricate system of canals called bile ducts that carry bile (a fluid that allows us to digest fats). The liver is provided with blood from two distinct sources: blood from the heart via the hepatic artery (not shown) and blood from the gastrointestinal tract via the portal vein.

Source: 123rf.com image ID 29113002.

the predominant cell type in the liver, there are numerous other cell types that also live in the liver; we won't try to enumerate them. But one other intricate feature of the liver, in addition to all its arteries and veins, is its system of canals called bile ducts. These canals carry the substance called bile, logically enough. Bile is created by the liver and secreted into our small intestine enabling us to digest fats.[27] All this is just a very small fraction of the labyrinth that is the liver, just enough to give you an idea of why it's so challenging to mimic liver tissue using bioprinting.

Focusing on the liver's vasculature, at any given moment the liver holds about one pint or 13% of the body's blood supply.[28] It's been rightly said that

> many of the liver's unusual features are linked to its intimate association with blood. ... The liver likes its bloodlines leaky. In contrast to the well-sealed vessels that prevent direct contact between blood and most tissues of the body, the arteries and veins that snake through the liver are stippled with holes, which means they drizzle blood right onto the hepatocytes.[29]

Liver specialist Dr. Markus Heim of the University of Basel put it this way: "Hepatocytes are swimming in blood. That's what makes them so incredibly efficient at taking up substances from the blood."[30]

Your Liver Is Like a Plush Bath Towel

Heim also said that fingerlike protrusions "massively enlarge" the surface area of liver cells in contact with blood.[31] That's why your liver is like a fancy hotel bath towel. A nice bath towel has a surprisingly large surface area. "Just how large?" you might ask. The approximate answer is simple to calculate.

Assume the towel is a large rectangle measuring 1 meter by 2 meters (roughly 3 feet by 6 feet) such as might be found in a very nice hotel. By including both sides of the towel, the surface area of the towel is 4 square meters or 36 square feet. Next, we guess that the number of fibers per square centimeter is more than 10 and less than 1,000. We suppose 100 fibers per square centimeter or about 650 fibers per square inch.

Now model each fiber as a cylinder and take each fiber to be about 0.5 centimeters long with a diameter of 1 millimeter. The surface area of a cylinder (neglecting the end caps) is the product of the length of the cylinder multiplied by the circumference of the cylinder. The circumference of the cylinder is given by $2\pi \times$ radius of the cylinder.

So the surface area, A_{fiber}, of a single fiber is

$$A_{fiber} = \text{length} \times 2\pi \times \text{radius}$$
$$A_{fiber} = 0.5 \text{ cm} \times 6.3 \times 0.5 \text{ mm}$$
$$A_{fiber} = 0.16 \text{ cm}^2$$

Then, to get the total surface area of the towel,

Total surface area = (area of towel) × (number of fibers per unit area) × (area of one fiber)

Total surface area = (4 m^2) × $(100 \text{ fibers/cm}^2)$ × (0.16 cm^2)

Total surface area = 64 m^2, which is approximately 576 square feet

Thus, the total surface area of the plush bath towel is 576 square feet—about the size of a studio apartment in Manhattan.

The point is that the area of the towel (including both sides) is 36 square feet if you don't have all the little fibers to help take up the moisture. But when you do have the fibers, that increases the surface area available to take up moisture by a factor of $576 \text{ ft}^2/36 \text{ ft}^2 = 16$. That's a huge increase in surface area. So it's not hard to believe that the fingerlike protrusions of the liver cells "massively enlarge" the surface area of liver cells that are in contact with blood.

Vascularization is vexing for the bioprinting of any tissue or organ. Now we know why this is particularly true for the liver. To approach the goal of transplantable liver tissue we must start modestly. Now we turn to those who have risen to the challenge of bioprinted liver.

Bioprinting of Human Pluripotent Stem Cells for the Generation of Liver-Like Cells

A team led by biomedical engineer Will Wenmiao Shu, then at Heriot-Watt University in Edinburgh, Scotland, described the first bioprinting investigation of human induced pluripotent stem cells that were directed to differentiate into liver-like cells.[32] Shu's team custom-built their own four-nozzle bioprinter designed to dispense cells in droplets.[33] To summarize Shu's main findings: His group demonstrated that stem cells could be printed without adversely affecting their biological functions, including viability and pluripotency. Importantly, the team bioprinted cells during the differentiation process and found no differences in what is called hepatocyte marker expression and morphology compared to non-printed controls.

Marker expression refers to a subject we've touched on before. We know that genes produce proteins. Specialized proteins coat the surface of every cell in the body. They're called receptors and they have the ability to selectively bind to other signaling molecules.[34] Receptors vary widely in their structure and their affinity for the signaling molecules. When receptors and signaling molecules are at work in your body, cells use them as a way of communicating with other cells and to carry out the untold number of functions that keep you going.

The stem cells that Shu's team used are no different than any other type of cell insofar as they know a convenient means of communication when they see one. Both mature liver cells and liver stem cells have a combination of receptors on their surface that makes them distinguishable from other kinds of cells. Shu's group took advantage of the biological uniqueness of stem cell receptors along with the chemical properties of particular compounds to tag or mark the stem cells.[35]

In Chapter 3, we introduced and illustrated the term phenotype by evoking a lovely passage from the writing of P. G. Wodehouse. In a less ethereal but more scientifically relevant fashion, cell surface markers are intimately tied to phenotype. If you want to see if your bioprinted cells have maintained or wandered from their phenotype, go looking for cell surface markers.

Searching for Cell Markers

Shu's group tested for cell markers using a popular technique to measure and analyze multiple physical characteristics of individual cells. It's called flow cytometry. In flow cytometry, thousands of cells per second in a fluid stream move through a laser beam in an ordered fashion. The light that's scattered in several directions is captured and analyzed by software to uncover cellular properties such as size and internal morphology.[36]

In fact, Shu used a form of flow cytometry that also incorporated fluorescence. Researchers often tag particles with a fluorescent chemical compound that will emit light of a particular wavelength when excited by a laser. In Shu's instrument, he was able to sort cells by means of their fluorescent labels as well as their scattering direction.

One protein Shu was chasing should be present if the stem cells have started to differentiate into liver cells. Generically, this protein is known as a transcription factor. That means it binds to the cellular DNA and participates in the process of making a type of RNA (namely, messenger RNA) from a DNA template.* Shu's group succeeded in their pursuit of the liver-specific transcription factor and identified this hepatocyte marker.

In addition to the transcription factor protein, as the cells mature they should begin to express the protein albumin. Albumin is synthesized in large amounts by mature hepatocytes; it's one of the liver's important assignments. If the stem cells are not very far along in the differentiation process (they're said to be immature), they will express albumin at much lower levels. The good news was that, compared to unprinted controls, Shu's induced pluripotent stem cells showed all the hepatic markers (including albumin) they were looking for. This is important because induced pluripotent stem cells are fragile and Shu showed they could use their custom bioprinter to print these cells even while they're differentiating.

Many of the new terms we encounter in this primer on bioprinting don't make our hearts race. It's true that albumin is not much of an icebreaker at social gatherings. Be

---

* In molecular biology, transcription occurs when one type of RNA uses DNA as a template and copies its genetic information. This process is one step in the sequence that creates proteins from genes in your DNA. In fact, a long-standing mantra in molecular biology is: DNA makes RNA makes protein.

that as it may, serum albumin is the chief protein in human blood plasma. It is a tax-icab driver, shuttling nutrients such as fatty acids to your nutrient-hungry cells. Fatty acids are building materials for lipids—molecules that make up that lipid bilayer in the cell membrane that we spoke about way back when. Fatty acids are a rich source of energy. Fat cells release fatty acids into the blood, and it's your albumin that delivers them to distant parts of your body.[37,38]

Incidentally, albumin's shuttle service shouldn't be taken lightly. Consider how handy it is to eat. We polish off our food with pleasure and with scant attention to the details of how our body will use the nourishment to provide fuel for our trillions of cells. Molecules like glucose are soluble in water so they float in our watery blood-stream. Cells can easily pick up such water-soluble molecules as they move past. But fatty acids (and many other molecules) are not soluble in water. They require special molecular carriers to get them to our hungry cells.[39]

Besides being a cab driver, albumin is a prognosticator. If you happen to go in for a physical exam, your doctor will likely run a complete blood count and ask you for a urine sample. If the urine in that cup has more than a very tiny amount of albumin, your doctor may request further tests to rule out any lurking kidney issues. Healthy kidneys don't leak albumin into your urine. Like raindrops showing up on the side-walk, albumin may herald a mere sun-shower or a downpour.

To find optimal printing conditions, Shu's team appraised printing nozzles of dif-fering lengths to see what effect that would have on cell viability. The viability of the cells was much lower with a longer print nozzle than with a short nozzle. This isn't unexpected given our discussion a while ago concerning the shear forces cells expe-rience when running the gauntlet through a bioprinting nozzle. But it was a helpful result in this exploratory study to confirm what their intuition would tell them and to decide on the best printing parameters.

One last consideration: Shu checked the pluripotency of printed stem cells against non-printed controls and verified that the bioprinting process itself did not trigger the cells to differentiate. Testing stem cells that were not directed to differentiate by means of a biochemical protocol, Shu's group found that they were morphologically unchanged by the printing process. That's good news: It meant that these stem cells remained pluripotent. They did these tests by looking for multiple markers on the stem cells that are known to reflect pluripotency.

To recap, Shu's work led the way to bioprinting hepatocyte-like cells starting with human induced pluripotent stem cells.

Patterning of Cell-Laden Hydrogels in the Form of Liver Tissue

We've seen variations on the themes of bioprinting. If we take canonical bioprint-ing to include the layer-by-layer placement of living cells as well as extracellular ma-trix and possibly scaffolds or growth factors, we've also seen bioprinting that does

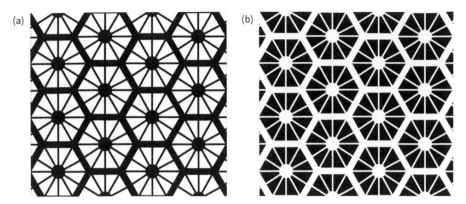

Figure 7.2 Optical (digital) masks for patterning liver cells (a) and extracellular matrix (b).

Source: Reproduced by permission from Xuanyi Ma et al., "Deterministically Patterned Biomimetic Human IPSC-Derived Hepatic Model via Rapid 3D Bioprinting," *Proceedings of the National Academy of Sciences of the USA* 113, no. 8 (February 8, 2016): 2206–11, https://doi.org/10.1073/pnas.1524510113.

not include cells or in which cells are seeded onto a printed construct. A further departure is to find a novel way to pattern the components of your bioconstruct. For example, you could turn to a process called digital light processing. Mechanical engineer Shaochen Chen of the University of California, San Diego did just that.[40] Chen's group used living cells, and the cells were encapsulated in suitable hydrogels. So far that is familiar. The twist comes in how he placed the components in the construct.

First we need to build on what we saw of the liver in Figure 7.1. The takeaway point from the figure is that the basic unit of the liver, the lobule, is a hexagonal prism. That's a three-dimensional object with a hexagonally shaped top and bottom and six sides that are rectangular. Let's imagine that we're looking down at the top surface of a section of liver tissue consisting of an assembly of multiple lobules. In all tissues, we know there are both cells and supporting structure—the extracellular matrix.

For the experiments we will describe, Chen made two optical masks. Think of these masks as opaque sheets with a pattern of through-holes. In one mask the pattern of holes defines the places where Chen wants to place the cells. In the other mask the pattern of holes defines where he wants to place the supporting structure. We can see the two masks in Figure 7.2. In panel (a), the white areas are those that allow (ultraviolet) light to pass through and they define the positions where the liver cell-laden hydrogel will be.[†] In panel (b), the white areas define the positions where the hydrogel containing the supporting matrix will be.

[†] They used ultraviolet light because it cross-links the hydrogel by forming chemical bonds between polymers within the hydrogel, and thus hardens the gel. This is not unlike the curing light dentists sometimes use to polymerize composite fillings.

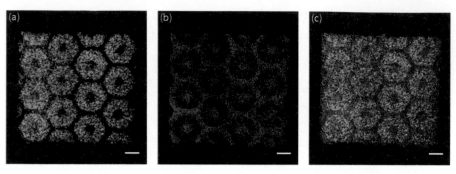

Figure 7.3 Images taken under fluorescent light showing patterns of fluorescently labeled human induced pluripotent stem cells differentiated to become liver progenitor cells (green) and supporting cells (red). Scale bars = 500 microns. The image in panel (c) shows the first two images combined to show the complete construct.

Source: Reproduced by permission from Xuanyi Ma et al., "Deterministically Patterned Biomimetic Human IPSC-Derived Hepatic Model via Rapid 3D Bioprinting," *Proceedings of the National Academy of Sciences of the USA* 113, no. 8 (February 8, 2016): 2206–11, https://doi.org/10.1073/pnas.1524510113.

Figure 7.3 shows fluorescently stained images of the cell-laden hydrogel (green), the matrix-laden hydrogel (red), and a combined image of the two after the bioprinting process. Now we need to make sense of how Chen's bioprinting procedure got us from the optical masks of Figure 7.2 to the printed constructs of Figure 7.3.

Figure 7.4 shows the two-step process Chen used to create his hydrogel-based liver construct. The way the process works is that instead of using a print head nozzle to place the gel-encapsulated cells at precise positions, the team sends ultraviolet light through a photomask such as we saw in Figure 7.2. However, there's a technical sleight of hand at work here: These aren't actual physical masks but, rather, virtual masks made by computer-aided design and stored as digital files, so they're referred to as *digital masks.*[‡] Nonetheless, the masks transmit the pattern onto a computer chip with a special name: a digital micromirror device.[§]

A digital micromirror device is a sophisticated term for a conceptually simple, albeit slick, trick. The device consists of over one half million tiny aluminum mirrors; a single mirror is about 16 microns in diameter. Each mirror is mounted on tiny hinges so that they can be individually tilted to two possible positions. In one position a given mirror is tilted toward the source of illumination. That particular mirror is now in the *on* state. In the second position a given mirror is not tilted toward the source of illumination. That particular mirror is now in the *off* state. All the mirrors that are in

---

[‡] Because no physical masks are used, this technology is also one that is characterized as a "maskless" technique.

[§] At this point, the careful reader may be saying, "Whoa, Nellie!" How is it that we're first referring to "optical masks" and then speaking of "photomasks"? The answer is that Shaochen Chen did this too. The terms are synonymous and the important point to remember is that they both call out masks generated by patterns produced by the digital micromirror device.

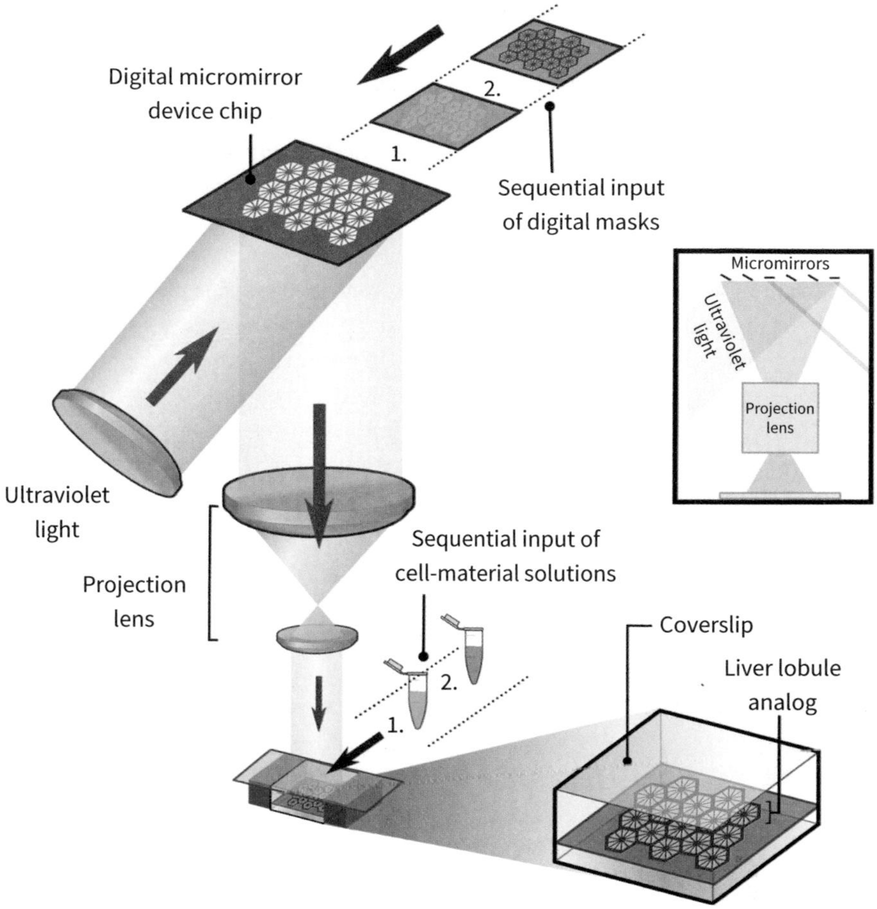

Figure 7.4 Two-step bioprinting of hydrogel-based liver construct. The first digital mask patterns human induced pluripotent stem cells that will differentiate into liver progenitor cells. The second mask patterns supporting cells. The inset shows how a mask is set with only six micromirrors represented, four in the *on* state and two in the *off* state.

Source: Reproduced by permission from Xuanyi Ma et al., "Deterministically Patterned Biomimetic Human IPSC-Derived Hepatic Model via Rapid 3D Bioprinting," *Proceedings of the National Academy of Sciences of the USA* 113, no. 8 (February 8, 2016): 2206–11, https://doi.org/10.1073/pnas.1524510113.

on states are able to reflect the image of the virtual mask toward the target, which is a cell-laden hydrogel.[41] The mirrors that are not tilted toward the source of illumination will not reflect the image of the mask toward the target.

Let's break down the last pithy sentence. The pattern for each layer to be built is encoded in photomasks that are read by the digital micromirror device chip. The experimenter presents the first mask to the device. This selects which mirrors are in the *on* position and which are in the *off* position. At that point the experimenter directs ultraviolet light onto the micromirror device. The mirrors of the device that are in the *on* position will reflect the light and deliver it to the hydrogel that is a photocurable material. Photocurable means that the hydrogel is made of components that can be hardened or cross-linked by either visible or ultraviolet light. Next, the experimenter selects the second photomask to feed to the micromirror device and the process is repeated. In the case of this experiment, there are only two photomasks. (The inset in Figure 7.4 attempts to show how a mask is set with only six micromirrors represented—four in the *on* state and two in the *off* state.)[42]

Continuing with Chen's experiment, the photomask pattern is now carried by the ultraviolet light bouncing off the tilted mirror array. The light goes through some optical components to massage it on its way to the ultimate target, the cell-laden gel. Mixed in with the gel is a chemical with a ridiculously long name but with the saving grace that it is cell-friendly and can be used to initiate polymerization of the gel material.** For each of the two printing steps—that is, for each of the two photomasks—a team member manually pipettes the gel mixture onto the plane where the ultraviolet light is projected.[43] The mixture is exposed to the light and polymerizes in the chosen pattern. The areas of the gel that are not hit by the light, as determined by the details of the photomask pattern, remain unpolymerized and are washed away.

It's All Done with Mirrors

The digital mirror device (invented at Texas Instruments in 1987) is at the heart of this bioprinting technique.[44] It's been said that there are two remarkable points about the device that Chen's group so expertly deployed: (1) that it works at all and (2) that it works so reliably.[45]

The more than one half million mirrors are packed together in a space no larger than a fingernail, and each micromirror weighs a few millionths of a gram. The mirrors rotate through 20 degrees on their individual hinges but in 5 microseconds (5 millionths of a second) the mirror's tip moves only two microns. Thus, the average velocity of a micromirror is only 40 centimeters/second. Compare this to the velocity of an oak leaf falling from a tree, which is about 100 centimeters/second.

The point here is the ingenuity of the design: The speed the micromirrors achieve is so modest that it doesn't induce breakage. Yet the durability of the micromirrors is still remarkable since each mirror can be switched on or off thousands of times per second.

The digital mirror device is a form of spatial light modulator—gadgets that modify some aspect of the incoming light such as its intensity. You use another form of spatial light modulator every day as part of your computer or smartphone display, namely, a

---

** The name is lithium phenyl-2,4,6-trimethylbenzoylphosphinate, so it's predictable that people immediately went in search of an acronym with the result that everyone calls it LAP.

sophisticated array of liquid crystals. One of the astute aspects of Chen's work is just that he chose to use a digital micromirror device in his bioprinting system.

Physiological Relevance

Chen's team designed the mask in panel (a) of Figure 7.2 to pattern human induced pluripotent stem cells that will differentiate into liver progenitor cells. Revisiting the flow of the experiment: The experimenter loads this mask onto the chip that controls the digital micromirror device, and a team member pipettes the gel solution containing these cells at the location in Figure 7.4 labeled "Sequential input of cell-material solutions." That's step one.

After washing and drying, step two begins. The experimenter loads the complementary mask (panel (b) of Figure 7.2) onto the chip, and a team member pipettes the second gel solution that contains two other cell types that are collectively referred to as supporting cells. The idea is to provide a biomimetic microenvironment in which the liver progenitor cells can thrive and functionally mature. The two supporting cell types are human umbilical vein endothelial cells and human fat-derived stem cells— that is, stem cells derived from adult fat tissue. Chen's group built their liver patterns in a matter of seconds, though they subsequently cultured the constructs for weeks in order to realize mature gene expression profiles (as explained shortly).

Chen chose endothelial cells based on their potential for forming vasculature. He chose the human fat-derived stem cells because of their high yield in fat tissue, their ability to differentiate into multiple lineages, and because they're considered a mediator of tissue regeneration by virtue of their secretion of valuable soluble factors.[46,47]

The team made a final construct that measured 3 mm × 3 mm with a thickness of about 200 microns. They compared hydrogel encapsulation of their liver cells at the progenitor stage with cells that had achieved higher degrees of differentiation. To do this, they assessed the expression of four key liver-specific genes by looking for the proteins these genes produced. A principal finding was the sustained higher level of albumin production when the cells were encapsulated at the progenitor stage rather than at a more mature stage. This is a result we've seen previously: Progenitor cells, rather than fully specialized cells, may carry advantages for bioprinted tissues.

The group also wanted to understand the activities of cells after bioprinting and studied cell viability, cell migration, and intercellular interactions. Using techniques we've become familiar with, they found that when they co-printed the liver progenitor cells along with human umbilical vein endothelial cells and fat-derived stem cells (what they called a triculture) they had better outcomes than if the supporting cells were not present. They saw improved structural organization and maintenance of the patterns they had worked so hard to impress on the cell-laden gels using their sophisticated micromirror array.

In further comparisons between the triculture and the controls, Chen's team employed the same marker expression techniques we described earlier in Shu's study. In the triculture, Chen detected higher liver-specific gene expression (proteins made by the liver). In addition to albumin, they found enhancement of other key proteins—for example, a transport protein the liver makes that carries both an important thyroid hormone and vitamin A to your body's appreciative tissues.

They also checked out the presence of crucial enzymes that the liver is responsible for producing and that enable it to metabolize drugs. Recall that an enzyme is a protein that accelerates the rate of chemical reaction, without the enzyme being altered in the process. If it were not for all manner of enzymes hard at work 24/7, the metabolism that keeps us going, wouldn't. An average cell is about one one-billionth the size of a drop of water and generally contains something like 3,000 enzymes.[48] Once again, Chen's group found that their triculture approach was superior to the controls (lacking supporting cells) in the expression of liver enzymes.

This experiment succeeded well at its goal of biomimicry. Chen's group showed that the elaborate liver cytoarchitecture and cell composition can be studied in a physiologically relevant model with the ability to enable maturation and functional maintenance of human induced pluripotent hepatic cells. The construct might well be used for early drug screening and disease modeling.

Cell viability in this work was only 76% as measured two hours after bioprinting, and after one week viability stood at 65%. As we said early on in this chapter, human hepatocytes are notoriously difficult to perpetuate in vitro. The thickness of the bioprinted liver model is only 200 microns and that means the cells receive their nutrients by diffusion rather than from a network of blood vessels. Chen's team did include human umbilical vein endothelial cells in their constructs, and these cells have the potential to aid in liver-specific vascular formation. But that's down the road.

Chen took his BS degree from Tsinghua University in Beijing, China. He joked to me that his career summary is the story of "how a boiler engineer turned himself into a tissue engineer." He was trained as a mechanical engineer to design and operate industrial boilers and turbines. These boilers for power generation are over 100 meters tall, 50 meters wide, and purposed to handle extremely high temperatures (over 600°C) and very high pressures. He didn't like the job. He said, "In my heart I have always wanted to do engineering in medicine." He earned his Ph.D. in laser optics and laser micromachining at the University of California, Berkeley in 1999. He was determined to be a university professor to be able to define his own research interests.

Chen said, "After a long soul search, I figured the best impact area of a 3D printing technique is for medicine." We know that the application of any 3D printing system to delicate biological materials necessitates a huge departure from high temperatures, high pressures, and harsh chemicals. Chen's odyssey, from grappling with mammoth power generation boilers to finessing the rapid bioprinting of human cells, is one of epic proportions.[49]

Bioprinting Liver Tissue by Fusing Liver Bud Spheroids

We got to know Dr. Koichi Nakayama in Chapter 1 and saw tissue spheroids skewered onto needles where they fused to form a tube. Here we'll discuss how Nakayama, co-corresponding author Dr. Eiji Kobayashi of Keio University School of Medicine in Tokyo, and their team applied his technique to bioprint liver tissue.[50]

In addition to the novelty of their bioprinting strategy, their liver research offers an alternative to traditional whole-organ transplantation. Nakayama, Kobayashi, and their team attempted to couple bioprinting to a new surgical approach. Regardless of experimental shortcomings, it's encouraging to see scientists and surgeons working together to try to translate bold ideas into practice. And to publish their findings with flaws in plain view—like wearing red pants.

Clinical trials of hepatocyte transplantation have shown therapeutic effects, and this treatment is less invasive than liver organ transplantation. But there are considerable obstacles, such as the lack of long-term efficacy.[51] However, Nakayama, Kobayashi, and colleagues drew attention to a recent study in a mouse model where the transplantation of in vitro-generated liver buds demonstrated outstanding success.[52]

Liver bud refers to the early embryonic development of tissue that will eventually become the *parenchyma* of your liver. Parenchyma denotes functional tissue as opposed to the connective and supporting tissue.[††] More generally, an *organ* bud—a condensed tissue mass of specific cell types prior to blood perfusion—can now be created in vitro.[53] In Nakayama's work, the team had a dual focus. They wanted to develop a method for growing a liver bud into a *scalable* organoid—that is, a method in which they could readily change the size of the organoid. They also wanted to develop a surgical technique for the transplantation of liver buds that directly connects the transplant to the recipient's native liver. These are profoundly ambitious goals.

The team worked with both mice and rats and encountered many setbacks. For example, the liver is prone to bleeding during surgery so they tried several ways to create a transplant site to overcome this concern. Suffice it to say that they finally succeeded. Another snag was to find the correct culture conditions for creating a large number of liver bud-like spheroids. Eventually they were able to find the right combination of cells, including human mature hepatocytes, human umbilical vein endothelial cells, and human mesenchymal stem cells, as well as an appropriate culture medium.

As their story continues to unfold, tuck away in a corner of your mind that they used *mature* hepatocytes, in contrast to earlier studies we've considered that used stem cells or progenitor cells.

[††] *Parenchyma* comes from the ancient Greek meaning *something poured in beside*. The idea is that the liver, lungs, and so on were purportedly formed of blood strained through blood vessels which then coagulated.

Caution: Road Work Ahead

The group faced additional obstacles. The fusion of unarranged spheroids didn't form controlled or uniform structures. And they were troubled with the formation of numerous small aggregates rather than single, appropriately sized spheroids. This too they overcame. They used their needle array and robotic handlers to fashion spheroids that fused as desired. After removal from the needle array, the holes from the needles closed up during two days of continuous culture. As we've seen in previous studies, Nakayama, Kobayashi, and team used immunofluorescence microscopy to check for items such as albumin production and microscopically examined the tissue they created. They were able to conclude with confidence that the three-dimensional bioprinted structures were growing into liver-like tissues through ex vivo tissue self-organization.

Ex vivo cultures are a model between in vitro and in vivo; a method in which whole tissue slices are cultured. In ex vivo models based on tissue slices, the cytoarchitecture remains intact, as do many of the intercellular connections.[54] So there is minimum alteration of the native conditions that would be present in vivo. By contrast, in vitro studies use components of an organism that have been isolated from their native environment. These latter studies are sometimes loosely referred to as *test tube experiments*.

Nakayama, Kobayashi, and collaborators achieved success, albeit with caveats. Their method of bioprinting enabled the immediate application of culture media circulation after printing and thus avoided cell death due to lack of nutrients reaching the construct after it was printed but before transplantation. And their surgical technique allowed for direct connection between the transplanted tissue and the native liver tissue: They observed growth of the transplanted liver bud and verified direct interactions between the hepatic cells in the graft and the hepatic cells in the recipient's liver.

However, one month following transplantation the team found the transplanted liver-like tissue wasn't healthy and was covered with scar tissue. They attributed this to poor survival of blood vessels in the grafted tissue. One conclusion was that a better strategy would be to fabricate vascularized tissue *before* transplantation. Another issue was low marker expression—for example, human albumin levels after transplantation were very low.

The group surmised that the mixed cells they used (human mature hepatocytes, human umbilical vein endothelial cells, and human mesenchymal stem cells) may have been problematic insofar as the cells were at separate stages of differentiation. The hepatocytes were mature, but the human umbilical vein endothelial cells and human mesenchymal stem cells were in what is called an early lineage stage— they were immature. The authors pointed out that the formation of cell aggregates and self-organization requires lineage dependent interactions and cooperative maturation. Thus in future work, they may use hepatic progenitor cells rather than mature hepatocytes.

Human Liver Tissue Seeds

Sangeeta Bhatia and Christopher S. Chen were the senior authors on our final liver study. We've been introduced to Bhatia. Chen is a physician and biomedical engineer at Boston University and, like Bhatia, is part of the 3D Organ Engineering Initiative at the Wyss Institute for Biologically Inspired Engineering at Harvard University and an Associate Wyss faculty member. Bhatia, Chen, and colleagues combined human hepatocytes, human dermal fibroblasts, and human umbilical vein endothelial cells to form a bioengineered tissue unit. Encapsulated in a fibrin hydrogel, they called this a "liver tissue seed."[55]

The group tested contrasting geometrical organizations of the three cell components in the seeds to assay the importance of tissue architecture and determined an optimized geometry. Note, however, that the seed patterning was not achieved with bioprinting but, rather, with a micromolding technique invented by the authors.[56] Vascularization was done by means of a hybrid printing/casting method (also invented by the authors). We'll say more about their vascularization approach in a moment.

The team implanted the liver seed into a mouse with a liver injury. They placed the seed some distance from the injured liver and found that the seed could perform many functions of the human liver, including the ability to grow and self-organize by responding to systemic regenerative cues from the host. In an interview, lead author Kelly Stevens said, "The hypothesis is that mouse liver injury will produce factors that will travel through the bloodstream and tell the human liver tissue to regenerate."[57]

Their strategy differs from work in which hepatocytes were directly engrafted into the host liver. Bhatia chose to circumvent the issue of limited engraftment success by developing a therapeutic satellite liver that could provide functional support to the failing liver. The liver seed expanded in size by 50-fold over the course of 11 weeks. This work provides additional evidence that in vitro-generated liver tissue can grow and function in vivo and might potentially help reduce the waiting list for transplantable livers.

Coming back to the method for vascularization, Chen, Bhatia, and colleagues invented the process years before the advent of their liver seed (Figure 7.5).[58] They used a 3D printer to make a biocompatible sacrificial material from glass fibers made of a mixture of sucrose, glucose, and a glucose biopolymer called dextran.[‡‡] The group built a lattice of these glass fibers for casting a three-dimensional vascular network. In

---

[‡‡] In everyday use, the word *glass* refers to an amorphous (non-crystalline) substance made from the chemical compound silicon dioxide that has special temperature-dependent properties. In science, *glass* refers to many different amorphous substances, not just those derived from silicon dioxide.

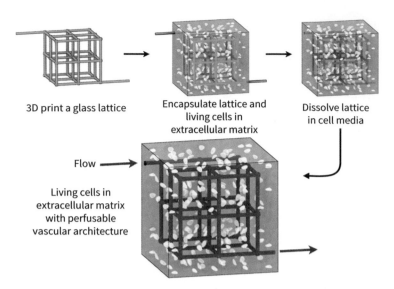

3D print a glass lattice

Encapsulate lattice and
living cells in
extracellular matrix

Dissolve lattice
in cell media

Flow

Living cells in
extracellular matrix
with perfusable
vascular architecture

Figure 7.5 3D printing and casting approach for creating vasculature.

Source: Reproduced by permission from Jordan S. Miller et al., "Rapid Casting of Patterned Vascular
Networks for Perfusable Engineered Three-Dimensional Tissues," *Nature Materials* 11, no. 9 (July 1,
2012): 768–74, https://doi.org/10.1038/nmat3357.

the casting step, they encapsulated the lattice in a thick extracellular matrix along
with living cells (hepatocytes). After cross-linking the matrix, they dissolved the lattice using cell culture media. What was left after they washed out the glass fibers was
an interconnected set of open channels. Since Chen and Bhatia were aiming to create
vasculature, they pumped a solution of endothelial cells through the channels. The
endothelial cells anchored to the channels' inside walls forming a confluent layer—
that is, a continuous layer—*confluent* from the Latin *flowing together*. Following this,
the group perfused the channels with blood. Blood diffused through the endothelial barrier and provided oxygen and nutrients for the encapsulated hepatocytes that
were within several hundred microns of a channel.

Notice that we said a *thick* extracellular matrix in the preceding paragraph. Chen
and Bhatia had invented a way to vascularize thick tissue. This method could have
appeared in Chapter 4 on vascularization; however, the work of Lewis' team was more
generalizable, so Lewis' study found a home in Chapter 4.

A Historical Event: Organovo

In 2007, Organovo became the first company in the bioprinting field. They were also
the first publicly traded bioprinting company and were listed on the NASDAQ stock
exchange. Headquartered in San Diego, California, Organovo was a pioneer in the
bioprinting of commercially available human tissues, notably their lead product, liver

tissue. In 2019, the company announced that the scientific data they had generated on their liver tissue was not decisive. Not long afterward Organovo entered into a merger agreement with a privately held company called Tarveda Therapeutics, Inc., focused on oncology (cancer) medicines.[59]

There are now other companies in the bioprinting field. As a matter of historical record, Organovo was the first.

We've seen that the barrier to bioprinting liver-like tissue is formidable. Yet it's a fact that among all human drug-induced organ toxicities, liver leads the pack.[60] This is to be expected because one of the principal functions of the liver is to keep you safe by metabolizing dangerous substances. A specialist in hepatology, Dr. Markus Grompe of the Oregon Health and Science University, said, "Our ancestors didn't have healthy refrigerated food. They ate a lot of crap, probably literally, and the liver in prehistoric times was continuously bombarded with toxins. You need every mechanism there is to adapt to that."[61] Nowadays, drug-induced liver injury (hepatotoxicity) is the number one cause of removing a drug from the market, failed clinical trials, and drug label change restrictions.[62]

In Chapter 4 we met the Greek physician and philosopher Galen, whose teachings set the course for Western medicine for well over a millennium. Galen believed that the liver was the principal organ of the human body. He argued that the liver emerged first of all the organs in the formation of the fetus.[63] Galen maintained that the heart was secondary to the liver in its importance because the heart was not the site of the production of the humors.

The ancient Greeks, notably Hippocrates in the fifth century BC, believed that the human body consisted of what they called four humors. These were four bodily fluids: yellow bile, black bile, blood, and phlegm. These four humors had to be kept in balance for proper health. Galen inherited and endorsed this idea, and his influence was such that the theory of the four humors endured until roughly the beginning of the eighteenth century.[64]

When you reflect on the earlier observations that the heart is mostly a pump and there's no machine that can replace the liver, perhaps the ancient Greeks and Galen had their organ ranking right, despite the dubious humors theory. Anyway, there's no doubt that ailments of both the liver and the heart are societal health burdens and constructing implantable bioprinted tissue for either of these VIP organs is daunting. We've had a look at the current state of affairs with respect to the bioprinting of liver tissue. Time to turn to affairs of the heart.

What do you do if you have a great idea and you know you're right? It has nothing to do with making money. It's all about helping others. But the most renowned and influential people to whom you present the idea look at you like you're nuts.

Sometimes it takes dramatic measures. A death. An autopsy. And a smoking gun.

8
The Heart

Cardiac Patches, Calcium, and Contraction Forces

Dr. James B. Herrick walked to the podium with a spring in his step to address the 27th Annual Meeting of the Association of American Physicians in Atlantic City, New Jersey.[1] It was 2:40 p.m. on May 14, 1912, and 86 members of the Association had elbowed their way into the Marlborough–Blenheim Hotel for the second presentation of the afternoon session.[2] Herrick approached the rostrum with the feeling that he would be outlining a thesis of significance to his peers:

> In 1912 when I arose to read my paper at the Association I was elated, for I knew I had a substantial contribution to present. I read it, and it fell flat as a pancake. ... I was sunk in disappointment and despair."[3,4]

But Herrick believed in his findings and put his vision into practice.[*] In fact, this began before his letdown in Atlantic City. He had worked out a simple model for coronary artery disease. The coronary arteries provide blood flow to the heart. If flow is obstructed (e.g., by atherosclerosis—fatty deposits on arterial walls), the heart muscle's demand for oxygenated blood will exceed its supply. It's this shortfall that produces the chest pain we commonly call angina, "the wailing of an anguished heart during that period of stress when it is getting inadequate perfusion."[5] If the imbalance between blood supply and blood demand is severe enough, the heart muscle cells start to die, producing a myocardial infarction, more commonly known as a heart attack.[6]

Herrick began to engage in what he called "missionary work, preaching the gospel of the pathology and clinical symptoms of acute coronary obstruction. ... I talked coronary occlusion almost ad nauseum. A few listened attentively, more, incredulously, the majority, indifferently." He began making the diagnosis of myocardial infarction on patients he saw in consultation before his disheartening 1912 presentation and spoke extensively of his insight at medical meetings in the Midwest in 1910 and 1911.[7] At a meeting of surgeons in the amphitheater of Rush Medical College in Chicago, "I can still see the quizzical look on Charlie Mayo's face as, from a front seat, he listened to, but was evidently not converted by, my sermon."[8,†] Finally, by

---

[*] His 1912 paper was also published as James B. Herrick, "Clinical Features of Sudden Obstruction of the Coronary Arteries," *JAMA* 59 (December 1912), 2015–20.

[†] "Charlie Mayo" refers to Dr. Charles Mayo, one of the founding brothers of the Mayo Clinic.

1919, Herrick's concepts were gaining traction. He published a paper that included an electrocardiogram from a 42-year-old physician Herrick diagnosed as having had a coronary thrombosis. When the patient died of pneumonia several months later, an autopsy revealed a healed myocardial infarction and an organized thrombus—a blood clot—in the left anterior descending coronary artery.[9,10] The electrocardiogram of the patient was taken in Herrick's office 41 days after the physician had first shown symptoms of coronary obstruction.[11,12] The healed myocardial infarction was a smoking gun.

Herrick's description of the pathology and clinical symptoms of a heart attack have changed how medicine approaches coronary disease, though with characteristic modesty he wrote, "These facts had been written about before. They had to be rediscovered."[13] And his early advocacy of electrocardiograms to help physicians diagnose acute myocardial infarctions was a paradigm shift. In his own words, "The electrocardiogram not only aided materially in diagnosis, but offered hope that it might reveal the location of the resulting myocardial damage, and, in a measure, its extent, and thus be helpful in the way of prognosis and treatment."[14]

Keeping a Sense of Perspective

Bioprinting aspires to change how medicine approaches coronary artery disease, too. We're going to check out three studies of bioprinted heart cells including the bioprinting of so-called cardiac tissue patches. Before we begin our next survey of bioprinted solid organs, this is a good time to recall just what we are and what we are not doing in this book.

We're trying to get a foothold on the current state of bioprinting and we're sampling morsels from the mountains of scientific knowledge that have enabled bioprinting to come into being. We're not comparing the bioprinting paradigm to other means of tackling problems such as reducing organ transplant waiting lists or developing new stem cell regenerative therapies.

On the subject of the heart, cardiac tissue patches, bioprinted and otherwise, are one of many potential solutions to the current impasse in achieving meaningful results using stem cells to help with heart disease. Other procedures that might be used for heart disease, such as genetic editing and trophic factor delivery, are outside our purview unless they also entail bioprinting.

Since we're committed to getting a foothold on bioprinting, we will only end up with, at best, a toehold on the bigger picture of the race to translate engineered tissues to the clinic—that is, to you.

The Sistine Chapel Ceiling Pre-Michelangelo

In 2015 biomedical engineer Robert Langer said,

> Right now, we can't make any organ or tissue to function properly in humans other than skin. 3D printing hasn't even been used successfully for that. Making an organ also requires blood vessels, nerves, and cells to behave properly. 3D printing can't do that.[15]

As we continue to learn about the current state of bioprinting, it's important to remember that "right now, only a human body can grow a complex human organ."[16]

The point of the last paragraph is not to be a party pooper. It's only to remind us that currently, bioprinting is like the ceiling of the Sistine Chapel before Michelangelo took several deep breaths, spit on his hands, and got the job done. At this point in time, bioprinting is full of strenuous possibilities.[17] More than that, who can say? It may happen that an alternative means of tissue engineering will be first past the post with transplanted tissue in humans. But if bioprinting fails to deliver, it won't be for lack of trying.

High-Resolution Bioprinted Heart Patch

Researchers are using a variety of methods to engineer tissues that might help to replace the cells that are lost during a heart attack. During an attack, as many as 1 billion contracting heart cells—cardiomyocytes—can be lost. These cells don't regenerate. Instead, the heart creates scar tissue that is a poor substitute. Cardiac scar tissue can't transmit electrical signals nor can it contract. As a result, a person is at risk of heart failure following a heart attack. One approach to healing the heart is to construct a patch that can be placed on the stricken heart, a patch that can contract and synchronize its rhythm with that of the heart.[18]

A team led by biomedical engineers Brenda Ogle of the University of Minnesota, Jianyi Zhang of the University of Alabama–Birmingham, and Paul J. Campagnola of the University of Wisconsin at Madison printed a cardiac muscle patch. Their construct was a scaffold in the form of a grid, designed to mimic in minute detail the extracellular matrix of heart cells. They seeded the grid with human induced pluripotent stem cells directed to differentiate into cardiomyocytes.[19]

One of the neat features of the work of Ogle and colleagues was the spatial resolution they achieved. To remind you, spatial resolution is a measure of how close two objects can be and still allow you to tell them apart.‡ In this work they had spatial resolution of less than 1 micron. That is, submicron resolution rather than the common

‡ So spatial resolution refers to the size of the smallest possible feature that can be detected.

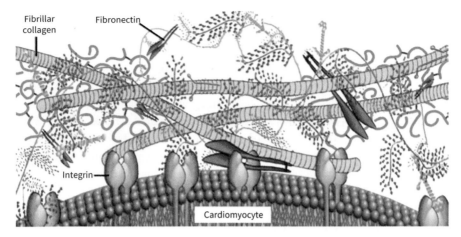

Figure 8.1 The extracellular matrix that surrounds cardiomyocytes. The visible portion of the cardiomyocyte cell is a portion of its lipid bilayer cell membrane with protruding transmembrane proteins called integrins.

Source: Reproduced by permission from Joseph M. Aamodt and David W. Grainger, "Extracellular Matrix-Based Biomaterial Scaffolds and the Host Response," *Biomaterials* 86 (April 2016): 68–82, https://doi.org/10.1016/j.biomaterials.2016.02.003.

1–10 micron resolution. That's spatial resolution of less than one-millionth of a meter. In everyday life it's like the tissues in your box of Kleenex. No, better still. It's like the two layers of an individual 2-ply tissue. Darn close together but you can still tell them apart.

To successfully print heart tissue you need to make the heart cells feel as comfortable as possible by giving them a homelike habitat. The extracellular matrix you provide them with needs to be just right. Figure 8.1 shows a cartoon of the extracellular matrix that surrounds cardiomyocytes.[20] With Ogle, Zhang, and Campagnola's printer, their spatial resolution is about the same as the feature size of the very components of that extracellular matrix. This allowed them to build a highly accurate matrix.

The team made a cardiac extracellular matrix architecture using a solution of photoactive gelatin polymers (gelMA) and cross-linked the polymers with the laser that's part of their printer. Their print capability was so exact that they could precisely control the position of the individual cross-links. In effect, what the team did was to make a tailor-made suit rather than settle for a store-bought one. That's how carefully they were able to design the in vitro extracellular matrix scaffold that they seeded with cardiomyocytes.

To accomplish this they used a rather technical strategy known as multiphoton-excited photochemistry and developed the printer that could pattern to their plan's high standards.[§] They used their instrument to make a three-dimensional

---

[§] Multiphoton-excited photochemistry is a fluorescence imaging technique that is similar to confocal microscopy that we've mentioned before. Multiphoton-excited photochemistry has an advantage in imaging optically thick specimens.

scan—a map—of the distribution of the structural protein fibronectin in the extracellular matrix of native mouse heart tissue and they converted this into a digital model.

As you can see in Figure 8.1, the protein fibronectin contains binding sites for other fibronectin molecules, for other components of the extracellular matrix, and also for cardiomyocyte cell surface receptors called integrins.[21] So fibronectin is prevalent throughout the extracellular matrix of cardiac tissue and a distribution map of this protein is a good way to reveal how cardiac tissue is structured. It's not unlike familiar thematic maps showing the variation of a topic across a geographic area; for example, a map showing the pattern of solar energy generating capacity in the United States (the southwestern part of the U.S. leads).[22]

Using immunofluorescent staining, the team found that fibronectin is roughly uniformly distributed around each cardiomyocyte in mouse heart tissue. Ogle and company used the fibronectin distribution to determine the dimensions of each individual compartment or channel in their gelMA scaffold that would serve as a surrogate for the native extracellular matrix, to be a niche for cardiomyocytes. Their analysis of the fibronectin distribution in native mouse heart tissue suggested that cardiomyocytes reside in channels that are about 15 microns wide by 100 microns long. They used this information to build their scaffold. Subsequently, they seeded the scaffold with cardiomyocytes to form a cardiac muscle patch, shown in Figure 8.2. The figure shows a fluorescent image of a human-induced pluripotent stem cell-derived cardiac muscle patch showing cardiomyocytes fitting snugly into the predesigned channels of the scaffold along with a higher magnification view.

Cardiac patches might significantly improve recovery from myocardial infarction if such patches could be readily made and applied soon enough. To understand why a cardiac muscle patch such as that of Ogle and friends could be valuable, let us reflect on what happens to the heart's extracellular matrix after a heart attack.

Heart Attack

A heart attack occurs because of a blockage in one of the coronary arteries that branch into the heart wall. We call them coronary arteries because they encircle the heart like a crown, and *coronary* derives from the Latin for *crown*.[23] The blockage is usually due to atherosclerosis and thrombosis that can lead to heart muscle damage and progress to heart failure.[24] *Atherosclerosis* refers to the buildup of fats, cholesterol, and other substances. The buildup is sometimes called plaque or hardening of the arteries and can restrict blood flow to the heart. Arterial *thrombosis* occurs when a blood clot blocks an artery. When a person has a heart attack, a portion of the heart muscle can die from oxygen deprivation. The heart does its best to remodel that part of its tissue that's been damaged. Yet this remodeling process results in big changes to the

Figure 8.2 Fluorescent image of a human induced pluripotent stem cell-derived cardiac muscle patch showing cardiomyocytes fitting snugly into the pre-designed channels of the scaffold along with a higher magnification view. The cardiomyocytes appear as pink; the blue stain shows the DNA within the cells' nuclei. Scale bars = 20 microns.

Source: Reproduced by permission from Ling Gao et al., "Myocardial Tissue Engineering with Cells Derived from Human-Induced Pluripotent Stem Cells and a Native-Like, High-Resolution, 3-Dimensionally Printed Scaffold," *Circulation Research* 120, no. 8 (April 14, 2017): 1318–25, https://doi.org/10.1161/circresaha.116.310277.

structure of the heart's extracellular matrix and results in the formation of a collagen-based scar.[25]

In a healthy heart, the composition and orientation of the extracellular matrix provides stability, helps transmit forces generated by the cardiomyocytes, and keeps the myocytes in the correct geometry.[26] Enzymes and other proteins that are present in the right ratio in a healthy heart strictly regulate the extracellular matrix composition and orientation. In Figure 8.1 you see that collagen fibers are part of the extracellular matrix. But be aware that the collagen isn't in a haphazard clutter like Mick Jagger's tousled locks on a bad hair day. Figure 8.1 is just a cartoon and the collagen fiber network in a healthy heart is well organized.[27]

In the first phase after a heart attack, collagen decays in the region of heart muscle that suffers from want of oxygen. Later there are further problems: The myocytes generate too much collagen in that region and in a haphazard orientation, and distorted collagen cross-linking occurs in the initial region of dead cells. These adverse outcomes spread to other regions of the heart and result in fibrotic scarring and more pathological remodeling.[28],** The cascade continues with reduced compliance—reduced ability of a heart chamber to expand when filled with blood—and increased stiffness of the heart muscle. Finally, heart failure sets in with swelling, shortness of breath, weakness, and fatigue. This is a grim reality and there is no cure for heart failure.

Back from the Abyss

Returning from our contemplation of the abyss, Ogle and team made a myocardial tissue equivalent using a cardiomyocyte-seeded scaffold and this they deemed a human cardiac muscle patch. The scaffold was made to mimic the native extracellular matrix: The photoactive gelatin polymers pinch-hit for the native extracellular matrix. Because of the submicron spatial resolution of their bioprinter, they were able to position cross-links in the gelMA with the precision of a tightrope walker with no safety net. The team seeded the scaffold in vitro with approximately 50,000 cardiomyocytes derived from human induced pluripotent stem cells.

To improve the ambience of their scaffold so that cardiomyocytes would feel more welcome, they added smooth muscle cells and endothelial cells to the seed mix. Within one day of seeding, the heart patch beat synchronously across its entire extent—measuring 2 mm × 2 mm × 100 microns thick—and generated calcium transients (more on this shortly). Note that the thickness is only 100 microns, so oxygen and nutrients are supplied by diffusion.

Over seven days in culture, Ogle and colleagues recorded significant increases in the beating speeds of contraction and relaxation, the peak strength of the calcium transients, and the expression of several relevant genes. The team had gone to great lengths to design a scaffold for the heart cells and fluorescence techniques rewarded them with images that showed the cells had aligned well with the long axis of the engineered channels. Not only was there proper orientation but the cells had also elongated to achieve a length-to-width ratio very close to that which is characteristic of native cardiomyocytes.

Next the team moved on to in vivo work in mice that had a myocardial infarction. Applying their heart patch to these mice, they found an engraftment rate of roughly 25% one week after transplantation—that is, 25% of the transplanted tissue cells remained engrafted in the hearts of the mice. *Engraftment* is the incorporation of grafted tissue onto the native tissue, and *engrafted* is often used in the literature as another way of saying *grafted*. The engraftment rate declined to only 11% by week four, though the researchers commented that this is still much higher than the rate reported in previous studies of cell-based myocardial therapy.[29]

** Fibrotic scarring describes the replacement of previously healthy tissue with tissue formed by an excess of fibers.

Goosebump Moment

I had an enjoyable conversation with Ogle, who shared with me the circumstances that led her to become a biomedical engineer.[30] She had thought to go into medicine but it didn't really click with her. After college and still uncertain of her career path, she became a nanny for a year. The father whose children she cared for was an orthopedic surgeon. She shadowed him at his workplace to gain insight into the profession and she still wasn't drawn in. However, he had a good friend with a Ph.D. in biomedical engineering and with whom he was starting a company to design spinal implants. Ogle discovered the field of biomedical engineering and knew that was what she wanted to do.

Her first academic position was at the University of Wisconsin–Madison. She joined the Wisconsin biomedical engineering faculty just a year before molecular biologist James Thomson discovered a way to make human induced pluripotent stem cells.[††] As Ogle told me, Thompson's discovery meant that she was in the right place at the right time for research using these new cell types:

> Jamie would go around to different groups—we hosted him at our group—showing people how to make these cells. That was the turning point for my lab when we started using induced pluripotent stem cells. It was a goosebump moment.

I asked Ogle about the prospect of a less invasive technique than open-heart surgery to introduce cardiac patches. She said this was an issue that must be addressed and mentioned the work of biomedical engineer Milica Radisic of the University of Toronto. Radisic is pursuing an injectable hydrogel that would promote survival and localization of the cardiomyocytes.[31,32]

I was also interested to learn more about the engraftment rate, a feature of her study that needs improvement.[33] In fact, this is part of the challenge for anyone trying to transplant cardiac patches by any method. Another barrier to success in transplanting stem cell-derived cardiac patches is the development of life-threatening arrhythmias.[34] The problem comes under the aegis of electromechanical coupling.

To explain the puzzle of electromechanical coupling: It's very tough to get the transplanted material to mechanically join the native tissue and remain viable; that's the *mechanical* part of electromechanical. Even if engraftment is successful, if the transplanted cells march to their own beat rather than the beat of the native tissue, that will produce dangerous rhythmic disturbances in the combination of native plus transplanted cells; that's the *electro* part of electromechanical.

Going forward, Ogle and team are in the process of improving engraftment by transitioning to a porcine model. Pigs and humans are more compatible for this work

---

[††] This was a year after Shinya Yamanaka had demonstrated induced pluripotent stem cells in mice. Thomson and Yamanaka demonstrated induced pluripotent stem cells in humans contemporaneously.

than are mice and humans. Ogle phrased it this way: Pigs are "more in sync with the human condition."

Who Is to Blame for the Pipe Water?

In Ogle's work we had occasion to call out calcium for yet a third time in our story as integral to the functioning of various tissues. Before moving on to our next cardiac bioprinting study, let's consider the unusual circumstances that led to calcium being fingered as the instigator of the beating heart.

Dr. Sydney Ringer was a physician at University College Hospital in London in the late nineteenth century. By his time, scientists already knew that if you took the heart out of a frog it would continue to beat on its own for some time. Ringer wanted to understand in detail why this was so. If you placed the heart in a dish of saline solution (water containing ordinary table salt), what chemical components were essential to keep the heart beating?

In 1882 Ringer reported his work in *The Journal of Physiology*, the official journal of The Physiological Society, saying that all you need for the heart to beat is a simple solution of sodium chloride.[35] However, in 1883 Ringer published what has been called "one of the most important retractions in the whole of physiology."[36] This withdrawal appeared in the same journal and the most telling sentence was: "I discovered, that the saline solution which I had used had not been prepared with distilled water, but with pipe water supplied by the New River Water Company."[‡‡] Ringer immediately repeated his experiments using distilled water—water that's been boiled and the steam collected, leaving impurities behind when the water evaporates. This distillation removes microbes and minerals such as calcium and magnesium.

When Ringer added sodium chloride to distilled water and placed the heart in this solution, the heart stopped beating immediately. Ringer correctly concluded that there was a contaminant in the pipe water that is required for the heart to beat. He commissioned an analysis of the pipe water and added back the contaminants one by one. Only when he added back the calcium did he find the heart started beating again. So Ringer discovered that calcium was required for the heart to beat.[§§]

Ringer's fortuitous discovery that calcium carried the signal that initiated heart contraction was a landmark observation. As we saw earlier for nerve transmission, it's actually the calcium ion (Ca^{2+}) rather than the elemental form of calcium to which we're indebted for our heartbeat. Ca^{2+} really gets around. It's the simplest and most versatile intercellular messenger that we know.[37]

---

‡‡   Ringer is saying that his assistant is to blame for the pipe water.

§§   The heart has its own inherent electrical system. In the heart's electrical system, an electrical stimulus is generated intrinsically by a small mass of specialized tissue in the right atrium (called the sinoatrial node). Both atria are then activated. The electrical stimulus travels down to another node (the atrioventricular node) and then further down through conduction pathways causing the ventricles to contract and pump out blood.

Pre-Vascularized Bioprinted Heart Patch

When doctors want to check your heart, a simple electrocardiogram shows if the heart's electrical firing pattern is normal, or too fast, or too slow, or erratic. The test can also show if there's an abnormality in the shape or the size of your heart. It's a good first test to find irregularity that would be associated with heart disease.[38] But an electrocardiogram isn't a very accurate gauge of your heart's pumping ability.

The next step up is an echocardiogram. This is an ultrasound of the heart and shows features of the internal structure and how blood flows through that structure. The ultrasound shows how well the heart valves are doing, how well the left and right sides of your heart are communicating with each other, the velocity of the blood exiting the heart, as well as the size and shape of the heart.[39]

Mechanical engineer Dong-Woo Cho of POSTECH in South Korea has developed a pre-vascularized bioprinted heart patch that uses decellularized extracellular matrix and cardiac progenitor cells in its preparation.[40] One of the striking features of this investigation was that Cho's team used multiple echocardiographic examinations at four and eight weeks after engrafting the patch onto the hearts of rats that had a myocardial infarction.

A Mini Noah's Ark

Cho used human cardiac progenitor cells and combined this with decellularized extracellular matrix from a pig's heart. There were a total of three separate bioink formulations used in this experiment: one with human cardiac progenitor cells, another with human mesenchymal stem cells plus vascular endothelial growth factors, and a third with both cardiac and mesenchymal cells plus vascular endothelial growth factors. (A vascular endothelial growth factor induces proliferation, migration, and permeability of endothelial cells.) The combination of cardiac and mesenchymal cells (plus vascular endothelial growth factors) when arranged in a patterned combination—rather than randomly mixed—resulted in the most successful cardiac patch.

Sometimes when you come across an experiment like this you feel as if you were on a mini Noah's Ark—cells from humans, extracellular matrix from pigs, echocardiograms on rats, and mice to examine vasculature formation. The mouse studies were encouraging. Four weeks after implanting the heart patch subcutaneously, the researchers found that blood vessels developed in the implanted patches. They appeared to be functional vessels because red blood cells were present.

Cho's team devised control groups with respect to the separate cell types and also controls to evaluate performance depending upon organization—whether the cells and growth factors were randomly mixed or carefully patterned. One distinct feature of the patterned structures was that they had the strongest development of vasculature: large numbers of vessels with diameters of around 50 microns, while the randomly mixed patches produced fewer blood vessels and their diameters were smaller than 22 microns.

In the context of a heart attack we spoke earlier about remodeling of the extracellular matrix. Cho's team found significantly less remodeling and fibrotic scarring after eight weeks in the rats (with a myocardial infarction) whose heart patch included the decellularized extracellular matrix compared to those that did not. And the synergistic effects of the combination of patterned cardiac and mesenchymal cells was apparent in their observation that the implanted patches were still engrafted—and tightly integrated—on the left ventricle wall of the rats' heart after eight weeks, again compared to controls that didn't flourish nearly as well.

As was true in the mice experiments, in the rat studies they also found mature blood vessel formation in the engrafted patterned patch. Rats being bigger than mice, these vessels were larger than 100 microns and contained red blood cells. In panel (a) of Figure 8.3, the printed patch you see is on the heart of a rat with a myocardial infarction.

Ejection Fraction and Echocardiograms

Each minute your heart pumps 1.5 gallons of blood. In one day or roughly 100,000 heartbeats, it shuttles 2,000 gallons of oxygen-rich blood through the approximately 100,000 miles of your entire circulatory system. Over a lifetime your heart pumps about a million barrels of blood—enough to fill three supertankers. And the little heart muscle that does this tips the scales in the neighborhood of 10 ounces.[41]

One of the most often cited measures of how well your heart is doing as determined by an echocardiogram is called the *ejection fraction*. The percentage of the blood within a ventricle that's pumped out in one heart muscle contraction is called the ejection fraction. The left ventricle of your heart is the main pumping chamber, and it's the left ventricle that's really at play in echocardiograms. The name ejection fraction stems from research done in 1962 by two physicians at the National Heart Institute in Bethesda, Maryland, who did a careful assessment of the "fraction of left ventricular volume ejected per beat."[42,43] In humans, ejection fractions of 55–70% are considered normal.

Cho's study used rats that had a reduced ejection fraction because of a heart attack. Cho's group aimed to restore function by means of a heart patch, and one measure of their success was the significant improvement in ejection fraction compared to control groups that didn't have the advantage of the heart patch. In vivo echocardiograms provided the data for panel (b) of Figure 8.3. We can see how much the ejection fraction was improved in the rat group that had the patterned patch with both cardiac and mesenchymal cells (plus vascular endothelial growth factors)—that's the patterned combination of Bioinks I + II shown in red in the legend on the graph.

Further confirming the value of ordered arrangement in bioprinting, in panel (b) the cell *compositions* were identical in the patterned Bioinks I + II patches and the

(a)

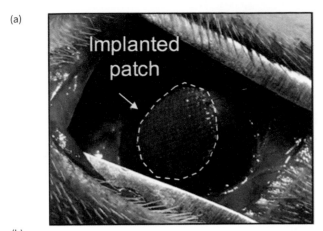

(b)

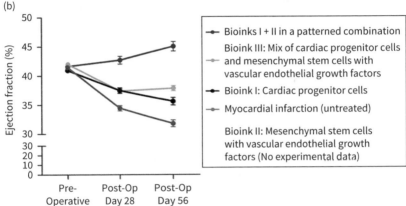

Figure 8.3 (a) Photograph of implanted stem cell patch in a rat. (b) Ejection fraction graphs as a function of time for patches of distinct composition and varying cell placement (patterned versus random mix).

Source: Reproduced by permission from Jinah Jang et al., "3D Printed Complex Tissue Construct Using Stem Cell-Laden Decellularized Extracellular Matrix Bioinks for Cardiac Repair," *Biomaterials* 112 (January 2017): 264–74, https://doi.org/10.1016/j.biomaterials.2016.10.026.

randomly mixed Bioink III patches, but the patterned patches had ejection fractions that were better by 18% than that of mix patches.

Fractional Shortening

Cho's team measured another feature of their heart patches with echocardiography. They determined the value of fractional shortening. Fractional shortening is the degree of shortening of the left ventricle diameter between the time when the heart relaxes and fills with blood and the time when the heart contracts to pump blood.

Fractional shortening, like ejection fraction, is a way of measuring the heart's performance. If you don't have sufficient fractional shortening, it suggests that the heart is not ejecting blood efficiently. Cho's work showed that the fractional shortening of Bioink I + II patches was improved by 23% compared to that of Bioink III patches as measured eight weeks after cardiac patch implantation.

Once again we see that in bioprinting, it's not only the ingredients you use in your stew; it's also how you add those ingredients together. Carefully patterned printed constructs performed better than printed constructs with randomly mixed constituents.

Bioprinted Functional and Contractile Cardiac Tissue

The healthy heart beats regularly 60–100 times per minute under normal resting conditions.[44] Each contraction of your ventricles represents one heartbeat. The contractile force of ventricular beating is an important measure of the health of your heart. En route to building cardiac tissues for implantation, chemical engineer Sang Jin Lee of the Wake Forest Institute for Regenerative Medicine led a team that bioprinted functional and contractile cardiac tissue and measured the contractile force of heart tissue in vitro. They used cardiomyocytes from rat hearts suspended in a bioink.[45]

The group built constructs 600 microns thick and with dimensions 1.8 cm × 1.6 cm. To maintain pattern alignment, they printed a polymer scaffold frame in the form of a rectangle and printed the cell-laden hydrogel to cross over the top and bottom edges of the frame. Following bioprinting, as the cardiac tissue self-organizes it shrinks for four weeks on the two sides not attached to the frame, after which it remains stable for at least nine weeks (panel (a) of Figure 8.4).

After three days in culture medium, the cardiac patch showed only weak, localized, and spontaneous contraction. One week later the tissue patch began spontaneous and synchronous beating in culture and became even more regular through weeks 3 and 4.

One of the novel elements of this work was the direct measurement of the contraction force of the beating heart tissue. The work was performed in vitro, and the team printed a micro-spring device attached to the frame that held the cardiac tissue to measure the contractile force when the cardiac tissue cells synchronously beat—shown in panel (b) of Figure 8.4.

Note that the labels accompanying the cartoons of the spring in panel (b) are familiar to us from blood pressure measurements. *Systole* is the period of contraction of the ventricles of your heart when that muscle ejects blood out to your body.[46] Blood leaves from each ventricle; a portion travels to the lungs for more oxygen, and blood that is already oxygen-rich travels to the rest of your body. *Diastole* is the period of relaxation of the heart muscle that's accompanied by the filling of the chambers with blood.[47]

(a)

Day 3

Week 1

Week 2

Week 3

Week 4

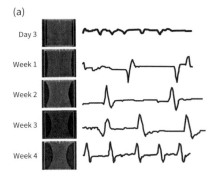

(b)

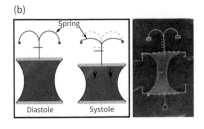

(c)

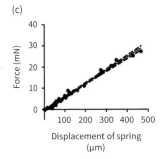

Figure 8.4 (a) Development of the cardiac tissue patch with time and spontaneous beating based on bright field video. (b) Contractile force measurement of bioprinted cardiac tissues by micro-spring integration illustrating micro-spring motion during diastolic and systolic phases of cardiac contraction cycle; image of micro-spring. (c) The displacement of the micro-spring is directly proportional to the force applied to it by the synchronous beating of the cardiac cells.

Source: Reproduced by permission of Zhan Wang et al., "3D Bioprinted Functional and Contractile Cardiac Tissue Constructs," *Acta Biomaterialia* 70 (April 2018): 48–56, https://doi.org/10.1016/j.actbio.2018.02.007.

When someone measures your blood pressure, the top number they report, the systolic blood pressure, is the pressure exerted on your arteries during systole which is normally about 100–120 mm of mercury. The bottom number they report, the diastolic blood pressure, is the residual pressure exerted on your arteries when the pumping part of the heart cycle is complete. This value is normally about 80 mm of mercury.

The systolic pressure is a way of assessing the *force* your heart exerts on the walls of your arteries each time it beats. The diastolic pressure assesses the force your heart exerts on the walls of your arteries in between beats.

Wait a minute. In Lee's experiment there were no arteries. And he couldn't put a blood pressure cuff on a bunch of cells. How did the team figure out the force from the attached micro-spring?

They didn't call this out by name, but the force determinations were a gift from the seventeenth century—English scientist and polymath Robert Hooke's explanation of elasticity known to us as Hooke's law.

Hooke's Law

Hooke interpreted his experimental results. Hooke's law is often written as $F = -kx$, where F is the restoring force a spring exerts on an object attached to it, k is the spring deformation stiffness, and x is the distance the spring is stretched or compressed.[***] The minus sign in the equation tells us that the restoring force of the spring is in the opposite direction to the spring's displacement. For example, during contraction, the beating cells in Lee's micro-spring experiment exert a force that pulls down on the spring. In response, the spring is stretched downward. The spring will then exert an upward restoring force that is equal to the force applied to it by the beating cardiac cells. Figure 8.4 (c) shows that Lee's micro-spring obeys Hooke's law.

Coming Back to Contractile Cardiac Tissue

Lee's team also measured the contraction force after administering various drugs and saw the forces increase or decrease depending on the drug they applied. In Figure 8.5, the team evaluated the dose effect of epinephrine on the contractile force from bioprinted cardiac tissue.

Epinephrine, also called adrenaline, is a hormone produced in your body that increases your cardiac output. This prepares us for the familiar fight–flight–freeze response to a perceived threat. The epinephrine dose in Lee's work ranged over a factor of 10,000 and was measured in units of micromoles (shown as µM in the graphs). The curves on the graph at the top of panel (a) show that at low doses nothing much happened to the measured contractile force and it remained in the neighborhood of 2.80 mN as quantitated in panel (b).[†††]

As shown in the bottom two graphs of panel (a), the team increased the dose to 20 µM and increased it again to 200 µM. And the peak contraction forces of the heart tissue shown in panel (b) increased to about 7 mN—raised by a factor of 2.5. The graphs of beat frequency and magnitude in the lower portion of panel (a) and the peak forces generated in panel (b) suggest that at those higher doses of epinephrine, if you chose flight over fight or freeze, your fleeing figure would be a blur. That is, higher doses of epinephrine made the heart cells beat faster and stronger. This is similar to what our own hearts do in response to a tense situation. When our adrenal glands secrete stress hormones including epinephrine, our hearts beat faster and stronger to prepare us for action in a nerve-racking situation.

[***] The spring deformation stiffness, k, is a property of the material the spring is made from and is a constant.
[†††] The units mN are milliNewtons. A Newton is an international unit of measure for force named after Sir Isaac Newton.

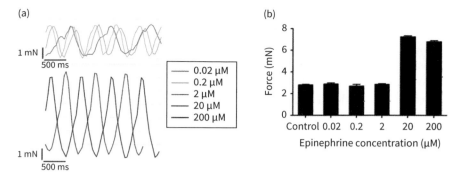

Figure 8.5 (a) Evaluation of the effect of epinephrine on the contractile force from bioprinted cardiac cells. The three graphs on the top portion show very little change in the magnitude of the force for low dosages of epinephrine, ranging from 0.02 micromoles to 2 micromoles. However, the two graphs on the bottom portion show a pronounced increase in the magnitude of the force at higher concentrations of epinephrine: 20 micromoles and 200 micromoles. (b) Quantification of peak force generated by bioprinted cardiac tissues in response to differing doses of epinephrine and including a control with no epinephrine.

Source: Reproduced by permission of Zhan Wang et al., "3D Bioprinted Functional and Contractile Cardiac Tissue Constructs," *Acta Biomaterialia* 70 (April 2018): 48–56, https://doi.org/10.1016/j.actbio.2018.02.007.

One last point about Lee's design concept. The team found that anchoring points were crucial for forming cardiac tissue. When tissues were not anchored to a polymer scaffold frame, they didn't observe orientation or maturation of the bioprinted cardiomyocytes. This is consistent with other studies that show that engineered cardiac tissues respond to mechanical load with improved growth rates and a greater degree of differentiation.[48] In their native environment, cardiomyocytes respond to mechanical interactions between the extracellular matrix and cell surface receptors as well as physical interactions between neighboring cells.

These are cellular mechanotransduction mechanisms: processes in which cells sense and respond to mechanical stimuli by converting them to biochemical signals that produce specific cellular responses.[49] So in Lee's research we can imagine the polymer scaffold frame as giving the cells a workout as a stand-in for the mechanical stimulation they experience within their extracellular matrix.

Monty Python's famous tag line was, "And now for something completely different." Before taking on another solid organ, suppose we switch gears and make an asset out of a liability. Translating that last bit, one pesky issue with some of the studies we've looked at (liver, as an example) has been the limitation in the thickness of the printed tissues. We know that we come up against the diffusion limit if we bioprint tissues more than about 100–200 microns thick and if we fail to supply them with a means of perfusion.

Turning this situation on its head, what can we do with thin tissue samples if we've gone to all the trouble of trying to mimic the native cytoarchitecture and had encouraging results in functional evaluations? One answer is to use the thin tissue samples for drug testing. Enter Eroom's law: Moore's law spelled backwards.

In Chapter 2, Jennifer Lewis called out Moore's law that describes the exponential increase in the number of transistors in a computer chip. The number doubled every two years for decades as the dimensions of transistors continued to shrink owing to more advanced fabrication techniques. An evaluation of Eroom's law in 2012 considered the trend in the number of new U.S. FDA-approved drugs per billion U.S. dollars of research and development spending in the drug industry. The study found that the number of new drugs has halved approximately every nine years since 1950.[50,‡‡‡] That's a bit like Moore's law in reverse gear and tells us that drug discovery models don't work too well.

To provide improved models for drug discovery, thin bioprinted, three-dimensional tissue samples have been partnered with continuously perfused, micron-sized chambers on a silicon chip for drug discovery and development and to provide model physiological functions of tissues and organs. The goal here is not to build an entire organ but to make minimal functional units that re-create tissue- and organ-level functions. Also, these so-called organs-on-a-chip might reduce reliance on animal testing, lessening the number of animals used for experimental procedures that can cause pain and suffering as well as the euthanizing of those animals.

Organs-on-a-chip are being developed for virtually every organ you can think of and for many combinations of organs on a single chip. An organ that is prominent in this growing endeavor is the lung. The reason for this isn't hard to understand.

Why do you breathe? You breathe because every part of your body needs oxygen to survive. Every bit of your daily life—digesting your food, moving your muscles, or even just thinking—requires oxygen. (Recall that cells use the oxygen during so-called cellular respiration to derive energy from the nutrients you ingest.) In all of these processes, the cells of your body produce carbon dioxide as a waste product. Your lungs have the responsibility of providing all your cells with oxygen and getting rid of the resulting waste gas, carbon dioxide.[51]

When breathing becomes an arduous task, your lungs will let you know they're having trouble doing their job. Walking up a steep hill, your leg muscles may tire, your heart rate will likely accelerate, you'll have to breathe more frequently, and depending on your level of fitness, you may become lightheaded. Even when not exercising, if the air you're breathing is contaminated, your lungs will react adversely. If the air pollution should reach toxic levels, you might become sick. Or you could die.

---

‡‡‡ Dollars are inflation-adjusted.

9
Organs-on-a-Chip
Tissues for Testing

In the mid-twentieth century, Donora was abnormally smoky. The borough of Donora is 28 miles south of Pittsburgh, Pennsylvania. The fumes from the mills tended to linger because bluffs and precipitous hills flank the area. On October 26, 1948, the fog became acute. There were eight physicians in Donora at that time, and their phones were ringing continuously. Everyone who called said the same thing. "Pain in abdomen. Splitting headache. Nausea and vomiting. Choking and couldn't get their breath. Coughing up blood."[1] A temperature inversion had gripped the borough: a reversal of the typical atmospheric conditions such that the air near the earth was cooler than the air higher up. As a result, the accumulation of smoke and fog was immobilized and the gases and fumes weren't carried upward away from the ground by convection currents. Days later, rain helped to dispel the noxious fumes. But 20 people died and nearly 6,000 of the population of 14,000 had been sickened.[2]

An investigation showed that the scourge was essentially an irritation of the respiratory tract. Its severity increased in proportion to the age of the victim and their predisposition to cardiorespiratory ailments. The ultimate cause of death was suffocation.[3]

A Lilliputian Landscape

Your lungs are the primary organs of your respiratory system and their healthy functioning cannot be overstated. The very first organ-on-a-chip (not bioprinted) was a lung-on-a-chip. An organ-on-a-chip is a Lilliputian landscape of living cells that goes beyond the self-assembling cellular microstructures—the organoids or spheroids—we've previously described. It's the action of shear stress and the flow of circulating blood on cells that are a distinguishing feature of an organ-on-a-chip and make it a valuable test bed.

When researchers began to bioprint multiple organs on the same chip, the lung was a natural choice for inclusion. In this chapter, we will learn how the organ-on-a-chip came into being and review several studies of bioprinted organs-on-a-chip that include the lung.

The word *chip* in organ-on-a-chip comes from the original fabrication method for organs-on-a-chip, a modified form of the semiconductor photolithography that's used to manufacture computer microchips. In semiconductor photolithography, the experimenter coats a surface with a light-sensitive polymer called a photoresist and applies a patterned mask to the surface; only unmasked areas of the polymer are exposed to light. Chemicals known as developers dissolve either the photoresist that was exposed or the photoresist that was not exposed, depending on the choice of photoresist. The end result is a transfer of the geometry of the patterned mask to form a three-dimensional relief image on the surface.

The organs-on-a-chip gang appropriated the word *chip*. And it can't be wondered at, since photolithography allows control of surface feature shapes and sizes on the same scale (from nanometers to microns) that living cells sense and respond to in their natural tissue environment.[4]

Before we describe the decades-long march to the realization of miniaturized, bioinspired, engineered organs, we provide two figures to get a sense of the brave new world of organs-on-a-chip. Figure 9.1 shows a lung-on-a-chip, the first-ever organ-on-a-chip. It was all contained within a piece of clear, hardened, rubber polymer smaller than a budget hotel soap bar. If you were to cut the pictured chip in half

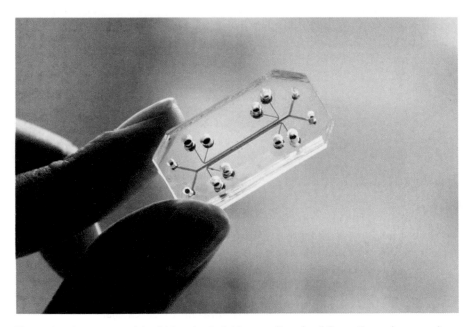

Figure 9.1 Lung-on-a-chip. A blood-mimicking medium (red) flows through a vascular channel and air (blue) flows through airway channels each lined by living human cells. A pump connected to the chip imitates the physical forces that breathing exerts on lung tissue.

Source: Reproduced by permission from the Wyss Institute for Biologically Inspired Engineering at Harvard University.

and look into the long channel lines, you would see what is shown in Figure 9.2. The blood-mimicking medium is carried in the passageway below the capillary cells and air flows through the passageway above the lung cells.

The group that built this chip came up with a design that reproduces the actual mechanical distortion of the lung tissue–capillary interface in our lungs that's caused by breathing movements. They incorporated pressure-driven stretching by including two large, side chambers in the device design. When the experimenters applied a vacuum to these chambers, they produced deformation of the thin wall that separates the cell-containing channels from the side chambers. This caused

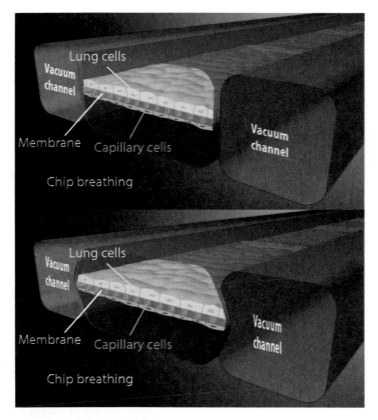

Figure 9.2 A blood-mimicking medium (red) flows through a vascular channel and air (blue) flows through airway channels each lined by living human cells. The lung mimic device uses compartmentalized channels to form a lung tissue–capillary barrier on a porous, flexible membrane coated with extracellular matrix. The device re-creates physiological breathing movements by applying a vacuum to the side chambers that causes mechanical stretching of the membrane forming the lung tissue––capillary barrier.

Source: Reproduced by permission from the Wyss Institute for Biologically Inspired Engineering at Harvard University.

stretching of the porous, flexible membrane coated with extracellular matrix and the tissue layers on either side of the membrane. When they released the vacuum, elastic recoil of the thin wall caused the membrane and attached cells to relax to their original size.

Donald Ingber's Voyage to Organs-on-a-Chip

Now we sally forth with an account of how physician, biologist, and bioengineer Donald Ingber put organs-on-a-chip on the map. Ingber is a professor at the Harvard Medical School and the Harvard School of Engineering and Applied Sciences, as well as the founding director of the Wyss Institute for Biologically Inspired Engineering at Harvard. It's a happy circumstance that at the end of the previous chapter we spoke about cellular mechanotransduction since Ingber started down the path to organs-on-a-chip because of his fascination with mechanical forces within cells. That goes clear back to his school days at Yale.

As a child, Ingber loved art but didn't find the occasion to express this passion. He entered Yale at age 17 majoring in molecular biophysics and biochemistry. The incredibly rich environment for arts and the humanities stimulated him. Because there were so many art majors compared to the number of art classes, he found it almost impossible—since he wasn't an art major—to get into an art class. He recalled having interviews through the night to gain acceptance to art classes. One interview was at 2:30 a.m. for a painting class. He didn't get in.[5]

It so happened that one day he saw students walking nearby with sculptures that drew his attention—they looked like viruses he'd seen in biology texts. He asked about these curiosities and the students said they had built them in an art class called Three Dimensional Design. Ingber's girlfriend was an art major and she was taking a class with the same professor who taught Three Dimensional Design. She introduced Ingber to the professor, and Ingber was able to talk his way into the class.[6]

Tensegrity—Say What?

The professor taught Ingber about a type of three-dimensional construction called *tensegrity*. (Sculptor Kenneth Snelson was the first person to demonstrate the principle of tensegrity and R. Buckminster Fuller coined the word tensegrity—a portmanteau combining tensile and integrity.[7,8])

Ingber described tensegrity as "an architectural system composed of opposing tension and compression elements that balance each other, thereby creating a resting tension or 'prestress' that stabilizes the entire structure."[9] He explained that tensegrity structures are mechanically stable because of the way the structure *as a whole* distributes and balances mechanical stresses, rather than the strength of individual members of the structure.

Prestress means that even before any external force is applied, all the structural members are already in tension or compression. Ingber and his colleagues demonstrated that cells are constructed like tents with prestressed cables and membranes that are winched in against internal struts and external tethers. These cellular components are analogous to tent poles, ground pegs, and ties to tree branches and allow prestress of the entire cell and thus give it mechanical stability.[10] His recognition that tensegrity provided a framework for cellular mechanotransduction gave him insight at the molecular level.[11]

As one example, Ingber predicted that because of their tensegrity structure, cells are able to sense mechanical forces through specific surface proteins that link the cell's external environment (especially the extracellular matrix) to the cell's internal cytoskeleton.[12] What's more, he subsequently performed experiments that identified these cell surface proteins.[*,13]

Ingber's work showed that mechanical forces are crucial to the function of cells, tissues, and organs.[14] So he had this wonderful revelation. How to leverage it for the greater good at a time when molecular growth factors, not mechanotransduction, were considered the nonpareil explanation of cell function and tissue development? Ingber joined forces with eminent Harvard chemist George Whitesides. In Ingber's words,

> It was the challenge of convincing other biologists that cell shape distortion is a physiologically relevant control element that led me to collaborate with George Whitesides, and to first apply his soft lithography-based microchip manufacturing approach to engineer culture substrates that can control cell shape and function.[15]

Geometric Control of Cell Life and Death

Whitesides had found an inexpensive way to make microchips and Ingber saw this as an opportunity. The microprinting technique that Whitesides invented begins with patterning a piece of silicon using standard photolithographic methods. After etching the patterns into the silicon, the next step is pouring on a rubber polymer. Adding a cross-linking agent at high temperature hardens the rubber and at that point it's taken off the silicon. The details of the surface topography of the etched silicon are now imprinted in the hardened rubber stamp.[16]

Ingber used Whitesides' rubber molds by inking them with proteins and stamping out cell-sized islands of extracellular matrix proteins in a medley of patterns.[†] This triggered several startling results: Cells grown on protein islands of contrasting shapes behaved differently even when cultured with identical growth factors and conditions.

---

* They turned out to be the integrins that poke out of the cell membrane in Figure 8.1.
† This was much less expensive than using standard photolithography for each experiment.

In one study, Ingber, Whitesides, and colleagues showed that hepatocytes, the most prevalent cells of the liver, could be made to fit the boundaries of the protein islands. They actually made hepatocytes in the shape of squares or rectangles—with corners that approximated 90°—when they grew the cells on protein matrix islands shaped as squares or rectangles.[17] Talk about a bestiary of cells, indeed!

In subsequent experiments, Ingber, Whitesides, and colleagues found even more evidence in support of Ingber's conviction that mechanical interactions are critical to both the function and the fate of cells and, by extension, to tissues and organs.[18] Experiments revealed that cells grown on and confined to divergently sized protein matrix islands were "growing when spread flat on big islands and turning on a suicide program when crowded on tiny islands."[19] This was more proof that mechanical forces have a dramatic influence on cell behavior. In fact, it doesn't get much more profound, as reflected in the title of their write-up, "Geometric Control of Cell Life and Death."[20]

Miniaturized Plumbing

For Ingber, there were two more breakthroughs before reaching the promised land of organs-on-a-chip. The first of these came from another collaboration with Whitesides, this time to bring *microfluidics* into play. Microfluidics is essentially a field dedicated to miniaturized plumbing and fluid manipulation.[21] Microfluidic systems manipulate *really* small amounts of fluid in the range of nanoliters to picoliters. A liter is about the size of a quart container of motor oil you would purchase to change the oil in your car. A nanoliter is one-billionth of a liter; a picoliter is one-trillionth of a liter.

To get a feeling for these small numbers, remember that bioprinting began when Thomas Boland modified a standard inkjet printer and used it to print proteins and cells. Such a printer has a droplet volume of in the region of 1–5 picoliters.[22] For comparison, a drip from a leaky faucet—as measured by the United States Geological Survey, though not a scientific definition—has been guesstimated to have a volume of ¼ milliliter.[23] This is equal to 250 million picoliters.

Microfluidic devices developed in the 1960s showed that fluid components could be miniaturized and linked together, inspired by the precedent of microelectronic circuits miniaturized and integrated on a chip.[24] In the case of electronics, the vision was to have an entire computer on a chip, while with microfluidics the vision was of an entire lab-on-a-chip. Ingber's group started putting cells into the microfluidic chips. By controlling fluid flow, they realized, you could control a cell's microenvironment.[25]

As Ingber explained, "You could place cells in different positions relative to each other and deliver different chemicals to different points on the surface."[26] So now he could manipulate cells in microenvironments that mimic those in the human body. Here's the final step, the second breakthrough we called out: apply all this in a physiology-centered context.

Ingber, Whitesides, and colleagues used Whitesides' rubber stamp method to create microfluidic channels and to culture cells under flow conditions inside these channels.[‡,27] Later, another group fabricated a microfluidic chip device having an air-filled channel that was similar in size and shape to a lung bronchiole—the smallest air passageways in the lung.[28] They focused on a specific symptom that presents in patients with pulmonary diseases such as asthma, pneumonia, cystic fibrosis, and other conditions.[29] The symptom is a respiratory crackling sound that can be heard through a stethoscope.

A breath of air that is opening airways clogged with thick fluid plugs causes the plugs to burst when air expands the lungs during breathing. Doctors have long considered the crackling sound from these bursts as a symptom of pulmonary disease. This new microfluidic study showed that the crackling is not only a symptom but also a serious event. A co-author said, "We've shown that these liquid plugs are injurious, particularly when they rupture. The rupture sends a very strong stress wave onto the cells."[30]

These researchers created a microfabricated plug generator and connected it to the cell culture chamber on the chip. The plug generator was a vial of liquid; scientists pumped air into the vial so that drops of liquid entered the simulated airways of the chip and in due course the drops burst.[31] They tested for periods of 10 minutes and found that at least 24% of the epithelial cells lining the simulated lung bronchiole had died after persistent exposure to bursting liquid plugs. Ingber knew of this work and recalled, "The sound that came out of this little rubber chip was exactly the same sound I was taught in medical school to listen for with the stethoscope when you suspected a patient might have pneumonia or 'fluid-on-the-lungs.' It was mind-blowing."[32]

One of the authors of this work applied to Ingber's lab at Boston Children's Hospital:

> When Dan Huh applied to be a postdoctoral fellow in my laboratory, I told him how impressed I was with this breakthrough. However, I asked, "Why don't we build a real living, breathing 'lung on a chip' lined by human lung cells and vascular endothelium interfaced in the center of two microfluidic channels?" And this is precisely what Dan did, which led to our lung alveolus chip publication.[33]

Ingber is referring to work that is widely acknowledged as the first true organ-on-a-chip.[34] Ingber's group devised a biomimetic microsystem that reconstituted the critical alveolar–capillary interface of the human lung. In discussing Figure 4.1 in Chapter 4, we mentioned alveoli as being those small sacs that are so vital to us for re-oxygenating blood in the capillary beds of our lungs.

---

‡ To create a channel from a rubber stamp, the process begins by etching a silicon wafer to form a raised feature. After pouring on a rubber polymer, cross-linking to harden the rubber polymer, and removing the rubber from the silicon, you're left with a channel in the polymer stamp corresponding to the raised feature on the silicon wafer. A plasma (ionized gas) treatment of the rubber stamp and a glass coverslip bonds the stamp and the glass via surface chemistry reactions to form a *closed* channel.

Ingber explained, "What makes an organ an organ isn't just tissue. It's having two or more different tissues that come together and form an interface, where new functions emerge."[35] His team went further to capture not only the lung's biological environment but also its mechanical environment: They added simulated breathing motions to mimic those of the lung, as we explained earlier.

The lung-on-a-chip was a watershed. The group showed that their creation could imitate the lung's response to infection and airborne particulates.[36] Just two years later, they showed that the chips could mimic the complicated human disease state of pulmonary edema—and demonstrated the ability of a new drug to prevent this life-threatening condition.[§,37,38] Other successes soon followed, such as human gut-on-a-chip.[39] As Ingber said,

> We get organ-level function in the chips because we put tissue/tissue interfaces, fluid flow and mechanical forces—breathing motions for the lung, peristaltic motions [coordinated waves of contraction and relaxation] for the gut—together. Without all that, you don't get in vivo physiology.[40]

We now have the story of the emergence of organs-on-a-chip in our cache of knowledge. Organs-on-a-chip might end up to be one of the bulwarks of bioprinting down the road. It's time now to examine a selection of bioprinting studies using organs-on-a-chip.

Bioprinted Three-Tissue Organ-on-a-Chip Platform

Dr. Anthony Atala and tissue engineer Thomas Shupe, both of the Wake Forest Institute for Regenerative Medicine, led a large multi-university collaboration in creating an assortment of bioengineered tissue organoids and tissue constructs integrated together in a closed circulatory perfusion system to facilitate inter-organ responses.[41] Tissues in the body do not exist in isolation but, rather, in a highly integrated and dynamically interactive environment. What happens in one tissue can affect other downstream tissues, just as in your neighborhood. If the people living two houses away have one of their trees fall on the main power line, you might experience a power outage.

Atala, Shupe, and colleagues built a three-tissue organ-on-a-chip system composed of liver, heart, and lung tissues and highlighted examples of inter-organ responses to drug administration. They found drug responses that depend on inter-tissue interaction—including one that took them completely by surprise. Their experiment illustrated the value of multiple tissue integration for in vitro assessment of

§ Pulmonary edema is a condition caused by excess fluid in the lungs, making it difficult to breathe.

both the efficacy and side effects associated with drug candidates. They incorporated supportive cells such as endothelial cells and fibroblasts as well as structural extracellular matrix proteins to re-create the dynamic interactions of cells with one another and with the cellular microenvironment. Figure 9.3 gives a schematic impression of their setup.

The team created each tissue—liver, heart, and lung—with the cell types present in the native human tissue, and with similar relative proportions where feasible. The group bioprinted customized tissue-specific bioinks for liver and cardiac organoids in a microfluidic design envelope and connected these to a lung organoid at the air–liquid interface.

The experimental blueprint was an adroit one that allowed the three tissue organoids to respond to a miscellany of external stimuli independently or in concert, which mimics organ dynamics in the human body. The group developed biosensors to provide for real-time data acquisition. To showcase this new platform they

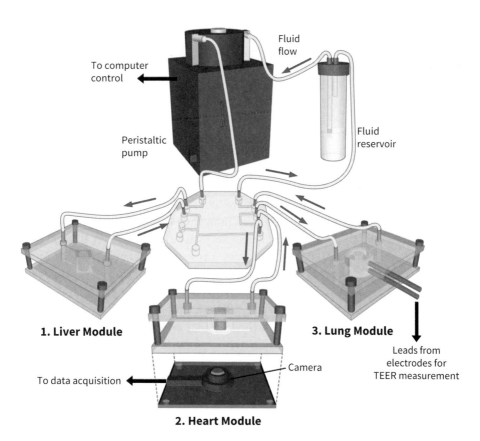

Figure 9.3 Schematic of the three-tissue organ-on-a-chip system.

Source: Reproduced by permission from Aleksander Skardal et al., "Multi-Tissue Interactions in an Integrated Three-Tissue Organ-on-a-Chip Platform," *Scientific Reports* 7, no. 1 (August 18, 2017), https://doi.org/10.1038/s41598-017-08879-x.

performed physiological characterization, functional testing, and drug and toxi-cology screening—and critically, they demonstrated integrated responses to agents that highlight the importance of inter-organ interactions. Before we charge into the fray, we comment on the label transepithelial/transendothelial electrical resistance and provide a primer on our lungs.

TEER is an acronym for transepithelial/transendothelial electrical resistance.** It's a widely used sensor for quantitatively measuring (and monitoring over time) the integrity of the junction between endothelial and epithelial layers.[42] TEER values are indicators of the sturdiness of the cellular barriers that Atala, Shupe, and friends were evaluating for drug transport.[††] Before venturing further into this intricate research we must take a breather and pay a very brief visit to our lungs to better understand these cellular barriers.

Our respiratory system has numerous components that humidify, warm, filter, and conduct the air we breathe until that air reaches its destination and gas exchange takes place.[43] At journey's end, the respiratory system basically consists of a branching set of air spaces (the alveoli) that are in close proximity to pulmonary capillaries.[44] These alveoli are the primary gas exchange units of the lungs.[45] There is an extremely thin gas–blood barrier between the alveoli and the pulmonary capillaries that allows for rapid gas exchange. In order for oxygen from the air we breathe to reach the blood, it must diffuse through this barrier that consists of a layer of epithelial cells (the alve-olar epithelium), a thin (about one micron) intervening space of extracellular matrix, and a layer of capillary endothelium. Carbon dioxide from the blood follows the re-verse course to reach the alveoli.[46] You can see a simulation of this three-layer barrier in Figure 9.2, where the lung cells provide the alveolar epithelium, the membrane is coated with extracellular matrix, and the capillary cells provide the capillary endothe-lium. By the way, one often quoted source estimates the number of alveoli in a human lung at 480 million with a range of 274–790 million.[47]

Rejoining Atala and Shupe, they housed individual tissue constructs in modular microreactors—small versions of the bioreactors we introduced at the beginning of this book. Liver, cardiac, and lung organoids have their own microreactors. These reactors are connected serially under fluid flow so the setup allows for a modular *plug and play* capability, as the authors write. Plug and play is a bit of jargon from the com-puter world. It means that devices work with a computer system as soon as they're connected rather than requiring any further messing about.

The collaborators formed spherical liver organoids with human hepatocytes and additional types of liver cells that are normally present along with hepatocytes. They also formed spherical cardiac organoids using induced pluripotent stem cells. For both the liver and the cardiac organoids encapsulated in hydrogels, they bio-printed straight into their respective microreactors. The team used a liver-specific

---

** Rather than pelt you like a hard rain, we've scrupulously abstained from inundating you with acro-nyms in this chronicle. TEER is the exception.

†† In our bodies, this cellular barrier acts to protect us against respiratory viruses in the airway epithelium.

extracellular matrix bioink for the liver organoids and a fibrin–gelatin combination bioink for the cardiac organoids. Both liver and cardiac organoids were close to 250 microns in diameter.

Atala approached the lung module by employing a semi-porous membrane and seeding it with layers of epithelial cells, lung fibroblasts, and lung endothelial cells. Self-assembly of the layered lung organoid produced a cellular organization similar to native airway tissue: an epithelial surface exposed to the air–liquid interface, lung fibroblasts for tissue structure, and an endothelium as a thin vascular barrier exposed to liquid media. The advantages of the layered technique of lung organoid formation are simplicity and the ability to rapidly establish an organized tissue.

Flexing Tissue Module Muscles One at a Time

Atala and Shupe first tested their liver, cardiac, and lung modules individually before going on to a two-module combination of liver and heart, and ultimately the three-module combination. To begin, they worked with their liver module and performed tests to measure urea (a metabolic waste product) and albumin outputs such as we've seen earlier. They also checked out the functionality of key liver enzymes by monitoring the metabolism of diazepam, perhaps more familiar to you by the brand name Valium. Their liver module passed the test by metabolizing that widely used medicine into its appropriate metabolites (products of metabolic breakdown).

Moving on, they first instigated a toxic attack on the liver and then a counter-measure. They exposed the liver organoids to acetaminophen. This pain medication provides palliative relief to many people but it can cause liver damage in higher doses. After running the liver module for five uneventful days, on day 6 they treated the liver organoids with one of the following: no drug whatsoever, increasing doses of acetaminophen, or high-dose acetaminophen along with a drug (called N-acetyl-L-cysteine) to counteract the anticipated liver toxicity of high-dose acetaminophen.

The results of these trial runs of the liver module were reassuring. The group evaluated cell viability and found the no-drug group did very well, the low-dose group showed reduced viability, and the high-dose group had very few viable cells remaining. Co-administering the acetaminophen and the countermeasure drug maintained the liver cells' viability. In fact, the organoid morphology appeared more like that of constructs that received the lower acetaminophen treatment.

The team also biochemically assessed the circulating, nutrient-filled media for every drug condition by measuring albumin and urea output of the liver organoids. Results showed a significant cytotoxic response—toxicity at the cellular level—of the organoids to acetaminophen and also their rescue by the countermeasure treatment, accurately reflecting clinical responses seen in human patients. This provided more confidence that the liver module was mimicking human reactions.

The cardiac organoids showed normal biomarker expression and long-term viability. Even so, cardiac toxicity may not necessarily show up as cell death. Toxicity can show up as a change in function. The most common form of cardiac toxicity is arrhythmia—an irregular or abnormal heart beat rhythm. So cardiac organoid beat rates are a good measure of cardiac organoid health. An onboard camera system customized to integrate with the cardiac microreactor housing provided this essential information.

As we saw in Chapter 8, epinephrine increases heart rate in humans. Prescription medications called beta blockers suppress the effects of epinephrine. When you take beta blockers, your heart beats more slowly and with less force and thus reduces your blood pressure.[48] The research team used epinephrine and the beta blocker propranolol to examine their efficiency at inducing and preventing cardiac organoid beating rate increases.

With the right software and real-time video, it's possible to manipulate images of cardiac organoids to zone in on organoid contraction in each video frame and get not only a visual representation of organoid beating but also a quantitative measure of beating rates. As with the acetaminophen liver tests, they used a range of doses of epinephrine to determine the lowest concentration that initiated a clearly discernable increase in beating rate.

With this knowledge in hand, they used a range of propranolol, incubated organoids at these doses for 20 minutes, and added the lowest concentration of epinephrine that had previously initiated rate increases. Increasing concentrations of propranolol more effectively prevented epinephrine-induced increases in beating rates, demonstrating an appropriate beta-blocking response in the presence of epinephrine. So far, so good.

Flexing Tissue Module Muscles Two at a Time: Liver- and Heart-on-a-Chip Interactive Tests

For their initial demonstration of multi-organoid interactions, the researchers performed experiments in which the functionality of a cardiac construct was dependent on the metabolic capabilities of an upstream liver construct. What this means is that first the liver gets to do its thing on whatever insult comes along, after which the heart will reveal the net effect of the insult. The team did this for single-drug response in the two-organoid system, followed by two-drug response in the two-organoid system.

To begin we have a cardiac-only system—no liver present. Propranolol by itself gave a small yet significant decrease in the beating rate of the heart organoids. When the liver module was plugged in upstream of the heart, there was no decrease in beating rate, indicating that the liver organoid had metabolized the propranolol before it reached the heart. Similarly, epinephrine by itself gave a significant increase in beating rate when only the cardiac organoid was present. When they plugged in the liver organoid, the liver component did not negate the epinephrine-induced beat rate increase but did reduce the effect to around a 30% beat rate increase. This also validated the integrated response of the two-organoid system with only one drug present.

Next, they assessed the interplay between both drugs in cardiac-only and two-organoid setups. So now we have two drugs evaluated with either the heart alone or the heart plus the liver. In a turnabout from our earlier explanations, this time they administered propranolol first and added epinephrine subsequently. Depending on whether liver organoids were present, epinephrine effects varied. With only cardiac organoids, propranolol remained active and successfully checked the effects of epinephrine. That's because there was no liver component to metabolize the blocking agent—the propranolol.

When both cardiac and liver were in the circuit, they found a 25% increase in beat rate after they added epinephrine as compared to a 50% increase in cardiac beat rate in control groups where they didn't administer any propranolol. This suggested that the liver organoids were able to metabolize a sufficient amount of propranolol so that epinephrine induced a beat rate increase of around half of the control (epinephrine-only) response.

The takeaway is that multiple organoid systems have advantages over single organoid systems in revealing drug responses.

Lung Module Characterization

Following the two-organoid interactive investigation, the experimenters directed their attention to the lung for integration into the test platform. The lung is a unique point of entry for either toxic particles in the air or, on the plus side, aerosolized drugs. Before proceeding with three-organoid interactive experiments, the team did electrophysiological sensing of the lung organoid to confirm that it responded in a functional, physiological manner to a variety of tests, including its TEER response to histamine administration.[‡‡] Histamine is known to cause permeability in lung endothelial tissue, and the team confirmed this was also true with their model system when TEER measurements showed a pronounced dip in electrical resistance after they introduced histamine.

Three at a Time: Liver, Heart, and Lung Interactive Tests—Enter Sherlock Holmes

With the lung module completing its preflight inspection, the team was ready to go for broke. One initial goal of this experiment was to use bleomycin to demonstrate a specific targeted insult to the lung while not harming the liver or cardiac organoids.

[‡‡] Histamines are made by your immune system and released during inflammatory and allergic reactions. The antihistamines we take for our allergies are designed to block histamines.

Bleomycin is a drug used against some cancers that causes significant lung fibrosis and inflammation. As such, it's commonly used in lung-specific studies. So the objective was to establish the capability of the platform to specifically target only one tissue in the system (the lung) with a drug, not all three tissues.

After two days of warm-up—running the platform with all three organoids in the loop—the group infused bleomycin into the circulating fluid on day 3. On day 9, the study was concluded and cell viability analysis of both the control group (no bleomycin) and the bleomycin-treated groups showed only low numbers of dead cells in all three organoids. On that basis, all three organoids seemed to have weathered the drug tolerably well.

However, the team did observe a change in cardiac organoid morphology under bleomycin exposure. The cardiac organoids appeared less tightly aggregated. They were no longer compact and healthy; their sphericity was compromised. Meanwhile, the cardiac organoids in the no-bleomycin control platform didn't show any morphological change. To see this, in Figure 9.4 compare the cardiac organoid in the rightmost figure in panel (b) to the cardiac organoid in the rightmost figure of panel (a).[§§] For example, look at the organoid close to the 1:00–2:00 o'clock position in both panels.

The group also did a quantitative evaluation. The untreated cardiac organoids had a consistent beat rate. Not so with those in the bleomycin platform. These cardiac organoids had completely ceased beating—center graph of Figure 9.4, panel (c). The cardiomyocytes were not dead cells, yet their collective beating had ceased. This is an (extreme) instance of arrhythmia, the cardiac toxicity that we called out earlier.

The result was completely unexpected. Bleomycin is not known to cause cardiotoxicity. So they repeated their work using a cardiac-only system. And bleomycin did not cause cessation of cardiac organoid beating. Intriguingly, these results *suggested* that rather than directly affecting the cardiac organoids, bleomycin may induce production of a secondary factor from one of the other tissues in the platform.

We now take on the role of Dr. Watson and follow the group's Holmesian sleuthing in clever pursuit of the culprit. In the best tradition of Sherlock Holmes, Atala and Shupe had not formed a firm conclusion—beyond the elementary facts that the miscreant was a Freemason, left-handed, chewed tobacco, and had traveled in the East.[49]

Bleomycin *is* known to cause *upregulation* of inflammatory product secretion (i.e., increased secretion) in lung tissue and lung epithelium. This enhances cell function and production of inflammatory mediators. Knowing this led the sleuths to assess lung inflammatory factors by sampling the circulating fluid media for interleukin-8.

Interleukins are a large family of messenger molecules in the form of secreted proteins. One of their jobs is to regulate immune responses such as inflammation. Interleukins aren't stored within cells but are instead secreted rapidly in response to a

---

[§§] As to the *liver* organoid, don't be fooled by the differences in staining between panels (a) and (b)— that's only a preparative artifact. The bleomycin did not change the liver organoid morphologies—that is, their sphericities (how they compare to a perfect sphere).

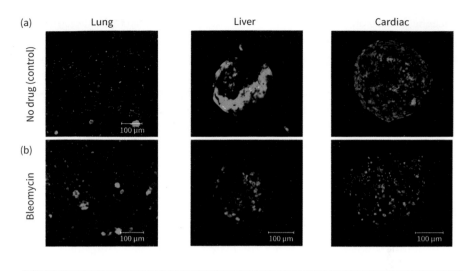

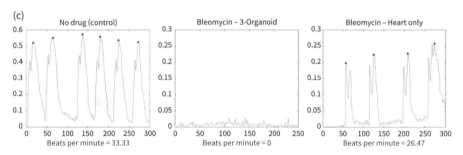

Figure 9.4 Unanticipated side effect in the three organoid liver, heart, and lung system—the drug bleomycin induces lung inflammatory factor-driven cardiotoxicity. (a) and (b) Viability of lung, liver, and heart organoids with no drug administration and with administration of bleomycin. The cardiac organoid morphology—the sphericity of the cardiac organoid with bleomycin—appears somewhat disaggregated. (c) Cardiac organoids ceased beating completely under bleomycin treatment. Note: Liver and cardiac organoids are close to 250 microns in diameter. All scale bars = 100 μm.

Source: Reproduced by permission from Aleksander Skardal et al., "Multi-Tissue Interactions in an Integrated Three-Tissue Organ-on-a-Chip Platform," *Scientific Reports* 7, no. 1 (August 18, 2017), https://doi.org/10.1038/s41598-017-08879-x.

stimulus. Once the body produces an interleukin, it travels to its target cell and binds to it via a receptor molecule on the cell's surface. This triggers a cascade of signals within the target cell that alters its behavior.[50]

Atala and Shupe found the interleukin-8 levels in the circulating fluid media (in the three-organoid system) were elevated. They also found elevated levels in tests on *isolated* bleomycin-exposed lung modules, verifying secretion from the lung constructs.

Since the interleukin-8 levels were elevated in the other bleomycin-treated organoids in the three-organoid system, they concluded that the lung constructs produced these inflammatory interlopers and the inflammatory agents quickly spread to the other organoids.

Not only that, the detectives also found a second inflammatory factor, interleukin-1. They quickly followed up on this additional clue by once more using lung modules in isolation and treating them with bleomycin—and these too produced interleukin-1. And the three-organoid system confirmed the spread of interleukin-1 to liver and cardiac organoids.

The Plot Thickens

Interleukin-8 is not regarded as a cardiotoxic factor. So their gumshoe work led them to the hypothesis that bleomycin-induced lung epithelium production of interleukin-1 contributed to inducing cardiotoxicity. And a literature search confirmed that interleukin-1 has been shown to cause cardiomyocyte toxicity previously and may be a side effect of bleomycin administration in vivo. The team wanted to get to the bottom of this and get convincing evidence that there were cardiotoxic effects from interleukin-1 but not interleukin-8. So they performed the following experiment.

They prepared cardiac organoids and verified that they had a consistent baseline beating rate throughout the batch. At this juncture the investigators added human *recombinant* interleukin-8 or interleukin-1 to the organoids in the cell culture media. Recombinant means that they used genetic engineering to create the interleukins that the body makes but doesn't store. They also maintained control organoids that received culture media with no interleukin-8 or interleukin-1. Over the course of six days they took videos and processed the videos to derive beat plots. These plots revealed that the cardiac organoids in the control group and interleukin-8 group had little, if any, significant changes in beating rate. Thus, there was no evidence that interleukin-8 was the guilty party.

By contrast, addition of interleukin-1 produced an increase in beating rate on the order of 60%. This rate increase endured for 96 hours. By the end of the experiment, the rate sharply decreased to levels significantly below those of the controls and interleukin-8 groups. Alas, by then the damage had been done. These results suggest that interleukin-1 has influence in the beating rates of cardiomyocytes.

We reiterate that cardiotoxicity is not typically attributed to the death of cells in the heart. Rather, in the case of many drugs that have been recalled from the market due to cardiotoxicity, the toxicity takes the insidious form of significant changes in beat rate (or beating kinetics as it's called) such as arrhythmias and other heart rate abnormalities. Importantly, these results demonstrate the detection of a toxic side effect that was not expected and would not be detectable without an integrated multi-organoid platform.

Atala and Shupe suspected from the outset that the butler (bleomycin) did it. He lulled their suspicions by acting incognito, but by elementary deduction they unmasked him. We must temper our zeal, however, because it's unclear whether this specific effect is present in humans who are given bleomycin, though the extant literature does link bleomycin and patient cardiac effects. What we can say is the work of Atala, Shupe, and team highlights the ability of multi-organ-on-a-chip platforms to uncover potential negative drug effects for further evaluation during the pre-clinical phases of drug development. If captured early in the drug development pipeline, such findings could save both lives and billions of dollars.

I had an opportunity to speak to the first author on this large collaborative project and was surprised to learn how deep his roots were.[51] Aleksander Skardal was a post-doc at the Wake Forest Institute for Regenerative Medicine when the project started and several years later became a faculty member at the Institute.[***] But earlier in his personal journey, he entered Johns Hopkins as an undergraduate chemistry major. He became interested in biomedical engineering and transferred into that program with a focus on biomaterials and tissue engineering. From there he became a graduate student in the bioengineering program at the University of Utah, in the lab of distinguished chemist Glenn Prestwich.

It was a good match for Skardal because Prestwich's lab was a subcontractor on a National Science Foundation Consortium that included Gabor Forgacs as principal investigator and Vladimir Mironov among others. It was referred to at the time as the *organ printing* consortium. Prestwich and Skardal were tasked with the development of what, in those days, was referred to as *biopaper* while Forgacs and Mironov worked on bioink. Biopaper was the early name for what we know as the *hydrogel* that provides lodgings for bioprinted organoids.[52,53] Skardal and Prestwich were really in at the ground level! Skardal remembers going back to his parents' house in New Hampshire in the summer previous to starting graduate school. They asked him, "Wait, uh, what is it you're going to be doing?" At the time it really was like science fiction.

While he was a student in Prestwich's lab, Skardal flew to Forgacs' lab at the University of Missouri every summer to spend several weeks testing his biopaper/hydrogel formulations with Forgacs' bioink/organoid constructs and Forgacs' state-of-the-art bioprinter.

Bioprinting a Functional Airway-on-a-Chip

The human airway is the primary site of exposure to noxious environmental agents and the consequent disorders instigated by these provocations, such as asthma, sinusitis, and chronic obstructive lung disease. Because of increasing air pollution, such

---

[***] Skardal is now on the faculty of the Department of Biomedical Engineering at The Ohio State University.

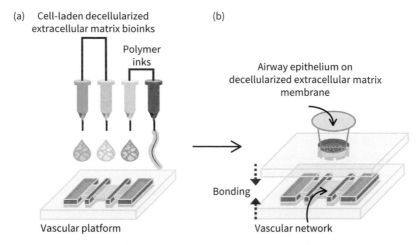

(a) Cell-laden decellularized extracellular matrix bioinks

Polymer inks

Vascular platform

(b)

Airway epithelium on decellularized extracellular matrix membrane

Bonding

Vascular network

Figure 9.5 Functional airway-on-a-chip: an airway model coupled with a bioprinted vascular platform. (a) Bioprinting of cell-laden decellularized extracellular matrix bioinks and two polymer inks (to create channels and to bond top and bottom frames). From left to right, the vascular platform contains a channel for cell growth media, a reservoir for lung fibroblast cells in decellularized extracellular matrix bioink, a cell growth media channel, endothelial cells bioink reservoir at the center, a cell growth media channel, a reservoir for lung fibroblast bioink, and a channel for cell growth media. (b) Assembly with airway epithelial cells.

Source: Reproduced by permission from Ju Young Park et al., "Development of a Functional Airway-on-a-Chip by 3D Cell Printing," *Biofabrication* 11, no. 1 (October 30, 2018): 015002, https://doi.org/10.1088/1758-5090/aae545.

respiratory problems are on the rise. There is need for an in vitro airway model to better characterize these inflammatory diseases and to vet potential medications for their treatment. Creating a model is challenging because human airways are so intricate with layers of dissimilar cells interacting with one another.

A team led by materials scientist Noo Li Jeon of Seoul National University and mechanical engineer Dong-Woo Cho of POSTECH in South Korea have risen to the occasion.[54] They have bioprinted a functional airway-on-a-chip coupled with a naturally derived blood vessel network. This work called for many cell types and adroit innovation. Jeon, Cho, and company created a vascular platform and investigated its function when integrated with an airway epithelium model. Using pathophysiological stimulation— that is, inducing disease states—they caused inflammatory responses in their model and quantitatively analyzed the effects of functional blood vessels to these offenses.

Panel (a) of Figure 9.5 shows the basis for their vascular platform. Two bioinks contained cells and two contained polymer scaffolding material. The bioinks that contained cells held stromal fibroblasts and endothelial cells.[†††] These cell-laden inks

---

[†††] Stromal refers to connective or supportive tissue, as opposed to functional tissue.

were encapsulated in a decellularized extracellular matrix hydrogel derived from mucous membrane in the trachea (windpipe) of pigs. The first of the polymer-scaffolding inks was used as a bonding material and the second, a biodegradable polyester, was the material for the frame of the airway-on-a-chip. This guide frame aligned an array of reservoirs and channels.

From left to right in panel (a), the bioprinted vascular platform consists of cell growth media, lung fibroblast cells in decellularized extracellular matrix bioink, more cell growth media, endothelial cell bioink reservoir at the center, cell growth media, more lung fibroblast bioink, and finally more cell growth media. To create a functional airway-on-a-chip, the team needed to couple the bioprinted vascular platform to an airway with a membrane intervening. They did this by inducing airway epithelium differentiation, accomplished by seeding primary human tracheal epithelial cells onto the pig tracheal mucous membrane. Yes, this airway-on-a-chip consisted of a mixture of components from two distinct animals: mucosal cells isolated from human trachea on top of a decellularized extracellular matrix isolated from pig trachea.

To complete the chip, the authors injected a bonding agent into the polymer frame to attach the upper chip, as shown in panel (b) of Figure 9.5.

Now we'll try to fill you in on how the vascular platform came to have a vascular network. For the endothelial cells, Jeon, Cho, and colleagues used human dermal microvascular endothelial cells that formed capillary-like structures on the hydrogel. To stimulate the growth of a vascular network, they used a potent *angiogenic* factor—a molecule for promoting the formation of blood vessels—called vascular endothelial growth factor. The natural vascular network formed after culturing for seven days. Their design enabled interactions between the separate cell types and triggered endothelial cells to assemble open structures and hence a three-dimensional vascular network shown as jagged red lines (panel (a) of Figure 9.6).

There's a subtlety to the vascularization story that lies in the geometry of the vascular platform and may not be simple to appreciate in panel (a) of Figure 9.6: For the lung fibroblast reservoirs, the polymer framework was printed only 100 microns high. By contrast, for those portions of the media channels not next to the lung fibroblast reservoirs, the polymer framework was printed 300 microns high (panel (b) of Figure 9.6). The media was introduced to the microchannels through a connecting tube. Cho's team adjusted the growth media to a level of 300 microns. This allowed the media flowing through the microchannels to submerge the fibroblast hydrogels and thus gather up angiogenic growth factors. So the growth factors secreted by the lung fibroblasts were able to diffuse into the endothelial cell reservoir through the openings between the polymer posts, thus promoting vascularization. The polymer posts are shown in blue in panel (a) of Figure 9.6. The vascular network the team formed was so stable it was maintained for three months in culture conditions in vitro.

(a)   (b)

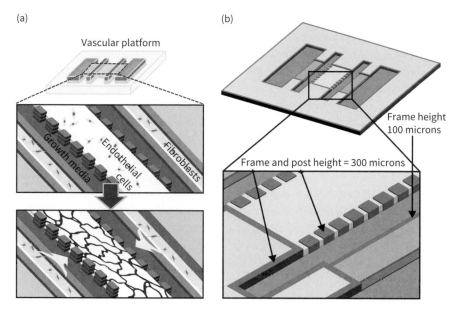

Figure 9.6 (a) Growth factors from the fibroblasts penetrate the center channel of endothelial cells through the openings between polymer posts. This drives endothelial cells to re-orient and create a network of blood vessels that are lined by a monolayer of endothelial cells. (b) It is the geometry of the vascular platform that allows growth media flow to submerge fibroblasts. That allows angiogenic (blood vessel forming) growth factors from fibroblasts to diffuse into the endothelial cell reservoir, as suggested in panel (a).

Source: Reproduced by permission from the Supplementary Material for Ju Young Park et al., "Development of a Functional Airway-on-a-Chip by 3D Cell Printing," *Biofabrication* 11, no. 1 (October 30, 2018): 015002, https://doi.org/10.1088/1758-5090/aae545.

Asthmatic Airway Epithelium on the Bioprinted Airway-on-a-Chip

In the previous investigation led by Atala and Shupe, we had our first exposure to interleukins. It's time for a second round. Before we dip into the role of interleukins in the Jeon and Cho work, it's best to learn just a little more about these mediators of inflammation.

Interleukins belong to a broader class of molecules called cytokines. Cytokines are a large group of molecules secreted by different types of cells throughout our body including cells of the immune system. They're a category of signaling molecules that mediate and regulate immunity, inflammation, and the process by which blood cells are formed. As signaling molecules, they act through cell surface receptors: Each cytokine has a matching cell surface receptor.

The binding event between a cytokine and its receptor molecule triggers a signaling cascade inside the cell to which it binds, and the signal ultimately reaches the cell's nucleus. Once at the nucleus, the effects of the cytokine are manifest in changes in gene transcription and protein expression—that is, certain genes may be turned on or off and consequently the gene's production of that protein may be stimulated or inhibited.[55] These phenomena will come into play in the Jeon/Cho study.

Jeon, Cho, and company wanted to challenge their epithelial/endothelial barrier with interleukins and see what happened. To set the stage, they began with a TEER measurement to check the structural integrity of their cellular barrier. Their evaluation showed that the airway epithelium in their vascular airway-on-a-chip was within the range of healthy human airway epithelium. This is a good indicator that the model they worked so hard to construct has relevance as a test bed for exploring pulmonary diseases and medications for these diseases.

Encouraged by these results, the team went further. They applied interleukin-13 for one week to the tissues at the bottom of the eye cup-like gizmo in panel (b) of Figure 9.5. Interleukin-13 is a cytokine that can induce features of allergic asthma. Sure enough, they observed pathological changes related to asthma. We won't show the immunofluorescence microscope images but simply an easily read graph. In Figure 9.7 you can see that the TEER value was reduced to less than 90% in the asthmatic case in which interleukin-13 was applied (so the cellular barrier was damaged), as compared with the control group that was spared the interleukin inquisition.

This is the second study we've looked at that uses this arcane TEER business. Here's why it continues to pop up. Clinical reference values—TEER numbers being a prime example—can quantitatively assess the epithelial barrier function. So TEER is a strong candidate for evaluating airway responses to new drugs or toxic substances using in vitro models and microfluidic organs-on-a-chip.[56] Despite its clinical significance, studies on conventional airway-on-a-chip models have heretofore failed

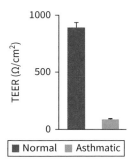

Figure 9.7 TEER values are reduced in the asthmatic tissue samples indicating the breakdown of the cellular barrier in the airway epithelium caused by interleukin-13.

Source: Reproduced by permission from Ju Young Park et al., "Development of a Functional Airway-on-a-Chip by 3D Cell Printing," *Biofabrication* 11, no. 1 (October 30, 2018): 015002, https://doi.org/10.1088/1758-5090/aae545.

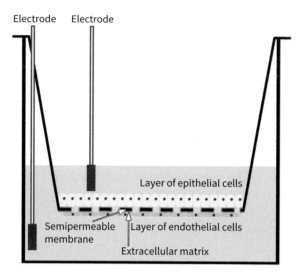

Figure 9.8 A classic TEER measurement. Note the thin layer of extracellular matrix (green) coating the semipermeable membrane between the layer of epithelial cells and the layer of endothelial cells.

Source: Reproduced by permission from Balaji Srinivasan et al., "TEER Measurement Techniques for in Vitro Barrier Model Systems," *Journal of Laboratory Automation* 20, no. 2 (April 2015): 107–26, https://doi.org/10.1177/2211068214561025.

to provide TEER values or showed very low values outside the TEER range of the healthy human airway epithelium.

Not so with the research of Jeon, Cho, and colleagues. Their bioprinted airway-on-a-chip may well have advanced the state-of-the-art by delivering a platform in which TEER values can offer meaningful measures. Figure 9.8 may help you to visualize the configuration of a classic TEER measurement.[57] Note, however, that although we've been speaking about TEER measurements across an air–liquid interface, in practice the TEER electrodes maintain electrical connection via fluid (e.g., cell culture media or a buffer solution). If there were not fluid on the top of the layer of cells, the electrical connection would be disrupted.[58]

Mucus

We move with an abrupt transition from the high-tech world of TEER to the low-tech world of mucus. Mucus secretion is the body's first resort against respiratory irritants. Mucus secretion is generally hard to observe in vitro and especially difficult in an asthmatic airway epithelium, yet the team did just that in their vascular airway-on-a-chip model. You can see this in Figure 9.9. They observed excessive mucus secretion from asthmatic airway epithelium that was more than 12 times normal. They also found the viscosity of the mucus secreted in the normal tissue was watery while that

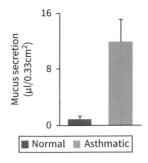

Figure 9.9 Mucus secretion was greatly increased in the asthmatic samples.

Source: Reproduced by permission from Ju Young Park et al., "Development of a Functional Airway-on-a-Chip by 3D Cell Printing," *Biofabrication* 11, no. 1 (October 30, 2018): 015002, https://doi.org/10.1088/1758-5090/aae545.

secreted by the vascular airway-on-a-chip was gelatinous. So asthmatic airway epithelium secreted thicker and more plentiful mucus.

The group strengthened its case. The interaction of the airway epithelium with the vascular network has a major function in inducing pathological responses from airway mucosa—and their in vitro model responded with in vivo-like characteristics. This is exactly what you would hope the in vitro experiment would reveal: Normal mucus secretion is modest and thin, while mucus secretion in asthmatic tissue is excessive and thick, just as in real life. We've described some messy findings, but it's no joke when people have asthmatic attacks. Good science like this might eventually contribute to alleviating their symptoms and suffering.

Invasion of the Dust Mites

This subheading would have been more technically correct if it had read, "Asthmatic Exacerbation by *Dermatophagoides* Mix on the Bioprinted Airway-on-a-Chip." Jeon and Cho used house dust mite *extract*, specifically, *Dermatophagoides* mix, on their airway-on-a-chip. (Derived from the ancient Greek, *dermis* means *skin* of course, *phagos* means *refers to feeding*, and the suffix *-oides* means *to look like*. Putting this together, the word *Dermatophagoides* roughly translates as *thing that looks like those that eat skin*.)[59]

An increased sensitivity and constrictive response of the airways to inhaled irritants is one of the symptoms of asthma. It's directly related to the severity of a person's asthma. It's also a complicated phenomenon. Airway mucosal inflammation can be induced and exacerbated by dust mites.[‡‡‡] There aren't a great many studies on the underlying mechanisms, and replicating these responses in vitro has been a

---

[‡‡‡] According to the Asthma and Allergy Foundation of America, dust mites may be the most common trigger of year-round allergies and asthma. They are on every continent except Antarctica.

predicament. So Jeon, Cho, and colleagues set out to test how exposure of their vascular airway-on-a-chip to dust mite extract could model hyperresponsiveness to house dust mites. Specifically, they wanted to see whether their chip would show expression of cell surface receptors on the vascular endothelium under fluid flow.

Using the dust mite extract, they stimulated both normal and asthmatic airway epithelium in their vascular airway-on-a-chip. What they were keen to find was the expression of a particular cell surface receptor. In the normal epithelium this protein was scarcely expressed. But the dust mite extract significantly increased the presence of the protein receptor on cell surfaces.

The protein receptor belongs to a family called cell adhesion molecules. Inflammatory cytokines stimulate the expression of adhesion molecules on the endothelium, which then enhances recruitment of immune cells into inflammatory sites. The adhesion molecules bring white blood cells in numbers to the cell surfaces. When cell adhesion molecules appear, it's like someone opening a large bag of potato chips; it ropes in the whole household.

The participation of cell adhesion molecules is consistent with the fact that most acute asthma patients have severe symptoms characterized by sensitive allergic inflammation, whereas non-asthmatic people usually don't respond to environmental allergens.[§§§]

Stimulation of the airway epithelium for four hours with dust mite extract triggered production of two inflammatory cytokines from both normal and asthmatic airway epithelia. The overall secretion of each of these cytokines was roughly double in the asthmatic case. You can see this in Figure 9.10. Jeon and Cho concluded that there was so-called cross-talk between epithelium and endothelium (signals from one tissue that modify the other tissue) during house dust mite-induced airway hyperresponsiveness. They suggested that the endothelium in their asthmatic vascularized airway-on-a-chip acted as a regulator of what is known as a *cytokine storm* (the phenomenon that took many lives in the COVID-19 (coronavirus) pandemic).

Return Engagement

Chuck Hull had a lot of fun inventing 3D printing. He could see that those busy with bioprinting were having lots of fun, too. Rather than be a spectator, he opted to be a participant and play a return engagement. Here's what he told me:

> I forgot, do you know I'm very involved with bioprinting? I have a project with the
> pharmaceutical company United Therapeutics to produce transplant lungs. Why
> lungs, the most difficult organ? Because United Therapeutics is very much the

---

[§§§] An allergen is a substance that's capable of triggering a response that starts in the immune system and results in an allergic reaction. Your immune system responds by releasing chemicals that typically cause symptoms in the nose, throat, eyes, ears, skin, or roof of the mouth.

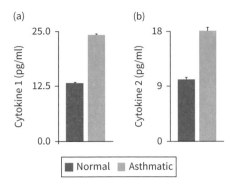

Figure 9.10 Stimulation of the airway epithelium with dust mite extract triggered production of inflammatory cytokines. The reaction in asthmatic tissue samples was twice that in normal tissue.

Source: Reproduced by permission from Ju Young Park et al., "Development of a Functional Airway-on-a-Chip by 3D Cell Printing," *Biofabrication* 11, no. 1 (October 30, 2018): 015002, https://doi.org/10.1088/1758-5090/aae545.

authority on lung health and they are motivated to solve the shortage of lung transplants. They do the biology work on the project and my team does the 3D printing work and we both work on all the things in between. As a poor non-biologist engineer/physicist I've had to get familiar with the topics you touch on [in the book], plus the detailed anatomy of these organs. This is obviously not a short-term project. I have several other bioprinting projects, less grandiose than this.[60]

Bioprinting Versus Manual Seeding

Some years before the work of Atala/Shupe and Jeon/Cho, cell biologist Barbara Rothen-Rutishauser of the Merkle Institute at the University of Fribourg in Switzerland conducted what may be the first bioprinted examination of an in vitro lung tissue analogue.[61] Rothen-Rutishauser and her colleagues reported on a three-dimensional air–blood tissue barrier analogue en route to an in vitro alveolar model. They used alveolar epithelial cells and endothelial cells (derived from human umbilical vein endothelial cells) on a layer of commercially available extracellular matrix material (Matrigel).

We will not present the complete study, only one notable result. Rothen-Rutishauser's team found that if they seeded cells by the conventional manual method on Matrigel, the cells grew in patches, often overgrowing one another and forming multi-layered clusters. (The study provided evidence that in the manual method, the cells were not seeded in clumps.)[62] In contrast to the manually seeded cells, bioprinted cells spread over the available growth surface to form confluent, thin monolayers.

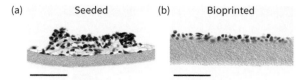

Figure 9.11 Bioprinting resulted in more uniform and homogeneous deposition than manual seeding. This induced optimal growth of the epithelial and endothelial cells. Scale bars = 100 microns.

Source: Reproduced by permission from Lenke Horváth et al., "Engineering an in Vitro Air–Blood Barrier by 3D Bioprinting," *Scientific Reports* 5, no. 1 (January 22, 2015), https://doi.org/10.1038/srep07974.

Figure 9.11 shows cross sections of both manually seeded and bioprinted thin tissue samples imaged in simple bright field microscopy. The staining shows cell nuclei as dark brown, cell cytoplasm as maroon, and collagen fibers of the Matrigel extracellular matrix material as light blue. In these side views, the bioprinted results are conspicuously thinner and more homogeneous than the seeded results.

Rothen-Rutishauser made the point that since conventional manual seeding resulted in the formation of multi-layered, often discrete, cell clusters embedded in thick Matrigel layers, this might inhibit direct cell–cell interactions. But with bioprinting, they generated thin layers of Matrigel that induced optimal growth of the epithelial and endothelial cells. This is more mimetic of actual physiological conditions and could be expected to promote greater cell–cell communication of dissimilar cell types and greater functionality.[63] There were no differences in cell viability.

Animal Testing

There has been increasing public awareness of the use of animals for testing drugs, antibiotics, and vaccines, for example, to measure both safety and efficacy. Many in the bioprinting community hope that organs-on-a-chip might reduce reliance on animal testing. Indeed, they also hope that by bioprinting partial or complete organ replacements using a human patient's own cells and maturing them in a bioreactor, they can lessen the number of animals used for experimental procedures that can cause pain and suffering as well as the euthanizing of those animals.

Since 2012, a company called Lush Cosmetics Limited has awarded what are called the Lush Prizes. Lush manufactures and markets handmade cosmetic products and is committed to opposing animal testing.[64] The annual Lush prize fund is £250,000 (about $340,000 as of the end of 2020) and the company awards this to scientists and others across five categories, including the Science Prize, the Public Awareness Prize, and so on. In 2017, the company awarded the Lush Science Prize to Jennifer Lewis and the Lewis bioprinting team. The Lush Science Prize is "for individuals, research teams, or institutions working on developing toxicity pathways, which are often better at predicting human reactions than mice, rats or rabbits."[65] The citation for the Lewis lab read as follows:

> Jennifer Lewis and her team have created organ-specific human tissues on a chip that would be suitable for drug screening, toxicology and disease modeling, as well as cosmetics testing. It was highlighted as one of the top 10 breakthrough technologies by MIT Technology Review.

Ironically, we've seen many animal experiments in this book that have been used in the attempt to bring bioprinting techniques to a level of reproducibility and sophistication where the benefits of reduced animal testing may also be realized. In Copenhagen in 2017, Dr. Donald Ingber gave a talk titled "Human Organs-on-Chips as Replacements for Animal Testing." During that talk, he said,

> Animal testing, which is required for regulatory approval and can take years to complete, often does not predict clinical responses in humans. Thus, when we founded the Wyss Institute, I focused one of our major platforms on this problem of developing alternatives to animal testing in order to accelerate drug development, and potentially to create a new approach to personalized medicine as well.

Ingber went on to say,

> While we cannot expect these devices to replace animal testing in the near future, if all goes well, we will see one animal model replaced at a time as data are generated that confirm chips lined with human cells can be equally robust as animal models, but produce even more relevant and predictive results.[66]

We left the bioprinting of solid organs in the lurch when we veered to bioprinted organs-on-a-chip. But we didn't abandon solid organs altogether, and to prove the point we now embark on an examination of the formidable task of bioprinting kidney tissue.

Whether it is bioprinting or some other method to help people with failing kidneys, those who would provide help must know what constitutes proper functioning. They must first have mastered how the organ is supposed to work. This requires investigative tools, commitment, and imagination.

As regards observational tools, the first microscopes (with more than one lens) emerged in the Netherlands in the late sixteenth century and provided a means to examine animals and plants in greater detail than had ever been possible.[67] Curious people began studies in many countries, chronicling and illustrating what they saw. Analytical tools took time to evolve and much of biology was descriptive. Nevertheless, that didn't preclude speculation on the functions of what these scientists observed. Not surprisingly, opinions differed and there was often a lack of consensus on what was happening on the microscopic scale. Some of these debates lasted for centuries.

10
The Kidney

The Ne Plus Ultra of Filters

The year is 1666. At the Academy of Messina, Italian physician and physiologist Marcello Malpighi is bent over his microscope examining the kidney anatomy of a frog. After injecting a mixture of black fluid and wine into the renal (kidney) artery, he notices distinct features in the kidney. He describes them as "dark vascular structures resembling fruits suspended on a branch."[1,2] He has made the first microscopic observation of what will come to be known as glomeruli.*

Malpighi goes further. He believes these structures link blood vessels to what are called fibers (to be renamed as tubules) that a colleague had recognized in earlier work.[3,4] Malpighi postulates that the glomeruli separate urine from blood and the urine somehow finds its way to the little canals known as fibers.[5,6]

Not everyone agrees with Malpighi's interpretation that the dark vascular structures he has observed are essentially filters. A Dutch anatomist has a competing hypothesis that involves secretion rather than filtration.[7,8] This sets off a scientific dispute.

For the next 260 years, scientists could not agree on whether the glomerulus was basically a filter or whether it was a part of the kidney's secretory system. The controversy was at the very crux of how the kidney works and revolved around filtration, reabsorption, and secretion.

Your kidneys have multiple functions: They *filter* your blood; they engage with blood vessels to *reabsorb* water, minerals, and nutrients; and they remove wastes and extra water to *secrete* urine. How do your kidneys accomplish their tasks? In the 260 years following Malpighi's account, scientists had made progress in unraveling the riddle of the kidney. But they still hadn't reached agreement on the physiology of urine formation. The issue centered on two theories: a theory that urine formation begins with glomerular filtration followed by tubule reabsorption versus a theory that urine formation begins with glomerular secretion.[9] No one had been able to devise a crucial experiment to determine the answer. Then a young physician from Harvard, Dr. Joseph Wearn, joined the academic medicine laboratory of Professor Alfred Newton Richards at the University of Pennsylvania. The year was 1921.[10,11]

---

* Malpighi is celebrated as one of the founders of the science of microscopic anatomy.

Wearn proposed that they use a newly developed micromanipulation technique to withdraw fluid directly from the glomerular structure of a living (anesthetized) frog's kidney for chemical analysis.[12] It was an exacting experiment. Wearn wrote,

> When I got into the problem of trying to withdraw glomerular fluid, the work proved to be so delicate that we could not have interruptions or even heavy footsteps that shook the apparatus. Therefore, I began to work at night when the only visitors to my laboratory were a friendly mouse and an occasional cockroach. Richards came in one evening and, seeing the light in my laboratory, looked in and from that time on throughout the time I worked with him, he came into the laboratory frequently for long hours of uninterrupted work in the evenings.[13]

Their experiments were a spectacular success. "The result was one of those simple, direct, unambiguous experiments that researchers dream of but seldom attain."[14]

Wearn collected glomerular fluid directly from the kidney and compared the chemical analysis of that fluid to bladder urine. His results showed that the normal constituents of blood—such as chloride and sugar—were present in the glomerular fluid but were reabsorbed between the glomerulus and the bladder. The report they released in 1924 was the first direct and definitive evidence of tubular reabsorption—providing strong evidence for glomerular filtration.[15] It showed that the kidney was a filter extraordinaire.

Together, Wearn and Richards solved the kidney controversy and ushered in the modern era of kidney physiology.[16] Their experiment established that urine formation begins with the passage of an ultrafiltrate of the blood across the glomerular capillaries, followed by the reabsorption of filtrate by the tubules. *Ultrafiltration* is so named because fluid filtered by the glomerulus is virtually free of large proteins and blood cells. Following glomerular filtration, the kidney tubules reabsorb what the body finds valuable for reuse from the glomerular filtrate—items such as glucose, protein building blocks, and water.[†] The kidney sends excess water and soluble cellular metabolic waste products such as urea and creatinine to the ureters that convey this to the bladder for excretion.

The Nephron Is the Kidney's Workhorse

When you urinate, you seldom give it a second thought. Although you're emptying your bladder, it's your kidneys that have done all the grunt work beforehand. And if your kidneys fail to filter properly, you're in a heap o' trouble. Before we get involved

---

† We speak of *re*absorption rather than absorption because the digestive tract initially absorbed the nutrients following ingestion of food.

in kidney tissue bioprinting, we need to expand on the pithy information we've provided so far on how our kidneys work.

The kidneys are usually described as two bean-shaped organs, each about the size of a fist. They're located just below the rib cage on either side of your spine.[17] The leftmost part of Figure 10.1 is a simplification borne of necessity since each of your kidneys has about 1 million nephron filtering units. Your kidneys take full advantage of this large number and filter about a half-cup of blood every minute, removing wastes and extra water to make urine. The entirety of your blood circulates through your kidneys many times a day. In a single day, your kidneys filter about 150 quarts of blood, though your body only contains about 4.8–6 quarts of blood.[18]

Figure 10.1 may seem complicated at first glance, but it isn't. When looking at the most detailed part of Figure 10.1 on the right-hand side, keep this in mind: *A nephron is basically an extremely fine tube.* At one end, the tube is folded into a cuplike structure (Bowman's capsule) that encloses a cluster of microscopic blood vessels called the glomerulus. Here, fluid filters out of the blood and into the tubule. The tubule

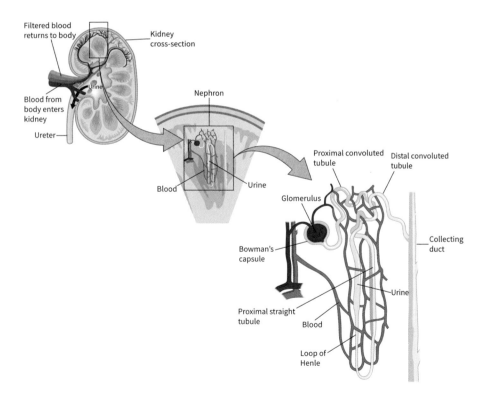

Figure 10.1 Cross-sectional view of a kidney; a single nephron; and the glomerulus, tubules, and associated capillary network.

Source: Reproduced by permission from Charles Molnar and Jane Gair, *Concepts of Biology: 1st Canadian Edition* (BCcampus, 2019), https://opentextbc.ca/biology.

(with individual names for its various sections) recycles items of value and leaves the rest as urine to be excreted.[19]

We've stressed the function of the kidney in removing waste products from the blood. This is true and it's one function that dialysis accomplishes when your kidneys can no longer fend for themselves. However, your kidneys do much more than filter out wastes. They help to keep your body on an even keel: Your kidneys help maintain balanced electrolytes and sustain fairly stable conditions necessary for survival.

Keeping on an Even Keel: Maintaining Electrolyte and Fluid Homeostasis

Let's unpack the phrase *maintaining electrolyte and fluid homeostasis*. We begin at the end: *homeostasis*. Homeostasis is a pillar of human physiology. The term was coined in 1926 by Harvard physiologist Walter B. Cannon.[20,21] *Homeo-* is derived from the ancient Greek word meaning *like* or *similar*, and *stasis* is from the Greek for *standing still*. Physiologists use the term homeostasis to mean "maintenance of nearly constant conditions in the internal environment [of the human body]."[22] All bodily tissues have functions that help maintain these constant conditions. Cannon's term became very popular and was appropriated for use in disciplines other than physiology after he wrote his most famous book, *The Wisdom of the Body*.[23] We started with the phrase *maintaining electrolyte and fluid homeostasis*. Now we move on to the adjectival modifiers of homeostasis: *electrolyte* and *fluid* in the context of the kidney.

To explain this, we draw on the venerable *Merck Manual* that started life as a small reference book for physicians and pharmacists in 1899 and is still chugging along today.[24] Water makes up more than half of the body's weight. It's convenient to think about the body's water as being restricted to various spaces, called fluid compartments. Three main compartments are fluid within cells, fluid in the space around cells, and blood plasma. To maintain normal function, the body must keep fluid levels from varying too much in each of these areas. [25] Now hold that thought as we introduce minerals.

Minerals are solid crystalline elements or compounds—inorganic substances that occur in nature.[26] The body needs lots of minerals, often called *essential minerals*. Essential minerals are sometimes categorized as major minerals and trace minerals. These two groups are both necessary for the body to function properly. The difference is that humans need larger amounts of major minerals and only smaller amounts of trace minerals.[27]

Some minerals function as *electrolytes*. Electrolytes are minerals that carry an electric charge when dissolved in a liquid such as blood. The principal blood electrolytes—sodium, potassium, chloride, and bicarbonate—help regulate nerve and muscle function. They also maintain the balance between acids and bases.[‡] In addition,

---

‡ There are multiple definitions of an acid (Arrhenius acid, Brønsted–Lowry acid, Lewis acid). The popular Brønsted–Lowry explanation is: An acid is a substance that is able to donate a hydrogen ion, H^+ (this

blood electrolytes aid in maintaining your body's water balance: The amount you take in whether by eating, drinking, or intravenously should equal the amount you lose through perspiration, respiration, urination, and defecation.

Electrolytes (especially sodium) help the body maintain normal fluid levels in the various fluid compartments because the amount of fluid a compartment contains depends on the concentration of electrolytes in it. If the electrolyte concentration is high, fluid moves into that compartment. Or, if the electrolyte concentration is low, fluid moves out of that compartment. To adjust fluid levels, the body can actively move electrolytes in or out of cells. So having electrolytes in the right concentrations is important in maintaining fluid balance in the fluid compartments. The kidneys help maintain electrolyte concentrations by filtering electrolytes and water from blood, returning some to the blood, and excreting what remains into the urine.

Here and elsewhere we've emphasized sodium. Why? All electrolytes found in the blood are important, and too much or too little of any one of them has consequences. Perhaps the reason that sodium is frequently singled out in the literature is that too much sodium can increase the risk of high blood pressure and cardiovascular disease, and cardiovascular disease is the leading cause of death in the United States and worldwide.[28,29,30]

To sum up, the kidneys support electrolyte and fluid homeostasis by helping to balance your excretion and consumption of electrolytes and water.

Proximal Convoluted Tubules—A Cherished Goal for Bioprinting

The proximal convoluted tubule is one of the gems of the kidney. For example, it has the virtue that it reabsorbs nearly all the glucose filtered by the glomerulus. And where would we be without glucose? It's the fuel that powers most of the biosphere, and each of our cells uses a steady stream of glucose for its metabolic needs.

We know of Jennifer Lewis. She and biologist Annie Moisan of Roche Innovation Center in Basel, Switzerland, led an effort to bioprint human renal proximal convoluted tubules. They surrounded the tubules with an extracellular matrix and housed them in perfusable chips.[31] You can steal a look ahead at panel (a) of Figure 10.2 for a cartoon of the proximal convoluted tubule. To get a sense of its place in the hierarchy of the kidney, have a glimpse back at Figure 10.1.§ The tubule is called proximal because it's the tubule that is proximal—closest—to the place where all the action of

---

is the same as a proton), and hence increases the concentration of H^+ when it dissolves in a solvent. The higher the concentration of hydrogen ions produced by an acid placed in a solvent, the higher is the acidity of the solution. A Brønsted–Lowry base is a substance that is able to accept a hydrogen ion, H^+.

§ In Lewis' original figure, they reversed the order of the two words *convoluted* and *proximal*. We're on the same page as Lewis and Moisan. However, they wanted to stress that the proximal tubules they fabricated are convoluted since other studies have only produced straight tubules.

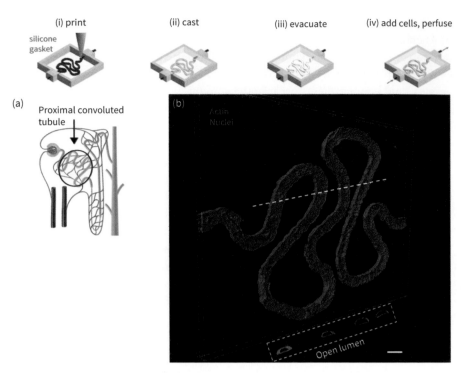

Figure 10.2 (i–iv) The steps in creating a kidney proximal convoluted tubule on a chip. In (i), Lewis, Moisan, and colleagues printed a silicone gasket, deposited extracellular matrix, and printed fugitive ink in a convoluted line. In (ii), they cast additional extracellular matrix to cover the line. In (iii), they evacuated the fugitive ink. In (iv), they added proximal tubule epithelial cells and then perfused. (a) Schematic of a nephron highlighting the proximal convoluted tubule. (b) A three-dimensional image captured by confocal microscopy of the printed proximal convoluted tubule. Actin is stained in red and nuclei are stained in blue, and the combination results in a purple rendering. The white dotted line denotes the location of the cross-sectional view shown below in which proximal tubule epithelial cells circumscribe the open lumen. Scale bar = 500 microns.

Source: Reproduced by permission from Kimberly A. Homan et al., "Bioprinting of 3D Convoluted Renal Proximal Tubules on Perfusable Chips," *Scientific Reports* 6, no. 1 (October 11, 2016), https://doi.org/10.1038/srep34845.

the nephron begins, namely the glomerulus surrounded by its epithelial cup known as Bowman's capsule. This is the site at which the process of urine formation begins with a filtrate of blood plasma. In each of the million or so nephrons in a kidney there is one glomerulus and an accompanying tubule. The tubule is composed of several sections. For purposes of understanding this piece of biological equipment, scientists call out individual segments, based on both morphological and functional distinctions. We'll run through them.

The proximal convoluted tubule drains filtrate away from the glomerulus. Next there's a portion called the proximal straight tubule. Following that, a segment called the loop of Henle descends, makes a hairpin turn, and comes back up. Then it segues into the *distal* convoluted tubule—the segment of all the convoluted structure that is furthest from the glomerulus. This segment makes a seamless transition to the last bit known as the collecting tube. The collecting tube makes its way to the enlarged upper end of the ureter and from there to your bladder to excrete urine. Quite a loop de loop ride.

Lewis, Moisan, and companions targeted the proximal convoluted tubule because it's the site most frequently damaged by drug-induced toxicity.[32] For example, it's a sad fact that many of the drugs we have in our arsenal to treat various cancers have nephrotoxic potential—the capacity to harm our kidneys. It's difficult to predict nephrotoxicity during preclinical drug development, and the nephrotoxic capacity of newly developed drugs is often underestimated. Renal toxicity accounts for 2% of failures in preclinical drug testing but it's responsible for almost 20% of failures in Phase III clinical trials.[33,**] Addressing this problem is one of the motivations of the work of Lewis, Moisan, and colleagues. (And they hope too that their proximal convoluted tubule can serve as a modular building block for the engineering of human nephrons and, perhaps ultimately, kidneys.)

The next question is: Why is the proximal convoluted tubule the part of the kidney that's most affected by drug toxicity? The answer is that the proximal convoluted tubule is responsible for 65–80% of nutrient absorption and transport from the renal filtrate—what the glomerulus produces—to the blood. As a result, circulating drugs and their metabolites often accumulate in the proximal convoluted tubule at high concentrations. Metabolites are the breakdown products the liver produces when acting on the drugs. Proximal tubule cells grown in traditional two-dimensional culture often lack, or rapidly lose, key characteristics such as phenotypic and functional aspects of the in vivo proximal tubule. In Chapter 3, we illustrated phenotype with P. G. Wodehouse's lovely description of his heroine. With the kidney's proximal tubule, the phenotype is not so lyrical. Still, it is what it is, and we'll elaborate on the phenotype in short order.

Apart from the work of Lewis, Moisan, and company, other researchers have developed kidney proximal tubule models including kidney organoids derived from pluripotent stem cells that have succeeded in representing various features of the nephron.[††] Even with these advanced techniques there remain problems: The kidney organoids are only roughly one millimeter in size. They also have no inlets and outlets

---

** FDA Phase III clinical trials are done to test efficacy and monitor adverse reactions. They usually involve several hundred to several thousand participants and test how well a new drug works compared with existing medications for the same condition. Phase III trials are larger and longer in duration than Phase II or Phase I clinical trials. Phase III studies are sometimes known as pivotal studies.

†† We note that Donald Ingber and colleagues have reported a human kidney proximal tubule-on-a-chip as well as a glomerular-capillary-wall-function on-a-chip. Also, the company Organovo that we mentioned in Chapter 7 reported their development of a bioprinted human proximal tubule model.

that the experimenter can individually access. So the proximal tubules within these miniscule organoids can't be directly probed and their perfusate—the fluid that's perfused through them—can't be easily collected and analyzed.

However, the study we've chosen to present is a method that combines bioprinting, three-dimensional cell culture, and organ-on-a-chip concepts to create a three-dimensional proximal convoluted tubule composed of a perfusable central cavity (the lumen in panel (b) of Figure 10.2). It also has a programmable architecture, which means that the architectural design can be changed by the experimenter or possibly by autonomous sensing.

The proximal convoluted tubule consists of an open cavity that is lined with proximal tubule epithelial cells. Lewis, Moisan, and colleagues embedded this in an extracellular matrix housed within a perfusable chip where they subjected the tubules to shear stresses similar to those found in the body. The unique combination of the three-dimensional geometry and controlled perfusion resulted in more differentiated proximal tubule epithelial cells than had ever been achieved before. In addition to reaching such a faithful replication of human proximal tubules, the team tested their creations with the nephrotoxin called cyclosporine A. They recorded the effects by direct imaging and also quantified the permeability of the epithelium in response to this toxic attack.

Starting out on the Road to Proximal Convoluted Tubules on Perfusable Chips

We start with a description of the steps that are shown in (i)–(iv) of Figure 10.2. In step (i), they printed a silicone gasket on a glass slide that demarcates the outer border of the three-dimensional tissue chip. They followed this by depositing an extracellular matrix composed of a gelatin–fibrin hydrogel within the gasket. Next they printed fugitive ink (shown in red) onto the extracellular matrix layer. In step (ii), they cast additional extracellular matrix around the printed feature.

After printing, they connected the fugitive ink to hollow metal pins interfaced through the gasket walls. In part (iii), we see that the group has evacuated the fugitive ink to create an open tubule. That's the counterintuitive step we commented on in Chapter 4—they cooled everything down to 4°C to liquefy the ink, and with its removal they're left with an open convoluted tubule channel embedded within the extracellular matrix.

Moving along to part (iv), they seeded the inside of the tubule with proximal tubule epithelial cells and perfused the seeded tubules. This sustained perfusion with culture media enabled the epithelial cells to grow inside the tubules to form a confluent layer. This takes about three weeks.

In panel (b), we see an image of a printed proximal convoluted tubule. Its purplish color comes from the use of two stains: Actin is stained in red and the nuclei of the epithelial cells are stained in blue. We're familiar with actin as the protein that forms filaments supporting the structure of virtually all cells, one of the primary proteins forming the cytoskeleton. Since both cell nuclei and the cell's cytoskeleton are ubiquitous, you can understand how the mingling of the two stains presents us with a purplish décor. The white dotted line denotes the location of the cross-sectional view shown below the main figure where we see the proximal tubule epithelial cells circumscribing the open lumen. They acquired the image using confocal microscopy.

Printed Proximal Tubules Form a Polarized Epithelium

There was much more to be done before the sun went down on the team's ambition, and they used a combination of light microscopy, scanning electron microscopy, and transmission electron microscopy to characterize the printed and perfused proximal tubules. We can build on our previous knowledge to fathom several of the results we'll present. To make sense of others, we need to pick up a few new terms.

The first one is polarity. Living organisms tend to develop with distinct anterior and posterior (or uppermost and lowermost) ends. To give an example from botany, when a plant sprouts from a seed, the shoot grows up toward the light and into the air while the root grows down into the soil.

Now consider cell polarity in our bodies. This refers to an intrinsic asymmetry that's observed in many types of cells, for example, in their shape, structure, function, or their organization of cellular components.[34] It's no coincidence that we introduce this term now because epithelial cells offer a great example of cell polarity. In the present context, epithelial cells line the inside of the proximal tubule. The part of the epithelial cell that faces the open passageway, the lumen of the tubule, is called the apical surface. The name is derived from the word apex, so the apical surface is the top of the cell. The bottom of epithelial cells is referred to as the basal surface, a name derived from the word base—think basement. So the polarization of the epithelial cells establishes an apical–basal polarity that reflects an asymmetrical distribution of lipids, protein complexes, and cytoskeletal components in the two portions of the cell that further reflects distinct functions.[35] Apical–basal polarity in cells isn't as familiar to us as shoot–root polarity in plants, but it amounts to the same thing.

What we're leading to is this: In the native epithelial cells that line the nephron's proximal tubule, there's a brush border of microvilli on the apical side facing the open lumen. (Brush border refers to closely packed little fingerlike projections that look like a brush under a microscope.)[‡‡] And this was what the research team found in their printed constructs—additional evidence that their in vitro model was faithful to in vivo reality.

[‡‡] Early anatomists working with simple microscopes saw a fuzzy structure that reminded them of the bristles of a paintbrush, so they deemed it a brush border.

Drug Toxicity Testing

Cyclosporine A is a drug commonly given to patients following transplant surgery to prevent rejection. Unfortunately, it's a known nephrotoxin that damages proximal tubule cells. The team exposed their perfused creations to various concentrations of the drug and monitored alterations of cell morphology and cytoskeletal organization by immunostaining of actin filaments. Microscope imaging revealed that the drug damaged the proximal tubule cells, including reorganizing the cell's actin; not a good outcome when you consider that actin forms the cell's cytoskeleton.

Besides microscopy, the researchers evaluated the drug-induced disruption of the epithelial barrier function by quantitatively measuring how leaky it became. One way they did this was by perfusing the tubules with cell media containing a fluorescing sugar polymer. Then they watched for leakage from the tubules to the surrounding extracellular matrix under live cell imaging with a fluorescent microscope. They found the barrier more and more permeable as the drug dose increased. It got so bad that the permeability increased by sixfold at higher drug doses.

However, the good news is that the results further indicated that their bioprinted constructs can be used to assess nephrotoxicity, and they did this both qualitatively with immunostaining and quantitatively by permeability measurements.

Vectorial Transport—A Directed Passage

Here is what the team was aiming for: a vascularized, proximal convoluted tubule model that would allow vectorial transport. This is movement of molecules in only one direction, from the epithelial-lined proximal tubule, through the intervening extracellular matrix, to the adjacent endothelial-lined vascular tubule: a molecular one-way street.[36] The proximal tubule and the adjacent vascular tubule are so close together that they're spoken of as being colocalized; in this experiment they were only around 70 microns apart.

Cell polarity is a fundamental feature needed for vectorial transport, and in the experiment we've been discussing, the team has amply demonstrated cell polarity. If you add vascularization to their renal tubule model, they will have at hand the endothelial-lined blood vessels needed to realize vectorial transport. In the kidney, vectorial transport means that nutrients are directed back to the bloodstream (and waste is sent to the urine).

After additional years of work they achieved that major milestone. Not perfect yet, but a huge accomplishment.[37] For epithelial cells in an absorptive structure like the proximal convoluted tubule, vectorial transport is the defining characteristic, the sine qua non that makes a proximal tubule a proximal tubule. To find brush borders and the like is great. To actually bear witness to vectorial transport is amazing. It's Tommy, the deaf, dumb, and blind Pinball Wizard, now able to hear and speak and see.

Their new exploit, showing renal reabsorption in a vascularized proximal tu-bule model, builds heavily on the accomplishments we've already presented. At the same time, they had to reinvent several key constituents. The two studies taken to-gether give us a well-documented example of how resourceful scientists reach their lofty goals.

To create vascularized models they had to modify the original extracellular matrix and fugitive ink used in their first experiment. To this end, they made an extracel-lular matrix from a gelatin–fibrin combination and labored to optimize its composi-tion. When they decreased the ratio of gelatin-to-fibrin from 7.5 to 0.4, they observed more than a fourfold reduction in the time required to achieve a confluent epithelium compared to proximal tubules embedded in their original extracellular matrix. It's like boiling a 20-minute egg in 5 minutes.

The group designed the modified extracellular matrix to have a solute diffusivity roughly 95% of pure water. A solute is a substance that creates a solution when dis-solved in a solvent, and diffusivity is a measure of how easily a solute will move in a solvent. You can readily imagine that because the new matrix is so permissive of movement, vectorial transport between adjacent proximal tubule and vascular chan-nels is as effortless as a water strider skimming across a pond.

The investigators also reformulated their fugitive ink to be compatible with the composition of the new extracellular matrix. The new matrix had an approximately 10-fold lower viscosity than in their original experiment.[§§] So when they introduced the original fugitive ink into this matrix, *viscous fingering* occurred between the printed features and the modified extracellular matrix. This expressive name denotes the displacement of a resident fluid by a second fluid in a porous medium.[38] Just like the original ink, the reformulated fugitive ink can be removed from the fabricated tissue by cooling to 4°C and with the advantage that it suppressed viscous fingering and ensured that smooth features were retained after encapsulation within the modi-fied extracellular matrix.

Now we consider their fabrication process as shown in panel (a) of Figure 10.3. In (i), they printed a silicone gasket, deposited a base layer of gelatin/fibrin extracellular matrix within the gasket, and printed fugitive ink in two adjacent convoluted lines on the extracellular matrix. In (ii) they cast additional extracellular matrix to cover the lines and incubated this at body temperature for two hours. They then cooled the en-tire construct to liquefy the fugitive ink. In (iii), Lewis' team flushed the fugitive ink out of both lines using cell media. This left behind open lumen.

Next it was time to confer an identity upon these open passageways by lining them with specific cells that give their existence purpose and meaning. This is shown in (iv). The proximal tubule and vascular channels are seeded with proximal tubule epi-thelial cells and glomerular microvascular endothelial cells, respectively. The glomer-ular microvascular endothelial cells were isolated from human kidney glomeruli.

[§§] This is prior to enzymatic cross-linking.

(a)

(i) Print (ii) Cast (iii) Evacuate ink (iv) Add cells, perfuse

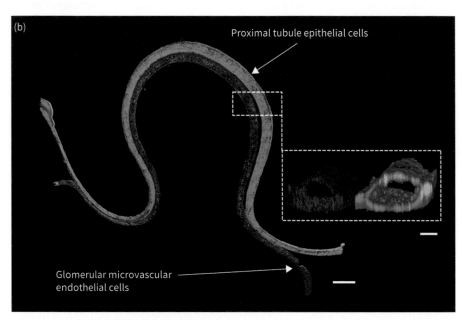

(b)

Proximal tubule epithelial cells

Glomerular microvascular endothelial cells

Figure 10.3 (a) The steps in creating two adjacent conduits, one lined with epithelial cells and one with endothelial cells. These colocalized channels will allow for transport from a proximal tubule to an adjacent blood vessel. In step (i), Lewis, Moisan, and colleagues printed a silicone gasket, deposited extracellular matrix within the gasket, and printed fugitive ink in two adjacent convoluted lines. In (ii), they cast additional extracellular matrix to cover the lines. In (iii), they evacuated the fugitive ink from both lines. In (iv), they added proximal tubule epithelial cells to one of the open channels and glomerular microvascular endothelial cells to the other channel and perfused the channels with epithelial/endothelial culture media, respectively. (b) Immunofluorescence staining. Scale bar = 1 mm. *Note:* The separation distance between the proximal tubule and vascular channels is approximately 70 μ. The inset shows cross-sectional images of the two open lumina. Scale bar = 100 μ.

Source: Reproduced by permission from Neil Y. C. Lin et al., "Renal Reabsorption in 3D Vascularized Proximal Tubule Models," *Proceedings of the National Academy of Sciences of the USA* 116, no. 12 (March 4, 2019): 5399–5404, https://doi.org/10.1073/pnas.1815208116.

In panel (b) we see immunofluorescently stained colocalized channels including high magnification images. The green, blue, and red stains highlight the plentiful sodium-potassium pumps, cell nuclei, and a specific cell receptor expressed by endothelial cells.[39]

Healthy and Mature Phenotypes

The group was careful to check the phenotypes of both the proximal tubule epithelial cells and the glomerular microvascular endothelial cells to see that they were maintaining their personalities. The ability of their bioprinted kidney tissue to replicate renal reabsorption depends strongly on cell function and cell maturity. Returning to the terms we introduced in their first study, the proximal tubule epithelial cells exhibited the appropriate cell polarity as determined by fluorescence microscopy. The team also evaluated barrier function. In the present case, they didn't use an electrical resistance method of testing, though they were in pursuit of the same information: the integrity of the cellular barriers.

What they did was to measure the barrier function of the epithelium and the endothelium within their model channels by perfusing them through their open lumen with culture media that contains a fluorescent marker. They literally watched what happened with a fluorescence microscope and quantified the results of the diffusional permeability. Just imagine you had a drinking straw and you wanted to know if there were any really small holes in it. If you forced a liquid through one end and observed bubbles or leaks before most of the fluid came out the other end, you'd know you had a leaky straw. This was the same idea but quite sophisticated since it was performed under live cell imaging with both the channels of interest and the surrounding extracellular matrix in the field of view.

Active, Selective Reabsorption of Key Solutes

The goal now was to test the barrier function of the proximal tubules and demonstrate their ability for active, selective reabsorption. Selective reabsorption happens in our bodies when vectorial transport moves precious nutrients from the proximal tubule channels into the adjacent vascular channels. In doing so, our kidneys recycle molecules that are vital to our health. To see if their construct had this capability, the group performed perfusion experiments using the closed-loop perfusion system you see in panel (a) of Figure 10.4. Lewis and Moisan's system enabled them to individually address tubular and vascular lumen, to circulate specific solutes and drugs, and to collect perfusates at intermittent time points during live imaging. Specifically, they made quantitative measurements of albumin and glucose reabsorption since both solutes have vital roles in renal physiology. We know of albumin's importance from

(a)

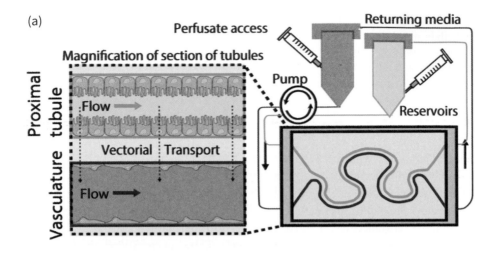

(b)

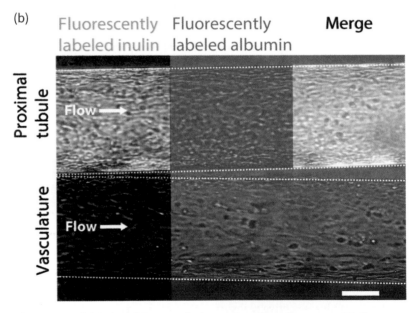

Figure 10.4 (a) A closed-loop perfusion system that allows quantitative measurements of kidney albumin reabsorption resulting from vectorial transport from the proximal tubule to the colocalized vasculature. (b) Confocal microscope images of colocalized proximal tubule and vascular channels—denoted by white dashed lines to guide the eye. The fluorescence microscope images reveal proximal tubule barrier function since inulin, fluorescently labeled to appear yellow, remains confined to the proximal tubule. The images also show selective transport of albumin from the proximal tubule channel to the vascular channel, since the albumin, fluorescently labeled to appear magenta, has appeared in the vasculature. Scale bar = 100 microns.

Source: Reproduced by permission from Neil Y. C. Lin et al., "Renal Reabsorption in 3D Vascularized Proximal Tubule Models," *Proceedings of the National Academy of Sciences of the USA* 116, no. 12 (March 4, 2019): 5399–5404, https://doi.org/10.1073/pnas.1815208116.

chapters past. Once in the blood, albumin assumes its identity as a molecular taxi driver, binding and carrying substances around in the blood.[40]

As to the importance of the key solute, glucose, consider the following. Studies have shown that a person weighing 70 kilograms has only four grams of circulating glucose in their blood. This amount of glucose will fill a teaspoon.[41] Although these four grams are a minute fraction of the person's weight, the cells of the body are highly attuned to its presence. If glucose levels fall significantly and the brain doesn't get sufficient glucose, this could result in seizures or even death.[42] Smaller shortfalls of glucose can have varied outcomes, such as dizziness, disorientation, jitteriness, and irregular heartbeat; symptoms vary depending on the individual.

Maintenance of steady levels of glucose is vital to preserve a constant source of this sugar to the brain, an organ that uses glucose as its principal metabolic fuel. And the brain has a voracious appetite for glucose: The brain consumes on the order of 60% of the blood glucose used in the sedentary, fasted person.[43] The kidney contributes to glucose homeostasis through multiple processes: glucose filtration, glucose reabsorption, glucose consumption, and glucose production.[44]

Right at the nephron's front door, the glomerulus filters glucose out of the bloodstream. This is in anticipation of the proximal convoluted tubule doing its job in reabsorbing the glucose from that ultrafiltrate—and how!

> "In normal individuals renal glomeruli [collectively] filter about 180 grams of glucose per day. This would result in an enormous loss of glucose through the ultrafiltrate if not recovered; thus, the main physiological undertaking of the kidney is to regain as much glucose as possible, and in normal situations almost all of it is reabsorbed in the proximal tubule by an insulin independent process.[45],***

We've now accounted for glucose filtration and glucose reabsorption. As to glucose consumption, your diet is typically your main source of glucose, although this isn't directly related to the kidneys. However, if you're fasting, whether by choice or by circumstances beyond your control, your kidney has two methods to keep you going. The first involves the breakdown of a more complicated form of glucose known as glycogen, which the kidney stores in modest amounts. Kidney experts categorize this as a glucose consumption contribution (rather than a production contribution), and although the amount is small, every little bit counts because the well-being of your brain is at stake.

The final way your kidney helps you to keep enough glucose on hand is glucose production. The kidney produces glucose from a variety of precursors. This occurs in the proximal convoluted tubule—the only part of the kidney with the appropriate enzymes for production of glucose from precursors (although the liver also has this

---

*** When you consider the balancing act that our kidneys perform for us every single day, it's remarkable that we're able to walk and chew gum at the same time—if we feel so inclined—for about 70–80 years.

capacity).[46] So we've established that the kidney has an important place in the orbit of glucose and that glucose is without question a key solute in our blood.

Measuring Albumin Reabsorption

Since the Lewis and Moisan-led team first measured albumin uptake, we follow their lead and initially consider albumin reabsorption. They performed this evaluation by perfusing not one but two fluorescently labeled probe molecules in a colorful and shrewd experiment. What they did was to bind differently colored fluorescent probes to the molecule of interest, albumin, and to a second molecule called inulin. The fluorescent probe attached to albumin has a magenta color, and the probe attached to inulin has a yellow color.

The use of inulin goes back to the early twentieth century. Pharmacologists at the University of Pennsylvania, led by the same Professor Alfred Newton Richards we met at the start of this chapter, examined properties of the renal tubules of amphibians. They found that inulin, a sugar found in many plant roots, *wasn't* reabsorbed in the kidney.[47] They extended their experiments to mammals and got the same null result. You may well wonder why this was worthy of attention.

The reason inulin isn't reabsorbed in the kidney is that it has minimal interaction with proximal tubule epithelial cells—so it serves as a hallmark for barrier function and has become a standard for kidney studies. One goal for Lewis, Moisan, and company: Proximal tubule epithelial cells should form near leak-tight barriers against the traffic of certain proteins such as inulin, when the epithelial cells are healthy and confluent.

The group perfused fluorescently labeled albumin and inulin through the proximal tubule using their closed-loop perfusion system. This permitted them to measure the concentration of both albumin and inulin in perfusate samples collected separately from the proximal tubule and vascular outlets (panel (a) of Figure 10.4). After one hour of exposure of the proximal tubule to fluorescently labeled albumin and inulin, they found significant albumin transport from the proximal tubule to the vascular lumen. The albumin uptake continued to increase over time, but the small amount of inulin in the proximal tubule remained unchanged throughout the experiment. *This is a demonstration of selective reabsorption.* Next, they used fluorescence microscopy and *directly visualized* vectorial transport between the proximal tubule and vasculature.

Panel (b) of Figure 10.4 presents a collection of actual images obtained by fluorescence microscopy. This vividly demonstrates the proximal tubule barrier function and the selective transport of albumin from the proximal tubule epithelial cells to the glomerular microvascular endothelial cells. Note that the white dashed lines have been added as a guide to the eyes in distinguishing the two colocalized channels.

Beneath the label *Fluorescently labeled inulin* at the far left of panel (b), you see fluorescent inulin (yellow) in the proximal tubule but virtually no inulin in the vasculature. Beneath the label *Fluorescently labeled albumin* in the middle, you see fluorescent magenta in both the proximal tubule and the vasculature. And at the far right beneath the label *Merge*, you see a third segment as it would appear if you added the red, green, and blue values from the first two images—the explanation follows.

More on the Merge: first the principles. White light is made up of three primary colors: red, green, and blue. Yellow (as seen in panel (b) in the upper far left) is primarily red + green. Magenta (as seen in panel (b) in the middle) is primarily red + blue. More generally, when light is formed by the addition of all three colors at maximum intensity—red, green, and blue—you get white light. And that's what we see in the upper part of the Merge column in panel (b): white (mixed with magenta). That's the concept behind the figure. Now we move from principles to practice.

To construct a display of what we've described, the team used software to break down the fluorescent colors in both the left and middle columns into their red, green, and blue constituents and then added them together.[†††] This gave the distinctive third column (Merge) where red plus green light (seen as the yellow fluorescence) was added to the red plus blue light (seen as the magenta fluorescence). The addition produced the white light in the portions of the proximal tubule where the yellow from the inulin's fluorescent marker was distributed, as seen beneath the vertical label Merge in the upper right of panel (b).

Thus, the Merge column helps to emphasize *selective* reabsorption of the fluorescently labeled albumin and also the barrier integrity of the proximal tubule that retains the lion's share of the fluorescently labeled inulin throughout the experiment. Whew!

Measuring Glucose Reabsorption

Next, to probe glucose reabsorption, the group directly measured the difference in glucose concentration between perfusates from proximal tubule epithelial cells and glomerular microvascular endothelial cells using a commercial glucose meter. They measured negligible glucose reabsorption for two groups of controls. In the first control group, both the proximal tubules and vascular channels were empty; they contained no cells. In the second control group, the proximal tubules were empty (no epithelial cells) and the vascular channels alongside them were lined with endothelial cells.

By contrast to the negligible glucose reabsorption in the controls, the team considered the following two other circumstances: The proximal tubules were epithelialized but the vascular channels were either empty or endothelium-lined. They found that regardless of whether the vasculature was empty or endothelium-lined, there were high glucose reabsorption levels. This highlights the crucial role of proximal tubule epithelial cells.

[†††] Just as if you wanted to divide up a dollar to give to three children, you would break the dollar down into coinage and give each child 33¢ while keeping a penny for yourself to avoid squabbling.

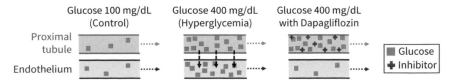

Figure 10.5 Schematics of hyperglycemic effects and their rescue. Left shows control. Middle shows hyperglycemia generated by perfusion of 400 mg/mL glucose into the proximal tubule channel. Right shows rescue from hyperglycemia by administration of dapagliflozin.

Source: Reproduced by permission from Neil Y. C. Lin et al., "Renal Reabsorption in 3D Vascularized Proximal Tubule Models," *Proceedings of the National Academy of Sciences of the USA* 116, no. 12 (March 4, 2019): 5399–5404, https://doi.org/10.1073/pnas.1815208116.

Before we present one final result, let's share the team's thoughts on where they're at and what they hope to do to improve. From tests we won't present, the team estimated that their proximal tubule epithelial cells are only about 20% as efficient in reabsorbing glucose compared to the proximal tubules in actual human kidneys. Lewis, Moisan, and colleagues think the reabsorptive properties of their engineered kidney tissues may be improved by reducing the proximal tubule lumen diameter as well as the separation distance between the proximal tubule and vascular channels. They wish to do better. Nonetheless, this is top-drawer bioengineering.

Hyperglycemia

Glucose reabsorption is also implicated in kidney disease. Hyperglycemia, an excess of glucose in the bloodstream (commonly called high blood sugar), is a signature of diabetes and a known risk factor for vascular disease. The group used their model to provide a well-characterized platform for probing the effects of hyperglycemia on cell function. To explore this disease state, they created a hyperglycemic condition by quadrupling the glucose level in the proximal tubule channel.[‡‡‡] They studied the effect of this glucose spike on the endothelium within the vascular channel (Figure 10.5).

After 36 hours of perfusion at these elevated glucose levels, there was hyperglycemia-induced endothelial dysfunction. The team determined this in part by fluorescent imaging that showed an elevated intensity of reactive oxygen-containing molecules. They're sometimes called reactive oxygen species, or you may have heard of them referred to as free radicals. These molecules can damage cells and did damage the endothelial cells in the vascular channel. The presence of elevated levels of glucose in the vascular channel is indicated in the middle cartoon in Figure 10.5.

[‡‡‡] With their independently accessed proximal tubule and vascular channel, it wasn't necessary for them to scarf down a gallon of Rocky Road ice cream to accomplish this.

At this point, they administered a drug—dapagliflozin—through the proximal tubule channel. Dapagliflozin is a prescription medication that acts to lower blood sugar by causing the kidneys to get rid of more glucose in the urine.[48] The drug acts by inhibiting the glucose transporter proteins at the surface of the tubular cells. As shown in the rightmost cartoon in Figure 10.5 and assayed by fluorescence microscopy and transmission electron microscopy, the adjacent vascular channel was now spared an assault by a glut of glucose. The drug effectively provides a barrier that protects the endothelium from exposure to excess glucose. Again, the design and functionality of their model tissue allowed for investigative studies of endothelium–epithelium crosstalk, this time under disease conditions.

So their biomimetic tissue is faithful to its calling in both sickness and health, or rather, sickness and homeostasis.

When will bioprinting be able to help those with kidney failure? When will bioprinted constructs be able to assist or replace any failing tissue or organ? When will bioprinted constructs be implanted in humans? When, when, when.

Chemist Nadrian Seeman of New York University is celebrated in the field of nanotechnology for his method of using strands of DNA as building material for nanostructures and nanodevices. When Seeman was interviewed via e-mail, the reporter peppered him with many questions. In time he got around to the inevitable question of *when* while making reference to the "many, well-hyped promises of nanotechnology."[49]

"In terms of technology and anything affected by technology, what will be different about our world in 5 years? In 10? In 50? What will have surprised us in 10 years, in 50?" Seeman didn't rise to the bait. His answer was, "I am a scientist, not a futurist. I don't make predictions."[50]

Scientists attempting to build the edifice of bioprinting, to make it viable and transition it to the clinic, are hopeful but they dislike prognosticating. This reticence toward speaking as futurists serves them particularly well when proof-of-concept experiments provide tantalizing hints at real-world applications. In this book, we can certainly do no better than to abstain from prognostication. In the next chapter, we do not prognosticate. Instead, we select subjects for contemplation from many possibilities. Even so, we run the risk that these choices will not be germane to the unfolding of bioprinting.

Inspired by the fact that bioprinting itself is a high-risk endeavor pursued by intrepid individuals, we'll take the plunge.

11
What's in the Offing?

At first glance Eddie looks like the kid next door. He's smiling. You can see that if you look in the mirror reflecting his lightly freckled face, reddish hair, and snub nose. It momentarily draws your attention away from the fact that you can't see anything else of him. Not even much of his neck, really—hidden by a rubber collar.

A few weeks ago he was playing at school recess. Warm weather had come early, well in advance of the summer break, and children were enjoying it with abandon. One of his classmates had used the toilet before recess and hadn't washed his hands properly. He and Eddie were roughhousing, just playing around. A day or so later Eddie felt like he was coming down with a cold. A week later things got worse and it was hard for him to stand up. Then he lost control of the muscles he would normally use to breathe on his own. He'd been infected with a virus but it didn't give him a head cold.

Now he's lying on his back in a cylindrical metal tank. Pumps rhythmically change the air pressure inside the tank and pull air in and out of his lungs for him. They're negative pressure ventilators designed to enable a person to breathe. Their mechanical wheezing and rasping is at once reassuring and unnerving. Everyone calls them iron lungs. Eddie has the paralyzing form of polio.

A Thumbnail History of Polio

Why bring up the subject of polio in a chapter purporting to look at the future of bioprinting? To answer the question, we first confess that in this book we've shamelessly pressed into service a *Ragtime* of notables from Mick Jagger to Mikhail Baryshnikov, from Seinfeld to Sinatra, from pinball's wiz to Charlotte's Web. The author of *Ragtime*, E. L. Doctorow, said something that we would like to apply to the future of bioprinting.

Journalist and author George Plimpton interviewed Doctorow in front of a live audience at the 92nd Street YMHA in New York City.[1] He questioned Doctorow about his approach to writing a book. "Do you have any idea how a project is going to end?" The author's response may describe the current journey that bioprinting is on. Doctorow said that in the main he did not. "It's hard to explain. I have found one explanation that seems to satisfy people. I tell them it's like driving a car at night: You never see further than your headlights, but you can make the whole trip that way."

What does this have to do with polio? Our first attempt to consider bioprinting's future will be to look in the rearview mirror of Doctorow's metaphorical car. Our chronicle has tried to present a snapshot of bioprinting's first two decades of existence. This is the work of a literary camera, not a Cassandra. There is no zoom lens at hand to gaze into the future. But the story of polio's past may give us a space in which to ruminate about the story of bioprinting's future. Two biomedical quests, each with its own sense of urgency. Can we modulate our expectations of bioprinting by considering how the effort to grapple with polio evolved?

Polio is the shortened name for a virus-borne disease called *poliomyelitis*, from the ancient Greek words *polios* for *grey* and *myelos* for *matter* (grey matter refers to the spinal cord) and the suffix *-itis*, meaning *inflammation*.[2] Poliovirus enters the body through the mouth, multiplying along the way to the digestive tract, where it multiplies even further. In paralytic polio, the virus leaves the digestive tract, enters the bloodstream, and seizes this opportunity to assault nerve cells. In severe cases, the throat and chest may be paralyzed and death may result if the patient isn't assisted by a means of artificial breathing support.[3] Polio seems to have circulated in human populations at low levels for most of the 1800s. Then the situation began to change.

Rutland County, Vermont, was the site of the first major documented polio outbreak in the United States in 1894. There were 18 deaths and 132 cases of permanent paralysis. One insight from this tragedy that went missing: the contagious nature of the disease.[4] It would take another decade for this to be discovered and the scene shifts to Sweden. After a series of polio epidemics in that country in 1905 affecting over 1,000 people, Swedish physician Dr. Ivar Wickman published two seminal findings.

He inferred from geographical field studies that polio was a contagious disease and that people who didn't appear to be severely ill could nonetheless have the disease— and could contribute to epidemic transmission.[5]

Back across the Atlantic and another decade later, we travel to New York City, c. 1916. During the summer months, health officials announced a polio epidemic in Brooklyn, New York. In New York City alone (primarily in the borough of Brooklyn), more than 2,000 people died. That figure rose to 6,000 across the United States, with thousands more left paralyzed. Summer epidemics were becoming common and lead to widespread closures of pools, amusement parks, and other places where children were present in large numbers.[6] The epidemic that swept New York City and the northeastern United States in 1916 was the event that marked the recognition of poliomyelitis as a major public health problem in the United States.[7]

Though the Black Death killed vastly more individuals, few diseases within contemporary history frightened parents more than polio did in the early part of the 20th century.

Philip Drinker was an assistant professor in the Department of Ventilation and Illumination at the emerging Harvard School of Public Health in the 1920s. Children's Hospital in Boston had called Drinker in for a consultation when the temperature dropped in one of the air-conditioned rooms for premature infants. While at the hospital, he caught sight of children in polio-induced respiratory paralysis. What he saw horrified him. "He could not forget the small blue faces, the terrible gasping for air." This led to the Drinker Respirator that was later dubbed the Iron Lung.[8] Drinker and Dr. Charles McKhann published a paper on the results in 1929 and ushered in the era of the Iron Lung.[*,9]

Eddie Cantor—comedian, dancer, singer, actor, and songwriter—was regarded almost as a family member by millions of radio listeners in the 1930s and 1940s. In 1938, Cantor suggested on the radio that people send dimes to President Franklin D. Roosevelt, himself a polio victim, to help fight polio. Within a few weeks, people had mailed 2,680,000 dimes to the President to help establish the March of Dimes to aid polio sufferers and fund efforts to find a vaccine against polio. Over the years this charity raised tens of millions of dollars, much of which went to the effort to find a vaccine.[10]

Moving forward one more decade, prior to 1949 polio research was restricted to facilities that could house large numbers of experimental animals. But in 1949 a paper appeared in the journal *Science* showing how polioviruses could be grown in non-nervous system tissue.[11] The breakthrough in culturing viruses in vitro led the way to simpler and less expensive methods for producing large quantities of virus for examination and eventually for vaccine production.[12]

In the same year as the innovation in culturing was published, another team contributed the crucial information that although there were 14 strains identified, they could be grouped into only three immunologically distinct types of poliovirus. Any vaccine would have to produce immunity to all three.[13,14]

In 1954 physician-turned-medical researcher Dr. Jonas Salk was head of the Virus Research Lab at the University of Pittsburgh working on developing a vaccine against poliovirus.[15] Encouraged by small-scale testing in 1952, in 1953 he injected himself, his wife, and their three sons with his experimental poliovirus vaccine. In 1954 a large trial began with over 1.3 million children vaccinated.[16]

In 1955 the results were announced: The Salk poliovirus vaccine was 80–90% effective against paralytic polio. The same day the results were announced, the U.S. government licensed Salk's vaccine, which led to its widespread distribution and use.[17] In the two years before the vaccine was widely available, the average number of polio cases in the United States was more than 45,000. By 1961 that number had fallen below 1,000.[18]

---

[*] The Iron Lung is a negative pressure ventilator, which means that the machine alternately decreases air pressure around the thorax and the abdomen (to stimulate spontaneous inhalation) and then returns the surrounding air pressure to normal atmospheric pressure.

Contemplating Bioprinting's Possible Future with a Sketch of Polio's History in Hand

It's not as if the search to protect people against polio is a template for the search to provide people with bioprinted tissues. There are surely many non-comparables in the two situations, but that doesn't make polio's history a nonstarter in contemplating bioprinting's possible future. Looking at our sparse matrix of the chronology of polio's timeline, we see that for approximately 20 years until the 1916 New York City epidemic, the solution to the menace of polio did not have the perceived importance that it assumed post-1916.

We know that bioprinting also had an incubation period beginning approximately at the turn of the twenty-first century. For a period of about a decade or more, there were very few research participants and there was almost a giddy sense of anticipation in making science out of science fiction.

In polio's timeline, the era of the Iron Lung beginning in the 1930s provided relief as it enabled children, who would otherwise have died, to live. The late 1930s also saw the inception of the March of Dimes, and money began to flow for efforts to study the poliovirus and to work toward a vaccine.

In bioprinting's second decade as a practiced art, proof-of-concept experiments galore and organs-on-a-chip have enabled bioprinting to gain traction in the scientific community. These accomplishments are meaningful in themselves. They're also seen as a bridge to a greater understanding of the biological obstacles and the hope for human implantation of bioprinted tissues. With the growing sophistication of proof-of-concept experiments has come a rush of funding from both the public and private sectors.

For example, the Department of Defense established the Armed Forces Institute of Regenerative Medicine in 2008, and in its first five years of existence it provided over $300 million for restoring the structure and function of damaged tissues and organs and curing previously untreatable injuries and diseases.[19] A chunk of this funding supported bioprinting research, much of it performed at the Wake Forest Institute for Regenerative Medicine directed by Dr. Anthony Atala. The first five-year Armed Forces award was very successful and was followed by a second five-year award and other initiatives.[20,21]

With the rise in the number of super-rich people, exceptionally large private funding has also occurred. An example is the founding of the Wyss Institute for Biologically Inspired Engineering at Harvard University. The Institute was founded in 2009 with a gift of $125 million from Hansjörg Wyss, who doubled his gift to $250 million in 2013.[22,23] In 2019, Wyss made an additional gift of $131 million, bringing his total Institute support to $381 million.[24] As with the Department of Defense funding, a chunk of the Wyss funding underpins bioprinting efforts. For example, Jennifer Lewis holds the position of Wyss Professor of Biologically Inspired Engineering.

Polio's chronology in the decade following the introduction of the Iron Lung and the rise in financial support of basic and applied research culminated in the 1949 discovery of the three distinct immunological types of the virus. This fundamental scientific achievement was essential for a rational plan to develop a vaccine against the poliomyelitis virus in all its distinct forms—and the first vaccine soon followed.

In the middle of bioprinting's second decade, vascularization was widely viewed as one of the most daunting obstacles to success. It was difficult to imagine scaling up constructs to relevant sizes. Voicing the concerns of many, Brian Derby, materials scientist from the University of Manchester, said, "We are still waiting for a breakthrough."[25] With the success of microchannels filled with fugitive inks, significant progress has been made in vascularization. With the benefit of hindsight, the success of the fugitive ink approach might come to be seen as a key enabling step, possibly analogous to nailing down the number of immunologically dissimilar polio viruses.

At present, however, in attempting to assign a comparable moment in bioprinting's timeline, we find that our crystal ball has transformed into a snow globe holding a figurine of Kermit the Frog with an inscrutable smile on his face and a scarf around his neck to protect him from the falling snow.

Unlike polio's three immunologically distinct forms, bioprinting aspires to tame a dozen demons—from articular and elastic cartilage to skin and bones and muscle and neuromuscular junctions, to a plethora of solid organs including liver, heart, and kidney to the vascularization and innervation that the majority of these targets require. We can pigeonhole some of these bioprinted tissue goals as two-dimensional, such as skin; hollow tubes, such as blood vessels; hollow non-tubular organs, such as the bladder; and solid organs, such as the kidney.[26] But there is no gainsaying the differing demands of bioprinting a liver versus a kidney. There is no single algorithm for finding the best bioink with its combination of cells and supportive biomaterial and training the cells via mechanical or electrical stimulation to prepare them for their specific in vivo environment.

Vascularization and Scaling up to Human Size

Two obstacles that we meet at almost every turn are vascularization and scale. Let's look at vascularization first. *Beat the Clock* was a 1950s television game show.[27] Contestants performed stunts while racing against a big ticking clock. The shenanigans required a good sense of humor as well as agility, balance, and good aim or possibly a willingness to be the target of whipped cream, pies, or getting soaked with water. Absent the messy stunts, bioprinting is presently a highly sophisticated form of beat the clock. In Jennifer Lewis' lab, team member Mark Skylar-Scott said,

We set a timer as soon as we lift the cells off the plate [to print them]. You watch the timer and you work fast. It takes more than one person to do this. That's why it's always been a team effort amongst us.[28]

And group member Kimberly Homan added, "You really don't want to bioprint alone."[29] It's fatal to allow the printed construct to become too large too slowly—you'll end up with a necrotic core. And yet there are attendant risks in trying to print the cells too quickly and subject them to damaging shear forces.

Then there's the issue of scale. Dr. Anthony Atala's team has used their custom integrated tissue and organ printer to fabricate human-scale jawbone.[30] Printing constructs like a jawbone could have enormous benefits to children born with facial anomalies and to those who suffer traumatic injury. To have scaled-up bioprinted bone available for human transplants is wonderful. However, if we're looking at scaling up the bioprinting of solid organs for human implantation, that is uncharted territory at present.

In an interview with physicist and science writer Margaret Harris, Jennifer Lewis had this to say about the questions of scale and vascularization:

Our thesis is that bioprinting alone is not the best way to create human tissues—we need to harness as much biology as we can. That could mean starting with stem cells and programming them along the same paths as your body does, or it could mean allowing blood vessels that have been printed at a large scale to start to form tips and sprouts—capillaries, if you will—through a process called angiogenesis. We want biology to do as much of the heavy lifting as possible, because if you think about what it would take to print a kidney or a liver, we're talking about tissue volumes that might range from a few hundred milliliters up to a liter. That's a lot to print, and then the tissue has to function, which is a tall order as well.[31]

One more point on the scaling up of bioprinting: While size is important, so is density. If you've managed to provide a thick tissue with vascularization, that's great. But the tissue also needs to have a high cellular density. The clever strategies for vascularizing thick tissues that we've previously discussed were important steps; however, they lacked the high cellular density that would allow them to fully mimic natural tissue.

In addition to scale and vascularization, Lewis added,

An even bigger challenge, as we move forward, is to integrate that tissue into a host without triggering rejection or other immune responses, and have it not only function but be sustained. These are grand challenges, but it's something that we're working toward, and both we and others (such as Sangeeta Bhatia at MIT and Chris Chen at Boston University, and also Mike McAlpine at the University of Minnesota) have got steps along the way that we're mapping out, one at a time.[32]

Aggregates of Pluripotent Stem Cells: Spheroids, Organoids, and Embryoid Bodies

Following our glance at polio, courtesy of the rearview mirror of Doctorow's car, we return our attention to what the headlights might reveal. We peer into the darkness, hoping to discern signage or distinguishing features of the landscape. One way to think about bioprinting's future is to look at new ideas that seem to hold promise. If the concept is not just speculative, if it has some demonstrable weight behind it, so much the better. Such a premise that has recently surfaced involves a kind of bootstrapping.

We're familiar with induced pluripotent stem cells and we've frequently mentioned spheroids and organoids. Jennifer Lewis' lab has proposed that patient-specific induced pluripotent stem cell-derived organoids can be assembled into what they call organ building blocks. That's one bootstrapping step. What's more, they've shown how to assemble hundreds of thousands of organ building blocks into constructs with high cellular density and with perfusable vascular channels, another bootstrapping initiative.

These constructs have cell densities in the region of 100 million cells per milliliter. This approaches the cell density of native tissues in the body.[33,34] Their resourceful biomanufacturing method introduces perfusable vascular channels via embedded printing. (Embedded printing is 3D printing in which the nozzle moves within a soft material that surrounds and supports the material being printed.)[35] Lewis and colleagues' overall experimental design comes to grips with the issues of scale and vascularization.[36] To begin, we'll add rigor to our previous mentions of spheroids and organoids while introducing a related variation called embryoid bodies.

Most cells in the human body acquire a three-dimensional organization through their interactions with neighboring cells and the extracellular matrix specific to the cell type.[37] Historically, studies in cell biology were performed using the cells of interest in a cell culture—that is, a favorable artificial environment—but in two-dimensions. In time, biologists realized that a three-dimensional cell culture is a more natural microenvironment and will better preserve the cells' phenotype. With three-dimensional culturing methods, cells can be cultured as aggregates.[38]

We mentioned well plates in Chapter 6. Using the shallow concavities in well plates is one means of forming the cellular aggregates known as spheroids.[39] In order for a spheroid to make it into the more exclusive club of organoids, it has to meet several criteria. When simply defined, an organoid is a cellular aggregate resembling an organ. Specifically, this means it must contain more than one cell type of the organ it models; it must display some function specific to that organ; and the cells should be organized in a way that is similar to those of the organ. Organoids are grown from human stem cells and from patient-derived induced pluripotent stem cells.[40,†]

---

† An echo from the beginning of our chronicle is Malcolm Steinberg's differential adhesion hypothesis that posits segregation of cells with similar adhesive properties. This seems to explain organoid self-assembly.

Now that we've distinguished between spheroids and organoids, we come to embryoid bodies. Embryoid bodies are three-dimensional aggregates of pluripotent stem cells that self-organize to develop tissues. As Lewis observed, most laboratory protocols for developing organoids begin by generating embryoid bodies that are composed of many induced pluripotent stem cells that can be placed into well plates. The experimenter can direct them to differentiate into kidney organoids, cardiac organoids, and so on.[41] (We note, however, that not all organoid methods make use of an initial embryoid body stage.)[42]

After creating hundreds of thousands of induced pluripotent stem cell-derived building blocks, Lewis' team placed them into a mold and compacted them by centrifugation. When they did this, they achieved a separation: The composition of the extracellular matrix components is less dense than that of the organ building blocks. So after centrifugation, the extracellular matrix material is on the top and can be easily removed, and the organ building block material is compacted on the bottom. It's at this juncture that the team rapidly patterned sacrificial ink within the block using embedded printing. After removing the sacrificial ink by warming, they were left with perfusable channels. Following this, Lewis demonstrated that these bulk vascularized tissues function and mature when perfused over long durations. We look further into their experiments.

High Cellular Density with Vasculature via Embedded Printing

Lewis' team grew induced pluripotent stem cells and transferred them into well plates to form spheroids or embryoid bodies (follow along in Figure 11.1). After harvesting, the cell aggregates can be used directly as organ building blocks or can be further differentiated by biochemical manipulation into specific organoids. For each desired bioconstruct, they cultured and harvested up to 400,000 organ building blocks, mixed them with extracellular matrix, and kept them at cold temperatures. Their tailored extracellular matrix was designed to retain fluid-like behavior at low temperatures and that enabled casting and embedded bioprinting to create perfusable vascular channels.

At this point, they used centrifugation on the cold slurry composed of organ building blocks and extracellular matrix. Centrifugation compacted the organ building blocks to produce a living tissue mass of high cell density (panel 6 of Figure 11.1). What's more, the cold slurry provided sufficient support for embedded printing of sacrificial gelatin ink that served as a template for the desired vascular channels. This is shown in panel 7 of Figure 11.1.

After patterning the channels with sacrificial ink, they placed the construct in an incubator set to human body temperature and the extracellular matrix gelled. This stiffening eased the way for the removal of the sacrificial ink. Once the channels were free of the ink, they connected the tissue construct to an external pump and perfused oxygenated media to maintain cell viability as shown in panel 8 of Figure 11.1.

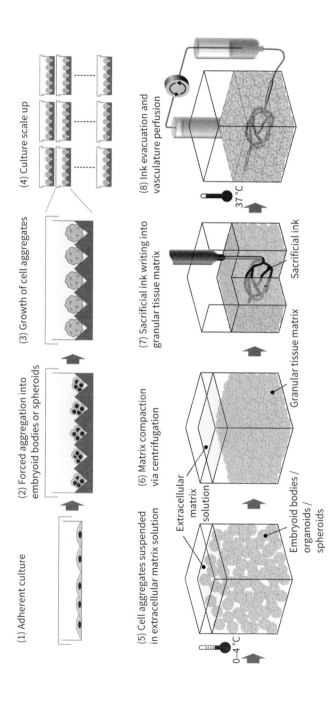

(1) Adherent culture

(2) Forced aggregation into embryoid bodies or spheroids

(3) Growth of cell aggregates

(4) Culture scale up

(5) Cell aggregates suspended in extracellular matrix solution

(6) Matrix compaction via centrifugation

(7) Sacrificial ink writing into granular tissue matrix

(8) Ink evacuation and vasculature perfusion

Extracellular matrix solution

Embryoid bodies / organoids / spheroids

Granular tissue matrix

Sacrificial ink

0–4 °C

37 °C

Figure 11.1 Fabrication of a bulk tissue of high cellular density containing a network of perfusable, hierarchically branched channels. After growing induced pluripotent stem cells in conventional culture conditions, Lewis' group transferred them into large-scale microwell arrays to create hundreds of thousands of so-called organ building blocks in the form of embryoid bodies, organoids, or multicellular spheroids. They mixed these cellular aggregates with an extracellular matrix solution, placed this combination into a mold, and compacted the organ building blocks using centrifugation. The majority of the less dense extracellular matrix collected above the compacted cellular aggregates and was removed. Then they rapidly patterned sacrificial ink within the compacted mass using 3D printing and warmed the tissue to body temperature to liquefy the fugitive ink. Following evacuation of the ink to create a network of channels, they perfused media through the channels. The highly dense tissue contained nearly half a billion cells.

Source: Reproduced with permission of AAAS from Mark A. Skylar-Scott et al., "Biomanufacturing of Organ-Specific Tissues with High Cellular Density and Embedded Vascular Channels," *Science Advances* 5, no. 9 (September 2019): eaaw2459, https://doi.org/10.1126/sciadv.aaw2459. © The Authors, some rights reserved; exclusive licensee American Association for the Advancement of Science.

To demonstrate the rapid vascular patterning they achieved, consider Figure 11.2. The images show a custom-shaped mold with inlet and outlet tubes for perfusion, filled with the compacted embryoid body/extracellular matrix slurry, and a branched, hierarchical channel network. To test the construct's durability, they perfused the structure for many hours and the vascular channels remained intact.

For cell viability testing, the group made a perfusable embryoid body-based tissue in a perfusion chamber that allowed circulation of oxygenated media through the

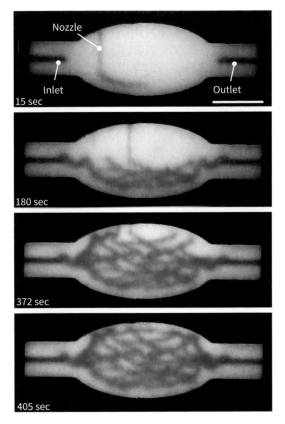

Figure 11.2 A timed image sequence showing rapid, embedded, 3D printing of a branched, hierarchical network of channels filled with fugitive ink. The printed channels were created within a custom-shaped mold containing a slurry of compacted, embryoid body-based, bulk tissue and extracellular matrix. Subsequent to the process shown, Lewis' group evacuated the fugitive ink, seeded the open channels with endothelial cells, and perfused the channel network with oxygenated media. Scale bar = 10 mm.

embedded vascular channels and around the periphery of the tissue. As a control, they cast a non-perfusable embryoid body-based tissue and only flowed media around its periphery in the same chamber. The control tissue developed a necrotic core within 12 hours; viable cells were restricted to a narrow band at the tissue periphery. By contrast, the tissue containing embedded vascular channels provided enhanced cell viability throughout the bulk tissue. So their vascularization scheme won its game of *Beat the Clock*.

Finally, Lewis' group created a functional, perfusable tissue containing human induced pluripotent stem cell-derived cardiac organ building blocks. Initially the cardiac construct beat asynchronously with calcium waves propagating sporadically and chaotically after a single day of perfusion. After seven days in culture, the tissue construct beat spontaneously and synchronously with calcium waves propagating rhythmically and rapidly throughout the tissue, showing functional contractility. (Cardiac contractility is essentially the strength of contraction of heart muscle fibers.)[43]

We've presented Lewis' design paradigm for vascularized, high cellular density, printed tissue because with additional advances it may offer a new direction for achieving the necessary increase in scale for therapeutic use. However, the authors of the study acknowledge current limitations. For example, their present protocol results in organ building blocks that lack sufficient cell maturity and microvascular network formation. As a result, they only observe modest contractility of myocardial fibers compared to the contractility observed in adult cardiac tissue. Consider this more closely.

Contractility, the strength of contraction of heart muscle tissue, is an important measure of cardiac function. The search for an optimal way to measure the contractile properties of heart tissue has been long and arduous.[44] In 1973, two researchers introduced the concept of strain to help understand the elastic stiffness of heart muscle. They defined *strain* as a dimensionless quantity that represented the percentage change in dimension from a resting state to a state after a force had been applied.[45] The idea is that healthy heart tissue should be flexible enough to extend in response to demands placed on it; stiffening can contribute to (and be a measure of) heart dysfunction. Unfortunately, for many years the techniques for measuring strain were invasive and involved a surgical procedure, so they weren't useful for the clinical setting.[46] The good news is that developments in echocardiography have made bedside use possible.

So nowadays we have a helpful quantity called strain, a measure of the deformation of the muscle tissue of the heart during contraction.[47,‡] What this discussion comes down to is: In adult cardiac tissue, contractility as measured by strain is in the neighborhood of 20%. But in the Lewis group study, they found contractility as measured by strain to be around 1%. The team is now focused on generating cardiac organ building blocks with enhanced tissue function, including contractility, by creating more mature and microvascularized organ building blocks.

‡ It's typical to apply this technique specifically to the left ventricle.

A Slice May Suffice

The headlights of Doctorow's car afford us another opportunity. We briefly flip on the headlights' high beams to get a different way of looking at the road that lies before us. If you were to search the internet for the term *bioprinting*, you'd have a hard time throwing a stone and not hitting a story focused on bioprinted whole organs. The thinking behind this isn't hard to understand. If a person's liver is failing, it would be great to be able to replace the liver. If a person's heart is failing, wouldn't it be sublime to replace their heart? But as scientists travel further down the bioprinting road, the roadbed seems to become rougher and rockier. Many researchers have spoken of the value of a partial organ replacement or organ adjunct.

For example, in the future, printed organs might be used outside the body as a supplement or more physiological alternative to dialysis.[48] It was in this context that Jennifer Lewis said, "Even though we might not reach the stage of a full organ transplant in my lifetime, there is still the possibility of substantial clinical benefit."[49]

Another voice weighing in on the subject is that of Brian Derby. Several years ago in a *Financial Times* news article titled "Printing Whole Organs Remains a Long Way Off," he said,

> After a heart attack you have a lot of scar tissue, so you might find ways of making patches that could augment organs. You are not going to be able to print a whole heart, kidney or liver at the moment.[50]

So in this chapter in which we are timorously trying to peek into the future, we recognize the option of fraction rather than full. We call out partial organs. And Dr. Anthony Atala. The emergence of his ideas on the theme is an interesting story, the story of the stone. It's a vignette that Atala has related on numerous occasions.

Atala was at Harvard for a long time before accepting the position of Director of the Wake Forest Institute for Regenerative Medicine. On a summer weekend in the late 1990s, Atala was striding along the sand on Boston's south shore.[51] He was walking with bowed head, passing seashells, seaweed, and sand, sand, sand. Then he spotted the stone. This particular stone attracted his attention because of its slightly curled, kidney-like shape. He reached down and removed the stone from its repose in the sand. Holding it up, he was delighted to find that along its entire length there ran a single ridge, its only distinct marking on an otherwise smooth surface caressed by the sea. This was no mere contour. It was a signature. It matched the Brödel line of the kidney.

Max Brödel was a major medical illustrator at the beginning of the twentieth century. He was also known as an anatomist and scientist, primarily for his description of an avascular area of the kidney that is now known as Brödel's line, or Brödel's bloodless line.[52] The Brödel line (which is actually a plane) delineates the place at which nephrotomy is done—that is, an essentially bloodless incision into the kidney. Even so, "although Brödel described in his original article some key anatomical landmarks, the identification of this avascular plane at the level of a lobulated normal kidney remains essentially related to the experience of a trained eye."[53]

Holding this sea-sculpted art in his hand, Atala had a moment of sudden insight. As a urologic surgeon, he realized that the Brödel line would be the perfect site to insert a thin slice—a wafer—of healthy kidney tissue and have it assimilate into the existing kidney. But his revelation went beyond the kidney and drew on a medical aphorism: the human body's 10-times reserve. "I just need 10%. ... Usually, a patient doesn't present with any serious symptoms till whatever organ is involved has lost 90% of its function."[54] Atala was proposing *renal sufficiency* as an alternative to renal replacement via bioprinting—or via a donor, for that matter. We must look into this organ reserve axiom.

Organ Reserve

Explanations often begin with the calculations of a gentleman named Benjamin Gompertz, a self-taught British mathematician employed as an actuary for an insurance company. In 1825 he published a paper on what he termed the law of mortality and which many now refer to as the Gompertz equation or Gompertz function. He focused on deaths due to the process of aging.[§] If you put aside deaths due to accident and infectious disease, and just zero in on age-dependent death, Gompertz found that age-specific mortality rates increased exponentially with age and he formulated an equation to describe this.[55,56] You know how automobile manuals always tell you to take your car in for a checkup based on mileage or time? Well, the human body checks out after a period of time, regardless of your mileage. It's called life span.

Early in the twentieth century, scientists began looking for biological explanations for Gompertz's law of mortality.[57] By the mid-twentieth century, scientists had tested various organ systems of many people through a range of ages. They estimated that most physiological functions of human beings decrease at a rate of between 0.5% and 1.3% per year after age 30.[58] The process of aging certainly has biochemical and cellular foundations. But our present understanding focuses on a linear decline (a roughly straight line with a downward slope) of the maximal function of our vital organs—that is, physiological changes in aging at the organ level.[59,60]

This would be true if you measured the maximum breathing capacity of your lungs, for example. The same is true for any other organ-system measurement, although the slopes of the lines won't be identical—the rates of decline will vary slightly.

Almost 40 years ago, Dr. James F. Fries, physician and professor of medicine at Stanford University, wrote on this subject. He explained that organ reserve potential is highest in early life. The decline as we age is in the so-called *reserve power* of

---

[§] As an actuary, he compiled and analyzed statistics relevant to mortality that insurance companies used to estimate premiums for life insurance.

the particular organ. The decline is apparent on measurement of maximum performance long before it's clinically visible as a limitation to activity. "Studies of this physiologic decline … uniformly indicate that the decline begins early in life in healthy individuals—well before it is reasonable to postulate any specific chronic disease effect."[61]

At play here is the medical maxim that Atala alluded to. So too is the concept of homeostasis that we learned about in the previous chapter. To give a specific example of the human body's 10-times reserve: The heart of a 20-year-old is capable of pumping about 10 times the amount of blood that is actually needed to keep the body alive. For those interested, a discussion of many organ-system-specific rates of functional decline—that is, neurological, renal, musculoskeletal, and so on—is available.[62]

In 1960, two researchers at the National Institutes of Health gerontology branch at Baltimore City Hospital detailed the relationship between the Gompertz function and physiological capacity measurements versus age.[63,**] Death occurs when the rate at which our bodies work to restore homeostasis is less than that demanded to overcome the effects of a given task. When our bodies' organ systems have declined to around 20% or less of their reserve capacity, we are unable to weather physical hardships and restore homeostasis.[64] This is consistent with Gompertz' experimental observations of mortality rates.

Our bodies have an innate expiration date.

A Portion Might Be the Prescription

On a more upbeat note, when I spoke to Atala, I asked him about kidney wafers. I knew he'd been thinking about this for decades, since that day on the beach. You don't really need a whole organ. You just need to give patients the function they require to live a productive life.

Atala told me that his proposal of aiming for renal sufficiency rather than renal replacement is already underway. Wake Forest has completed Phase I clinical trials, including a two-year follow-up in a handful of patients who were on the road to dialysis. Atala wanted to start with the easiest way first—a small hydrogel rather than a wafer. They're at the point now where they've gone to 75-year-old patients with end-stage kidney failure, done a needle biopsy of their kidneys, and gotten enough progenitor cells to grow the healthy kidney cells that are needed for therapy. Then they placed the cells back into the patients' bodies in a hydrogel scaffold using minimally invasive surgery with radiographic guidance. And the patients did well. The cell-laden hydrogel arrested the decline in their kidney function. After two years, one of the patients started to decline again. The FDA gave approval to reinject that

** Gerontology is the scientific study of the process of aging.

patient—and the patient restabilized. Now Phase II clinical trials have started in 10 centers throughout the United States. [65]

Atala looks forward to the next rendition where instead of a small hydrogel they would implant a wafer to hold the healthy kidney cells. The wafer might be a bio-compatible polymer material or decellularized kidney tissue, for instance. Perhaps that transplant would last a decade. He envisions progressing to larger structures that could provide 20 years of good health. That's the thinking.

As Atala explained in a magazine interview,

> The patient doesn't collapse climbing the stairs, short of breath, until his lungs are functioning at 10% capacity. The heart patient doesn't succumb to chest pain until her artery is 90% blocked. It's the same with the kidney or liver. If I can insert healthy tissue, equal to 10 or 20% of the size of the organ, I can keep that patient functioning and alive at a high quality of life. Ideally, with this technology [bioprint-ing], you screen people and you augment the existing organ when they've only lost 40 or 50% of their function. The patient never even reaches an emergency stage.[66]

Partial organs just might be the prescription—the ℞ in the future of bioprinting.

An Editorial Perspective

Two venerable bioprinting scientists, Anthony Atala and Gabor Forgacs, recently wrote a joint editorial titled "Three-Dimensional Bioprinting in Regenerative Medicine: Reality, Hype, and Future."[67] They pointed out that our organs are the results of millions of years of evolution during which time, in their words, "*nature experimented until it 'got it right'* [italics added]."[68]

When the general public started to become aware of bioprinting, the notion of living biological products captured their imagination. After all, 3D printing had come a long way from plastic samples for prototyping to manufacturing intricate tools, automotive parts, and entire buildings. Since bioprinting is a specific application of 3D printing, many people were under the impression that bioprinting would follow suit. If toys could be 3D printed at the push of a button, why not tissues and organs. As we've seen, it's not that easy.

Mark Twain memorably said, "The report of my death was an exaggeration."[69] If bioprinting could give tongue it would say, "The report of my success was an exaggeration." The coverage in the popular media of bioprinting's success is presently an exaggeration, but unlike Twain, bioprinting's outcome is not certain. Perhaps the most vexing undertaking in bioprinting is the *bio* part. Biology is just so intricate and mostly unknown.

Sir Peter Medawar was a British biologist and Nobel Prize winner in Physiology or Medicine for his work on tissue grafting—foundational to the practice of organ transplants.[70] He wrote,

Biologists work very close to the frontier between bewilderment and understanding. Biology is complex, messy, and richly various, like real life; it travels faster
nowadays than physics or chemistry (which is just as well, because it has so much
farther to go), and it travels nearer to the ground.[71]

However, Atala and Forgacs observed that only three decades ago, most human
cell types could not be grown and matured outside the body and human stem cells
had not yet been identified. Today, most types of cells can be grown in vitro, multiple types of human stem cells have been identified, and organs have been bioengineered in the laboratory. In small clinical trials, engineered organs such as skin,
urethra, blood vessel, and bladder have been implanted in patients. In fact, several
bioengineered tissues are presently advancing through the governmental regulatory
process on their way to commercialization and availability to patients. The authors
emphasized that most engineered tissues in the more advanced stages of regulatory
approval were *made by hand*.

Atala and Forgacs concluded their editorial by saying, "Bioprinting offers many
promising opportunities. However, patience and perseverance are needed to realize
the full potential of the technology."[72]

When we began our narrative, we spoke of monks printing by hand to reproduce
the scholarly jewels of civilization. Atala and Forgacs remind us that automated bioprinting was preceded by reproducing the jewels of the human body—functional
organs—by hand.

Perhaps the most famous example of this was Atala's own work in constructing
a bladder for a 10-year-old child named Luke. His malfunctioning bladder allowed
urine to leak and back up into his kidneys. The possibility of dialysis loomed. Atala
took cells from the boy's defective bladder and carefully nourished them in vitro so
that they multiplied. He designed a scaffold for a new bladder and manually pipetted
the cells onto it. Atala transplanted the bioengineered bladder into Luke's body in a
14-hour surgery. That was in 2001.[73] It's been working fine for two decades. The accomplishment has been heralded as a milestone in urologic tissue engineering.[74]

The advent of automated bioprinting, enabled by Chuck Hull's invention of 3D
printing, has been in practice for (a mere) 20 years. That has made it unnecessary to
build new organs by hand. Thanks to Hull and dozens of bioprinting research groups,
we are not waiting for a latter-day Gutenberg. The technology is here.

What we need to acquire in the coming decades is the additional knowledge to
successfully imitate nature in constructing complicated tissues and organs. With patience and persistence, it's possible that we too can "get it right."

Epilogue

The book you've been reading can only be a vignette, a brief description of an evolving field; life goes on. Most happily, so too has Nancy's life. Her kidney transplant was in May 2016, and she was able to come back quickly to her old job as full-time office manager at a thriving physical therapy clinic where she's highly esteemed by both staff and patients. She told me,

> The greatest gift of the kidney transplant is that I have much more energy to do what life requires of me. Where a lot of people want to celebrate the third anniversary of their kidney transplant by running a marathon or traveling to Australia or something they've always wanted to do, all I want to do is sit on the back porch and watch the birds and listen to the bees hum. That's what I treasure more than anything else—the ability to enjoy my own home and this beautiful town.[1]

Don't think she just sits on the back porch.

> For almost two years before my transplant I was feeling so fatigued that I couldn't do any gardening. But now I'm extremely grateful to be back on my hands and knees pulling weeds out of the flower beds and pruning flowers for a couple of hours at a time.

Because of her immunosuppressant regimen, she travels less than she used to. Yet,

> There are things that I thought I could never do again but I'm actually fine doing— such as ushering at the Colorado Shakespeare Festival. I worried that the immuno- suppressants would be a deal-breaker for me—that I would always be ill, unable to go into work, unable to attend concerts, movies, plays, or sit in crowds. But I found that as long as I avoid people who are hacking or coughing or look ill, it goes pretty well. I've become an avid hand-washer. And when people extend their hand to shake it, I will shake hands but [soon after] I'll put on a little hand sanitizer. I guard my life very carefully.

Nancy has acclimated to life with a third kidney—and since it's her third she some- times calls it Trina. Her daily immunosuppressant medications are a reminder that it's part of me but it's also not part of me. It's a giant piece of someone else's DNA in my body. One of the blood tests I need to have measures how much of the kidney's DNA is in my blood and how much of my DNA is in my blood.

The idea is that if the kidney starts to shed DNA then it's under attack from her immune system. That might require an adjustment to her immunosuppressant regimen.

Recently, Nancy met with her new nephrologist; her former doctor had left the practice. He spent some time reviewing her case history.

"Interesting, huh?" Nancy asked him.

"No," he replied. "Your case history is really, really boring. And that's just what we want to see."

Notes

A Note to the Reader

1. Groll, Jürgen, Thomas Boland, Torsten Blunk, Jason A. Burdick, Dong-Woo Cho, Paul D. Dalton, Brian Derby, Gabor Forgacs, Qing Li, Vladimir A. Mironov, Lorenzo Moroni, Makoto Nakamura, Wenmiao Shu, Shoji Takeuchi, Giovanni Vozzi, Tim B. F. Woodfield, Tao Xu, James J. Yoo, and Jos Malda. "Biofabrication: Reappraising the Definition of an Evolving Field." *Biofabrication* 8, no. 1 (January 8, 2016): 013001. https://doi.org/10.1088/1758-5090/8/1/013001.
2. Ibid., 2.
3. Groll, J., J. A. Burdick, D.-W. Cho, B. Derby, M. Gelinsky, S. C. Heilshorn, T. Jüngst, J. Malda, V. A. Mironov, K. Nakayama, A. Ovsianikov, W. Sun, S. Takeuchi, J. J. Yoo, and T. B. F. Woodfield. "A Definition of Bioinks and Their Distinction from Biomaterial Inks." *Biofabrication* 11, no. 1 (November 23, 2018): 013001. https://doi.org/10.1088/1758-5090/aaec52.
4. Ibid., 4.
5. Williams, David F. "On the Nature of Biomaterials." *Biomaterials* 30, no. 30 (October 2009): 5897–909. https://doi.org/10.1016/j.biomaterials.2009.07.027.
6. Augustine, N. "Essays on Science and Society: What We Don't Know Does Hurt Us. How Scientific Illiteracy Hobbles Society." *Science* 279, no. 5357 (March 13, 1998): 1640–41. https://doi.org/10.1126/science.279.5357.1640.

Preface

1. Inspired by Pelham Grenville Wodehouse. *Pigs Have Wings: A Blandings Story*. London: Penguin Books, 1957, 165.
2. National Kidney Foundation. "Organ Donation and Transplantation Statistics." January 11, 2016. https://www.kidney.org/news/newsroom/factsheets/Organ-Donation-and-Transplantation-Stats.
3. Forgacs, Gabor. Phone conversation with K.D., April 6, 2018.
4. Grüner, George, and James Tyrrell. "An Interview with Gabor Forgacs: From Theoretical Physics to the Business of 3D Bio-Printing." *Translational Materials Research* 2, no. 4 (January 15, 2016): 040203. https://doi.org/10.1088/2053-1613/2/4/040203.
5. Danaylov, Nikola. "Gabor Forgacs: We Live in a Time When It Is Really Difficult to Say 'This Is Impossible!'" *Singularity Weblog* (February 7, 2014). https://www.singularityweblog.com/gabor-forgacs-bioprinting.
6. Steinberg, M. S. "On the Mechanism of Tissue Reconstruction by Dissociated Cells, I. Population Kinetics, Differential Adhesiveness, and the Absence of Directed Migration." *Proceedings of the National Academy of Sciences of the USA* 48, no. 9 (September 1, 1962): 1577–82. https://doi.org/10.1073/pnas.48.9.1577.

7. Steinberg, M. S. "Mechanism of Tissue Reconstruction by Dissociated Cells, II: Time-Course of Events." *Science* 137, no. 3532 (September 7, 1962): 762–63. https://doi.org/10.1126/science.137.3532.762.

8. Steinberg, M. S. "On the Mechanism of Tissue Reconstruction by Dissociated Cells, III. Free Energy Relations and the Reorganization of Fused, Heteronomic Tissue Fragments." *Proceedings of the National Academy of Sciences of the USA* 48, no. 10 (October 1, 1962): 1769–76. https://doi.org/10.1073/pnas.48.10.1769.

9. Steinberg, M. S. "Reconstruction of Tissues by Dissociated Cells." *Science* 141, no. 3579 (August 2, 1963): 401–408. https://doi.org/10.1126/science.141.3579.401.

10. Foty, Ramsey A., and Malcolm S. Steinberg. "The Differential Adhesion Hypothesis: A Direct Evaluation." *Developmental Biology* 278, no. 1 (February 2005): 255–63. https://doi.org/10.1016/j.ydbio.2004.11.012.

11. Forgacs, Gabor, Ramsey A. Foty, Yinon Shafrir, and Malcolm S. Steinberg. "Viscoelastic Properties of Living Embryonic Tissues: A Quantitative Study." *Biophysical Journal* 74, no. 5 (May 1998): 2227–34. https://doi.org/10.1016/s0006-3495(98)77932-9.

12. Jakab, K., A. Neagu, V. Mironov, R. R. Markwald, and G. Forgacs. "Engineering Biological Structures of Prescribed Shape Using Self-Assembling Multicellular Systems." *Proceedings of the National Academy of Sciences of the USA* 101, no. 9 (February 23, 2004): 2864–69. https://doi.org/10.1073/pnas.0400164101.

13. Neagu, Adrian, Karoly Jakab, Richard Jamison, and Gabor Forgacs. "Role of Physical Mechanisms in Biological Self-Organization." *Physical Review Letters* 95, no. 17 (October 21, 2005). https://doi.org/10.1103/physrevlett.95.178104.

14. Forgacs, Gabor. E-mail to K.D., July 21, 2019.

15. Ringsdorf, Helmut. "A Moment of Reflection: Sixty Years After the Nobel Prize for Hermann Staudinger." *Hierarchical Macromolecular Structures: 60 Years After the Staudinger Nobel Prize I* 261 (2013): 1–19. https://doi.org/10.1007/12_2013_263.

16. Nancy. Conversation with K.D., March 1, 2017.

Introduction

1. The chapter subtitle is a corruption of "Frailty, thy name is woman," by William Shakespeare. Shakespeare, William. *Hamlet* (Act I, Scene 2, Line 350). In "Open Source Shakespeare: Search Shakespeare's Works, Read the Texts." Accessed December 8, 2019. https://www.opensourceshakespeare.org.

2. Shakespeare, William. *Hamlet* (Act III, Scene 1, Lines 1755–1756). In "Open Source Shakespeare: Search Shakespeare's Works, Read the Texts." Accessed December 8, 2019. https://www.opensourceshakespeare.org.

3. Shakespeare, William. *Hamlet* (Act I, Scene 2, Line 333). In "Open Source Shakespeare: Search Shakespeare's Works, Read the Texts." Accessed December 8, 2019. https://www.opensourceshakespeare.org/.

4. Shaer, Matthew. "Need a New Organ? Surgeon Anthony Atala Sees a Future Where You Can Simply Print It Out." *Smithsonian Magazine* (November 17, 2016). https://www.smithsonianmag.com/innovation/miracle-maker-anthony-atala-winner-smithsonian-ingenuity-awards-2016-life-sciences-180961121.

Chapter 1

1. Greer, Germaine. *The Obstacle Race: The Fortunes of Women Painters and Their Work*. London: Tauris Parke Paperbacks, 2001.
2. DeWitte, Sharon N. "Mortality Risk and Survival in the Aftermath of the Medieval Black Death." Edited by Andrew Noymer. *PLoS One* 9, no. 5 (May 7, 2014): e96513. https://doi.org/10.1371/journal.pone.0096513.
3. Burke, James. *Connections*. Boston: Little, Brown, 1978, 101.
4. Ibid., 101–102.
5. Man, John. *The Gutenberg Revolution: The Story of a Genius and an Invention That Changed the World*. London: Bantam, 2009, 26.
6. Ibid., 4–6.
7. Hull, Charles. "The Birth of 3D Printing." *Research-Technology Management* 58, no. 6 (December 23, 2015) : 25–29. https://www.tandfonline.com/doi/pdf/10.5437/08956308X5806067.
8. Ibid., 25.
9. Hull, Charles W. Apparatus for production of three-dimensional objects by stereolithography, issued March 1986. US Patent Number 4,575,330.
10. Wilson, W. Cris, and Thomas Boland. "Cell and Organ Printing 1: Protein and Cell Printers." *The Anatomical Record* 272A, no. 2 (May 6, 2003): 491–96. https://doi.org/10.1002/ar.a.10057.
11. Boland, Thomas, Vladimir Mironov, Anna Gutowska, Elisabeth A. Roth, and Roger R. Markwald. "Cell and Organ Printing 2: Fusion of Cell Aggregates in Three-Dimensional Gels." *The Anatomical Record* 272A, no. 2 (May 6, 2003): 497–502. https://doi.org/10.1002/ar.a.10059.
12. Mironov, Vladimir, Thomas Boland, Thomas Trusk, Gabor Forgacs, and Roger R. Markwald. "Organ Printing: Computer-Aided Jet-Based 3D Tissue Engineering." *Trends in Biotechnology* 21, no. 4 (April 2003): 157–61. https://doi.org/10.1016/s0167-7799(03)00033-7.
13. W. Cris Wilson, Jr., "Cell and Organ Printing 1."
14. Boland, Thomas. Phone conversation with K.D., March 15, 2018.
15. Vladimir Mironov, Vladimir. Phone conversation with K.D., April 3, 2018.
16. Thomas Boland, "Cell and Organ Printing 2."
17. Vladimir Mironov et al., "Organ Printing."
18. Mironov, Vladimir. E-mail to K.D., March 25, 2018.
19. Neagu, Adrian, Karoly Jakab, Richard Jamison, and Gabor Forgacs. "Role of Physical Mechanisms in Biological Self-Organization." *Physical Review Letters* 95, no. 17 (October 21, 2005): 178104. https://doi.org/10.1103/physrevlett.95.178104.
20. Norotte, Cyrille, Francois S. Marga, Laura E. Niklason, and Gabor Forgacs. "Scaffold-Free Vascular Tissue Engineering Using Bioprinting." *Biomaterials* 30, no. 30 (October 2009): 5910–17. https://doi.org/10.1016/j.biomaterials.2009.06.034.
21. Sun, Wei. Phone conversation with K.D., March 23, 2018.
22. Sun, Wei. E-mails to K.D., February 8, 2018; March 7, 2018; July 17, 2019.
23. Khalil, S., J. Nam, and W. Sun. "Multi-Nozzle Deposition for Construction of 3D Biopolymer Tissue Scaffolds." *Rapid Prototyping Journal* 11, no. 1 (February 2005): 9–17. https://doi.org/10.1108/13552540510573347.

24. Sun, Wei. E-mail to K.D., March 7, 2018.

25. Mironov, Vladimir, Nuno Reis, and Brian Derby. "Review: Bioprinting: A Beginning." *Tissue Engineering* 12, no. 4 (April 2006): 631–34. https://doi.org/10.1089/ten.2006.12.631.

26. Derby, Brian. Phone conversation with K.D., May 4, 2018.

27. Hutmacher, Dietmar W., Michael Sittinger, and Makarand V. Risbud. "Scaffold-Based Tissue Engineering: Rationale for Computer-Aided Design and Solid Free-Form Fabrication Systems." *Trends in Biotechnology* 22, no. 7 (July 2004): 354–62. https://doi.org/10.1016/j.tibtech.2004.05.005.

28. Hickey, Shane. "Chuck Hull: The Father of 3D Printing Who Shaped Technology." *The Guardian*, November 30, 2017. https://www.theguardian.com/business/2014/jun/22/chuck-hull-father-3d-printing-shaped-technology.

29. Scott, Clare. "Twin Babies No Longer Joined at the Brain Thanks to 3D Technology." 3DPrint.com, October 21, 2016. https://3dprint.com/153280/conjoined-twins-3d-surgery.

30. "3D printed splint saves the life of a baby." Biomedical Engineering at the University of Michigan, January 12, 2016. https://bme.umich.edu/3d-printed-splint-saves-the-life-of-a-baby.

31. Zopf, David A., Scott J. Hollister, Marc E. Nelson, Richard G. Ohye, and Glenn E. Green. "Bioresorbable Airway Splint Created with a Three-Dimensional Printer." *New England Journal of Medicine* 368, no. 21 (May 23, 2013): 2043–45. https://doi.org/10.1056/nejmc1206319.

32. "A Brighter Christmas for Hannah, First Teen to Be Saved by 3D Printed Airway Splint." Michigan Medicine, January 15, 2016. https://medicine.umich.edu/dept/otolaryngology/news/archive/201512/brighter-christmas-hannah-first-teen-be-saved-3d-printed-airway-splint.

33. Murphy, Sean V., and Anthony Atala. "3D Bioprinting of Tissues and Organs." *Nature Biotechnology* 32, no. 8 (2014): 773–85. https://doi.org/10.1038/nbt.2958.

34. Rouwkema, Jeroen, and Ali Khademhosseini. "Vascularization and Angiogenesis in Tissue Engineering: Beyond Creating Static Networks." *Trends in Biotechnology* 34, no. 9 (September 2016): 733–45. https://doi.org/10.1016/j.tibtech.2016.03.002.

35. Novosel, Esther C., Claudia Kleinhans, and Petra J. Kluger. "Vascularization Is the Key Challenge in Tissue Engineering." *Advanced Drug Delivery Reviews* 63, no. 4–5 (April 2011): 300–11. https://doi.org/10.1016/j.addr.2011.03.004.

36. Sender, Ron, Shai Fuchs, and Ron Milo. "Revised Estimates for the Number of Human and Bacteria Cells in the Body." *PLoS Biology* 14, no. 8 (August 19, 2016): e1002533. https://doi.org/10.1371/journal.pbio.1002533.

37. Vacanti, Joseph P. "Beyond Transplantation." *Archives of Surgery* 123, no. 5 (May 1, 1988): 545–49. https://doi.org/10.1001/archsurg.1988.01400290027003.

38. Ibid., 548.

39. Ibid.

40. Ibid., 549.

41. Ibid.

42. Hospodiuk, Monika, Madhuri Dey, Donna Sosnoski, and Ibrahim T. Ozbolat. "The Bioink: A Comprehensive Review on Bioprintable Materials." *Biotechnology Advances* 35, no. 2 (March 2017): 217–39. https://doi.org/10.1016/j.biotechadv.2016.12.006.

43. Magin, Chelsea M., Daniel L. Alge, and Kristi S. Anseth. "Bio-Inspired 3D Microenvironments: A New Dimension in Tissue Engineering." *Biomedical Materials* 11, no. 2 (March 4, 2016): 022001. https://doi.org/10.1088/1748-6041/11/2/022001.

44. Sean V. Murphy and Anthony Atala, "3D Bioprinting," 779.

45. Ibid., 782.

46. Monika Hospodiuk et al., "The Bioink."

47. Rozario, Tania, and Douglas W. DeSimone. "The Extracellular Matrix in Development and Morphogenesis: A Dynamic View." *Developmental Biology* 341, no. 1 (May 2010): 126–40. https://doi.org/10.1016/j.ydbio.2009.10.026.

48. Alberts, Bruce, Alexander Johnson, Julian Hart Lewis, David Morgan, Martin Raff, Keith Roberts, and Peter Walter. *Molecular Biology of the Cell.* New York: Garland, 2015, 1035.

49. Itoh, Manabu, Koichi Nakayama, Ryo Noguchi, Keiji Kamohara, Kojirou Furukawa, Kazuyoshi Uchihashi, Shuji Toda, Jun-ichi Oyama, Koichi Node, and Shigeki Morita. "Scaffold-Free Tubular Tissues Created by a Bio-3D Printer Undergo Remodeling and Endothelialization When Implanted in Rat Aortae." *PLoS One* 10, no. 9 (September 1, 2015): e0136681. https://doi.org/10.1371/journal.pone.0136681.

50. Kirschner, Chelsea M., Anthony Brennan, and James Schumacher. *Bio-Inspired Materials for Biomedical Engineering.* Edited by Chelsea M. Kirschner and Anthony Brennan. Hoboken, NJ: Wiley, 2014, 79–80.

51. "Shear Mystery." Nasa.gov, June 7, 2002. https://science.nasa.gov/science-news/science-at-nasa/2002/07jun_elastic_fluids.

52. Mironov, Vladimir, Richard P. Visconti, Vladimir Kasyanov, Gabor Forgacs, Christopher J. Drake, and Roger R. Markwald. "Organ Printing: Tissue Spheroids as Building Blocks." *Biomaterials* 30, no. 12 (April 2009): 2164–74. https://doi.org/10.1016/j.biomaterials.2008.12.084.

53. Landry, J., D. Bernier, C. Ouellet, R. Goyette, and N. Marceau. "Spheroidal Aggregate Culture of Rat Liver Cells: Histotypic Reorganization, Biomatrix Deposition, and Maintenance of Functional Activities." *Journal of Cell Biology* 101, no. 3 (September 1, 1985): 914–23. https://doi.org/10.1083/jcb.101.3.914.

54. Moldovan, Nicanor I., Narutoshi Hibino, and Koichi Nakayama. "Principles of the Kenzan Method for Robotic Cell Spheroid-Based Three-Dimensional Bioprinting." *Tissue Engineering Part B: Reviews* 23, no. 3 (June 2017): 237–44. https://doi.org/10.1089/ten.teb.2016.0322.

55. Manabu Itoh, "Scaffold-Free Tubular Tissues," 2.

56. Nakayama, Koichi. Phone conversation with K.D., March 28, 2018.

57. Fallows, James. "The 50 Greatest Breakthroughs Since the Wheel." *The Atlantic,* November 2013.

58. Raney, Jordan R., Brett G. Compton, Jochen Mueller, Thomas J. Ober, Kristina Shea, and Jennifer A. Lewis. "Rotational 3D Printing of Damage-Tolerant Composites with Programmable Mechanics." *Proceedings of the National Academy of Sciences of the USA* 115, no. 6 (January 18, 2018): 1198–203. https://doi.org/10.1073/pnas.1715157115.

59. Gocha, April. "A New Spin on Additive Manufacturing: Rotational 3-D Printing Controls Fiber Orientation to Print Stronger Functional Composites | The American Ceramic Society." The American Ceramic Society, January 18, 2018. http://ceramics.org/ceramic-tech-today/a-new-spin-on-additive-manufacturing-rotational-3-d-printing-controls-fiber-orientation-to-print-stronger-functional-composites.

60. Lewis, Jennier. E-mail to K.D., August 16, 2018.

61. James Fallows, "The 50 Greatest Breakthroughs."

62. Persaud, Natasha. "Bioprinting Human Tissue." *Renal and Urology News*, February 6, 2018. https://www.renalandurologynews.com/end-stage-renal-disease/bioprinting-human-tissue-anthony-atala/article/741779.

63. Powley, Tanya. "Printing Whole Organs Remains a Long Way Off." *Financial Times*, June 11, 2015. https://www.ft.com/content/2fb2fb5a-ffa5-11e4-bc30-00144feabdc0.

64. Findlay, Sam. "Can You 3D-Print Organs?" "ARC Centre of Excellence for Electromaterial Science, March 23, 2015. https://electromaterials.edu.au/2015/03/23/can-you-3d-print-organs.

65. Derby, B. "Printing and Prototyping of Tissues and Scaffolds." *Science* 338, no. 6109 (November 15, 2012): 921–26. https://doi.org/10.1126/science.1226340.

66. Sinha, Gunjan. "Cell Presses." *Nature Biotechnology* 32, no. 8 (August 2014): 716–19. https://doi.org/10.1038/nbt.2983.

67. Morber, Jenny. "Custom Organs, Printed to Order." PBS, March 18, 2015. http://www.pbs.org/wgbh/nova/next/body/3d-printed-organs.

68. Lewis, Phillip L., and Ramille N. Shah. "3D Printing for Liver Tissue Engineering: Current Approaches and Future Challenges." *Current Transplantation Reports* 3, no. 1 (January 22, 2016): 100–108. https://doi.org/10.1007/s40472-016-0084-y.

Chapter 2

1. De Duve, Christian. *A Guided Tour of the Living Cell.* New York: Scientific American Library, 1984, 18.

2. Ibid., 106

3. Wilhelm, Mathias, Judith Schlegl, Hannes Hahne1, Amin Moghaddas Gholami, Marcus Lieberenz, Mikhail M. Savitski, Emanuel Ziegler, Lars Butzmann, Siegfried Gessulat, Harald Marx, Toby Mathieson, Simone Lemeer, Karsten Schnatbaum, Ulf Reimer, Holger Wenschuh, Martin Mollenhauer, Julia Slotta-Huspenina, Joos-Hendrik Boese, Marcus Bantscheff, Anja Gerstmair, Franz Faerber, and Bernhard Kuster. "Mass-Spectrometry-Based Draft of the Human Proteome." *Nature* 509, no. 7502 (May 2014): 582–87. https://doi.org/10.1038/nature13319.

4. Christian de Duve, *A Guided Tour*, 28.

5. Ibid., 19.

6. Ibid., 21.

7. Ibid., 34–35.

8. Ibid., 41.

9. Ibid., 41

10. Milo, Ron, and Rob Phillips. *Cell Biology by the Numbers*. New York: Garland, 2016. http://book.bionumbers.org/what-is-the-thickness-of-the-cell-membrane.

11. Christian de Duve, *A Guided Tour*, 46.

12. Edidin, Michael. "Lipids on the Frontier: A Century of Cell-Membrane Bilayers." *Nature Reviews Molecular Cell Biology* 4, no. 5 (May 2003): 414–18. https://doi.org/10.1038/nrm1102.

13. Goodsell, David S. "Sodium–Potassium Pump." RCSB Protein Data Bank, October 1, 2009. https://doi.org/10.2210/rcsb_pdb/mom_2009_10.

14. Goodsell, David S. "Glucose Transporters." RCSB Protein Data Bank, April 1, 2017. https://doi.org/10.2210/rcsb_pdb/mom_2017_4.

15. Ibid.

16. Christian de Duve, *A Guided Tour*, 87.

17. Goodsell, David S. *The Machinery of Life*. New York: Copernicus, 2010, 78–79.

18. Christian de Duve, *A Guided Tour*, 93.

19. Ibid., 384.

20. Weintraub, Karen. "Stem Cells: Plenty of Hope, But Halting Progress." *The New York Times*, September 15, 2014. https://www.nytimes.com/2014/09/16/science/stem-cell-progress-begins-to-catch-up-to-promise.html.

21. National Institutes of Health. "Stem Cell Basics II." NIH Stem Cell Information Home Page, 2016. https://stemcells.nih.gov/info/basics/II.htm.

22. Ibid.

23. Ibid.

24. Murry, Charles E., Joseph Gold, Lil Pabon, and Lior Gepstein. *Heart Development and Regeneration*. Edited by Nadia Rosenthal and Richard P Harvey. Cambridge, MA: Elsevier, 2010, 878.

25. Scudellari, Megan. "How IPS Cells Changed the World." *Nature* 534, no. 7607 (June 15, 2016): 310–12. https://doi.org/10.1038/534310a.

26. Ibid.

27. "The Nobel Prize in Physiology or Medicine 2012." NobelPrize.org, 2012. https://www.nobelprize.org/prizes/medicine/2012/press-release.

28. "Cell Line." Lexico Dictionaries. Accessed December 22, 2019. https://www.lexico.com/definition/cell_line.

29. Karagiannis, Peter, Kazutoshi Takahashi, Megumu Saito, Yoshinori Yoshida, Keisuke Okita, Akira Watanabe, Haruhisa Inoue, Jun K. Yamashita, Masaya Todani, Masato Nakagawa, Mitsujiro Osawa, Yoshimi Yashiro, Shinya Yamanaka, and Kenji Osafune. "Induced Pluripotent Stem Cells and Their Use in Human Models of Disease and Development." *Physiological Reviews* 99, no. 1 (January 1, 2019): 79–114. https://doi.org/10.1152/physrev.00039.2017.

30. Attwood, Stephen, and Michael Edel. "IPS-Cell Technology and the Problem of Genetic Instability—Can It Ever Be Safe for Clinical Use?" *Journal of Clinical Medicine* 8, no. 3 (February 28, 2019): 288. https://doi.org/10.3390/jcm8030288.

31. De Coppi, Paolo, Georg Bartsch, Jr., M. Minhaj Siddiqui, Tao Xu, Cesar C. Santos, Laura Perin, Gustavo Mostoslavsky, Angéline C. Serre, Evan Y. Snyder, James J. Yoo, Mark E. Furth, Shay Soker, and Anthony Atala. "Isolation of Amniotic Stem Cell Lines with Potential for Therapy." *Nature Biotechnology* 25, no. 1 (January 2007): 100–106. https://doi.org/10.1038/nbt1274.

32. Hildreth, Cade. "Amniotic Stem Cells: An Extension of the Cord Blood Banking." BioInformant, March 17, 2018. https://bioinformant.com/amnion-foundation.

33. "What Are Progenitor Cells?" Boston Children's Hospital, 2009. http://stemcell.childrenshospital.org/about-stem-cells/adult-somatic-stem-cells-101/what-are-progenitor-cells.

34. Salzberg, Steven L. "Open Questions: How Many Genes Do We Have?" *BMC Biology* 16, no. 1 (August 20, 2018). https://doi.org/10.1186/s12915-018-0564-x.

35. Horning, David P., and Gerald F. Joyce. "Amplification of RNA by an RNA Polymerase Ribozyme." *Proceedings of the National Academy of Sciences of the USA* 113, no. 35 (August 15, 2016): 9786–91. https://doi.org/10.1073/pnas.1610103113.

36. Nissim, Sahar, Richard I. Sherwood, Julia Wucherpfennig, Diane Saunders, James M. Harris, Virginie Esain, Kelli J. Carroll, Gregory M. Frechette, Andrew J. Kim, Katie L. Hwang, Claire C. Cutting, Susanna Elledge, Trista E. North, and Wolfram Goessling. "Prostaglandin E2 Regulates Liver Versus Pancreas Cell-Fate Decisions and Endodermal Outgrowth." *Developmental Cell* 28, no. 4 (February 2014): 423–37. https://doi.org/10.1016/j.devcel.2014.01.006.

37. Sinha, Gunjan. "Cell Presses." *Nature Biotechnology* 32, no. 8 (August 2014): 716–19. https://doi.org/10.1038/nbt.2983.

38. Ibid., 717.

39. Ibid.

40. Marguerat, Samuel, and Jürg Bähler. "Coordinating Genome Expression with Cell Size." *Trends in Genetics* 28, no. 11 (November 2012): 560–65. https://doi.org/10.1016/j.tig.2012.07.003.

41. "What Is Your Conceptual Definition of 'Cell Type' in the Context of a Mature Organism?" *Cell Systems* 4, no. 3 (March 2017): 255–59. https://doi.org/10.1016/j.cels.2017.03.006.

42. Ibid.

43. "Home." Human Cell Atlas, 2016. https://www.humancellatlas.org.

44. Weule, Genelle. "Human Cell Atlas: The plan to map every cell in your body." *ABC News*, November 16, 2017. http://www.abc.net.au/news/science/2017-11-17/human-cell-atlas-the-plan-to-map-every-cell-in-your-body/9127096.

45. Daley, Jason. "Human Cell Atlas Releases First Major Data Set." *Smithsonian Magazine*, April 19, 2018. https://www.smithsonianmag.com/smart-news/human-cell-atlas-releases-first-major-data-set-180968831.

46. Gibson, Lydialyle. "Building Toward a Kidney." *Harvard Magazine*, December 7, 2016. https://harvardmagazine.com/2017/01/building-toward-a-kidney.

47. Fortier, Lisa A., Joseph U. Barker, Eric J. Strauss, Taralyn M. McCarrel, and Brian J. Cole. "The Role of Growth Factors in Cartilage Repair." *Clinical Orthopaedics and Related Research* 469, no. 10 (March 15, 2011): 2706–15. https://doi.org/10.1007/s11999-011-1857-3.

48. Laird, Dale W., Paul D. Lampe, and Ross G. Johnson. "Cellular Small Talk." *Scientific American* 312, no. 5 (April 14, 2015): 70–77. https://doi.org/10.1038/scientificamerican0515-70.

49. Gibson, "Building Toward a Kidney."

50. Ibid.

51. Ibid.

52. Ibid.

53. Ibid.

54. Ibid.

55. Gunjan Sinha, "Cell Presses," 717.

56. Ibid.

57. Dennis, Michael Aaron. "Gordon Moore." *Encyclopædia Britannica*, November 27, 2019. https://www.britannica.com/biography/Gordon-Moore.

58. Moore, Gordon E. "Cramming More Components onto Integrated Circuits." *Electronics* 38, no. 8 (April 19, 1965).

59. Tweney, Dylan. "April 19, 1965: How Do You Like It? Moore, Moore, Moore." *WIRED*, April 19, 2010. https://www.wired.com/2010/04/0419moores-law-published.

60. Arbesman, Samuel, and Gregory Laughlin. "A Scientometric Prediction of the Discovery of the First Potentially Habitable Planet with a Mass Similar to Earth." Edited by Enrico Scalas. *PLoS One* 5, no. 10 (October 4, 2010): e13061. https://doi.org/10.1371/journal. pone.0013061.

61. Farmer, J. Doyne, and François Lafond. "How Predictable Is Technological Progress?" *Research Policy* 45, no. 3 (April 2016): 647–65. https://doi.org/10.1016/j.respol.2015.11.001.

62. Lydialyle Gibson, "Building Toward a Kidney," 9.

63. Ibid.

Chapter 3

1. "Early Anatomical Theaters." Medicalalumni.org, 2019. https://www.medicalalumni.org/early-anatomical-theaters.

2. Brockbank, William. "Old Anatomical Theatres and What Took Place Therein." *Medical History* 12, no. 4 (October 1968): 371–84. https://doi.org/10.1017/s0025727300013648.

3. Abbott, Alison. "Hidden Treasures: Padua's Anatomy Theatre." *Nature* 454, no. 7205 (August 2008): 699–99. https://doi.org/10.1038/454699a.

4. Benedek, T. G. "A History of the Understanding of Cartilage." *Osteoarthritis and Cartilage* 14, no. 3 (March 2006): 203–209. https://doi.org/10.1016/j.joca.2005.08.014.

5. Wykes, Sara. "Incredible Cartilage." *Stanford Medicine*, July 2014. http://sm.stanford.edu/archive/stanmed/2014summer/SMSum14.pdf.

6. Ibid.

7. Malda, Jos, Janny C. de Grauw, Kim E. M. Benders, Marja J. L. Kik, Chris H. A. van de Lest, Laura B. Creemers, Wouter J. A. Dhert, and P. René van Weeren. "Of Mice, Men and Elephants: The Relation Between Articular Cartilage Thickness and Body Mass." Edited by Joseph P. R. O. Orgel. *PLoS One* 8, no. 2 (February 21, 2013): e57683. https://doi.org/10.1371/journal.pone.0057683.

8. Integrated Teaching and Learning Program, College of Engineering, University of Colorado Boulder. "Viscous Fluids—Lesson." TeachEngineering.org, March 6, 2018. https://www.teachengineering.org/lessons/view/cub_surg_lesson03.

9. Ibid.

10. Schuurman, Wouter, Peter A. Levett, Michiel W. Pot, Paul René van Weeren, Wouter J. A. Dhert, Dietmar W. Hutmacher, Ferry P. W. Melchels, Travis J. Klein, and Jos Malda. "Gelatin–Methacrylamide Hydrogels as Potential Biomaterials for Fabrication of Tissue-Engineered Cartilage Constructs." *Macromolecular Bioscience* 13, no. 5 (February 18, 2013): 551–61. https://doi.org/10.1002/mabi.201200471.

11. Toole, Bryan P. "Hyaluronan Is Not Just a Goo!" *Journal of Clinical Investigation* 106, no. 3 (August 1, 2000): 335–36. https://doi.org/10.1172/jci10706.

12. Hutmacher, Dietmar. Phone conversation with K.D., March 4, 2018.

13. Levato, Riccardo, William R. Webb, Iris A. Otto, Anneloes Mensinga, Yadan Zhang, Mattie van Rijen, René van Weeren, Ilyas M. Khan, and Jos Malda. "The Bio in the Ink: Cartilage Regeneration with Bioprintable Hydrogels and Articular Cartilage-Derived Progenitor Cells." *Acta Biomaterialia* 61 (October 2017): 41–53. https://doi.org/10.1016/j.actbio.2017.08.005.

14. Sula, Jai-Yoon, Chia-wen K. Wu, Fanyi Zeng, Jeanine Jochems, Miler T. Lee, Tae Kyung Kim, Tiina Peritz, Peter Buckley, David J. Cappelleri, Margaret Maronski, Minsun Kim,

Vijay Kumar, David Meaney, Junhyong Kim, and James Eberwine. "Transcriptome Transfer Produces a Predictable Cellular Phenotype." *Proceedings of the National Academy of Sciences of the USA* 106, no. 18 (April 20, 2009): 7624–29. https://doi.org/10.1073/pnas.0902161106.

15. Orgogozo, Virginie, Baptiste Morizot, and Arnaud Martin. "The Differential View of Genotype–Phenotype Relationships." *Frontiers in Genetics* 6, no. 179 (May 19, 2015). https://doi.org/10.3389/fgene.2015.00179.

16. Johannsen, W. "The Genotype Conception of Heredity." *The American Naturalist* 45, no. 531 (March 1911): 129–59. https://doi.org/10.1086/279202.

17. Pelham Grenville Wodehouse. *Something Fresh.* London: Arrow Books, 2008, 73–74.

18. "Genotype Versus Phenotype." Berkeley.edu. Accessed December 22, 2019. https://evolution.berkeley.edu/evolibrary/article/genovspheno_01.

19. Müller, Michael, Ece Öztürk, Øystein Arlov, Paul Gatenholm, and Marcy Zenobi-Wong. "Alginate Sulfate–Nanocellulose Bioinks for Cartilage Bioprinting Applications." *Annals of Biomedical Engineering* 45, no. 1 (August 8, 2016): 210–23. https://doi.org/10.1007/s10439-016-1704-5.

20. Riccardo Levato et al., "The Bio in the Ink," 42.

21. Malda, Jos. Phone conversation with K.D., May 7, 2018.

22. Glowacki, Julie. "In Vitro Engineering of Cartilage." *Journal of Rehabilitation Research and Development* 37, no. 2 (March 2000): 171–77. https://www.rehab.research.va.gov/jour/00/37/2/pdf/glowacki.pdf.

23. Ibid., 171, 175.

24. Alberts, Bruce, Alexander Johnson, Julian Hart Lewis, David Morgan, Martin Raff, Keith Roberts, and Peter Walter. *Molecular Biology of the Cell.* New York: Garland, 2015, 1061.

25. Oyen, Michelle. "Collagen Mechanics: Learning from Nature." University of Cambridge, September 2008. https://www.cam.ac.uk/research/news/collagen-mechanics-learning-from-nature.

26. Bruce Alberts, *Molecular Biology*, 125.

27. Michelle Oyen, "Collagen Mechanics."

28. Rhee, Stephanie, Jennifer L. Puetzer, Brooke N. Mason, Cynthia A. Reinhart-King, and Lawrence J. Bonassar. "3D Bioprinting of Spatially Heterogeneous Collagen Constructs for Cartilage Tissue Engineering." *ACS Biomaterials Science & Engineering* 2, no. 10 (August 4, 2016): 1800–805. https://doi.org/10.1021/acsbiomaterials.6b00288.

29. Feldkamp, Lee A., Steven A. Goldstein, Michael A. Parfitt, Gerald Jesion, and Michael Kleerekoper. "The Direct Examination of Three-Dimensional Bone Architecture in Vitro by Computed Tomography." *Journal of Bone and Mineral Research* 4, no. 1 (December 3, 2009): 3–11. https://doi.org/10.1002/jbmr.5650040103.

30. Bonassar, Larry. Phone conversation with K.D., March 23, 2018.

31. Kang, Hyun-Wook, Sang Jin Lee, In Kap Ko, Carlos Kengla, James J Yoo, and Anthony Atala. "A 3D Bioprinting System to Produce Human-Scale Tissue Constructs with Structural Integrity." *Nature Biotechnology* 34, no. 3 (February 15, 2016): 312–19. https://doi.org/10.1038/nbt.3413.

32. Emamaullee, Juliet A. "Profiles in Surgical Research: Anthony Atala, MD, FACS." *Bulletin of the American College of Surgeons*, May 2016. http://bulletin.facs.org/2016/05/profiles-in-surgical-research-anthony-atala-md-facs.

33. Ibid.

34. Pati, Falguni, Jinah Jang, Dong-Heon Ha, Sung Won Kim, Jong-Won Rhie, Jin-Hyung Shim, Deok-Ho Kim, and Dong-Woo Cho. "Printing Three-Dimensional Tissue Analogues with Decellularized Extracellular Matrix Bioink." *Nature Communications* 5, no. 1 (June 2, 2014). https://doi.org/10.1038/ncomms4935.

35. Cho, Dong-Woo. Phone conversation with Sangbok Kim; translated in e-mail to K.D., April 17, 2018.

36. Piez, Karl A. "History of Extracellular Matrix: A Personal View." *Matrix Biology* 16, no. 3 (August 1997): 85–92. https://doi.org/10.1016/s0945-053x(97)90037-8.

37. Maher, Brendan. "Tissue Engineering: How to Build a Heart." *Nature* 499, no. 7456 (July 2013): 20–22. https://doi.org/10.1038/499020a.

38. Ott, Harald C., Thomas S. Matthiesen, Saik-Kia Goh, Lauren D. Black, Stefan M. Kren, Theoden I. Netoff, and Doris A. Taylor. "Perfusion-Decellularized Matrix: Using Nature's Platform to Engineer a Bioartificial Heart." *Nature Medicine* 14, no. 2 (January 13, 2008): 213–21. https://doi.org/10.1038/nm1684.

39. Ibid., 218.

40. Ibid.

41. Pati Falguni, "Printing Three-Dimensional Tissue Analogues," 9.

42. Harald C. Ott, "Perfusion-Decellularized Matrix."

43. Brendan Maher, "Tissue Engineering," 22.

44. Ott, Harald C, Ben Clippinger, Claudius Conrad, Christian Schuetz, Irina Pomerantseva, Laertis Ikonomou, Darrell Kotton, and Joseph P. Vacanti. "Regeneration and Orthotopic Transplantation of a Bioartificial Lung." *Nature Medicine* 16, no. 8 (July 13, 2010): 927–33. https://doi.org/10.1038/nm.2193.

45. Song, Jeremy J., Jacques P. Guyette, Sarah E. Gilpin, Gabriel Gonzalez, Joseph P. Vacanti, and Harald C. Ott. "Regeneration and Experimental Orthotopic Transplantation of a Bioengineered Kidney." *Nature Medicine* 19, no. 5 (April 14, 2013): 646–51. https://doi.org/10.1038/nm.3154.

46. Dolgin, Elie. "Devices: Artificial Inspiration." *Nature* 489, no. 7417 (September 2012): S12–14. https://doi.org/10.1038/489s12a.

47. Gilpin, Sarah E., Philipp T. Moser, and Harald C. Ott. "Regenerative Medicine of the Respiratory Tract." In *Principles of Regenerative Medicine*. Edited by Anthony Atala, Robert Lanza, Antonios G. Micos, and Robert Nerem, 1059–72. New York: Elsevier, 2019. https://doi.org/10.1016/b978-0-12-809880-6.00060-6.

48. Skardal, Aleksander, Mahesh Devarasetty, Hyun-Wook Kang, Ivy Mead, Colin Bishop, Thomas Shupe, Sang Jin Lee, John Jackson, James Yoo, Shay Soker, and Anthony Atala. "A Hydrogel Bioink Toolkit for Mimicking Native Tissue Biochemical and Mechanical Properties in Bioprinted Tissue Constructs." *Acta Biomaterialia* 25 (October 2015): 24–34. https://doi.org/10.1016/j.actbio.2015.07.030.

49. Dzobo, Kevin, Keolebogile Shirley Caroline M. Motaung, and Adetola Adesida. "Recent Trends in Decellularized Extracellular Matrix Bioinks for 3D Printing: An Updated Review." *International Journal of Molecular Sciences* 20, no. 18 (September 18, 2019): 4628. https://doi.org/10.3390/ijms20184628.

50. Chen, Fang, James J. Yoo, and Anthony Atala. "Acellular Collagen Matrix as a Possible 'Off the Shelf' Biomaterial for Urethral Repair." *Urology* 54, no. 3 (September 1999): 407–10. https://doi.org/10.1016/s0090-4295(99)00179-x.

51. Rouwkema, Jeroen, Bart F. J. M. Koopman, Clemens A. Van Blitterswijk, Wouter J. A. Dhert, and Jos Malda. "Supply of Nutrients to Cells in Engineered Tissues." *Biotechnology and Genetic Engineering Reviews* 26, no. 1 (January 2009): 163–78. https://doi.org/10.5661/bger-26-163.

Chapter 4

1. Nelson, Michael. "Members Stories Archives." Cavernoma Alliance UK, April 19, 2017. https://www.cavernoma.org.uk/category/members-stories.
2. Ibid.
3. McCaig, Amy. "New Method for Analyzing Brain Activity Helps Understand Stroke Impact." Rice University News & Media, 2019. https://news.rice.edu/2017/02/23/new-method-for-analyzing-brain-activity-helps-understand-stroke-impact.
4. "Cavernous Angioma Fast Facts." Angioma.org, July 31, 2018. https://www.angioma.org/wp-content/uploads/2020/05/CavernousAngiomaFastFacts.pdf.
5. "Blood Vessels." The Franklin Institute, June 2, 2015. https://www.fi.edu/heart/blood-vessels.
6. Aird, W. C. "Discovery of the Cardiovascular System: From Galen to William Harvey." *Journal of Thrombosis and Haemostasis* 9, Special Issue 1 (July 2011): 118–29. https://doi.org/10.1111/j.1538-7836.2011.04312.x.
7. Ibid.
8. Schultz, Stanley G. "William Harvey and the Circulation of the Blood: The Birth of a Scientific Revolution and Modern Physiology." *Physiology* 17, no. 5 (October 2002): 175–80. https://doi.org/10.1152/nips.01391.2002.
9. Petruzzello, Melissa. "What's the Difference Between Veins and Arteries?" In *Encyclopædia Britannica*. Accessed December 22, 2019. https://www.britannica.com/story/whats-the-difference-between-veins-and-arteries.
10. Alberts, Bruce, Alexander Johnson, Julian Hart Lewis, David Morgan, Martin Raff, Keith Roberts, and Peter Walter. *Molecular Biology of the Cell.* New York: Garland, 2015, 1238.
11. Sender, Ron, Shai Fuchs, and Ron Milo. "Revised Estimates for the Number of Human and Bacteria Cells in the Body." *PLoS Biology* 14, no. 8 (August 19, 2016): e1002533. https://doi.org/10.1371/journal.pbio.1002533.s002.
12. Aird, William C. "Phenotypic Heterogeneity of the Endothelium." *Circulation Research* 100, no. 2 (February 2, 2007): 174–90. https://doi.org/10.1161/01.res.0000255690.03436.ae.
13. Alpert, Norman R., and David M. Warshaw. "Muscle." In *Encyclopædia Britannica*. Accessed August 8, 2019. https://www.britannica.com/science/muscle.
14. Hughes, Christopher C. W. "Endothelial–Stromal Interactions in Angiogenesis." *Current Opinion in Hematology* 15, no. 3 (May 2008): 204–209. https://doi.org/10.1097/moh.0b013e3282f97dbc.
15. Costa-Almeida, Raquel, Raquel Soares, and Pedro L. Granja. "Fibroblasts as Maestros Orchestrating Tissue Regeneration." *Journal of Tissue Engineering and Regenerative Medicine* 12, no. 1 (May 23, 2017): 240–51. https://doi.org/10.1002/term.2405.
16. Forgacs, Gabor. "Perfusable Vascular Networks." *Nature Materials* 11, no. 9 (August 23, 2012): 746–47. https://doi.org/10.1038/nmat3412.

17. Therriault, Daniel, Scott R. White, and Jennifer A. Lewis. "Chaotic Mixing in Three-Dimensional Microvascular Networks Fabricated by Direct-Write Assembly." *Nature Materials* 2, no. 4 (March 23, 2003): 265–71. https://doi.org/10.1038/nmat863.
18. Ibid., 270.
19. Wu, Willie, Adam DeConinck, and Jennifer A. Lewis. "Omnidirectional Printing of 3D Microvascular Networks." *Advanced Materials* 23, no. 24 (March 23, 2011): H178–83. https://doi.org/10.1002/adma.201004625.
20. Forgacs, Gabor. E-mail to K.D., July 31, 2019.
21. Hansen, Christopher J., Rajat Saksena, David B. Kolesky, John J. Vericella, Stephen J. Kranz, Gregory P. Muldowney, Kenneth T. Christensen, and Jennifer A. Lewis. "High-Throughput Printing via Microvascular Multinozzle Arrays." *Advanced Materials* 25, no. 1 (October 26, 2012): 96–102. https://doi.org/10.1002/adma.201203321.
22. Kolesky, David B., Ryan L. Truby, A. Sydney Gladman, Travis A. Busbee, Kimberly A. Homan, and Jennifer A. Lewis. "3D Bioprinting of Vascularized, Heterogeneous Cell-Laden Tissue Constructs." *Advanced Materials* 26, no. 19 (February 18, 2014): 3124–30. https://doi.org/10.1002/adma.201305506.
23. Gibson, Lydialyle. "Building Toward a Kidney." *Harvard Magazine*, December 7, 2016. https://harvardmagazine.com/2017/01/building-toward-a-kidney.
24. Pieribone, Vincent A., and David F. Gruber. *Aglow in the Dark : The Revolutionary Science of Biofluorescence*. Cambridge, UK: Belknap Press, 2007.
25. Ko, H. C. H., B. K. Milthorpe, and C. D. McFarland. "Engineering Thick Tissues—The Vascularisation Problem." *European Cells and Materials* 14 (July 25, 2007): 1–19. https://doi.org/10.22203/ecm.v014a01.
26. Radisic, Milica, Liming Yang, Jan Boublik, Richard J. Cohen, Robert Langer, Lisa E. Freed, and Gordana Vunjak-Novakovic. "Medium Perfusion Enables Engineering of Compact and Contractile Cardiac Tissue." *American Journal of Physiology—Heart and Circulatory Physiology* 286, no. 2 (February 2004): H507–16. https://doi.org/10.1152/ajpheart.00171.2003.
27. Kolesky, David B., Kimberly A. Homan, Mark A. Skylar-Scott, and Jennifer A. Lewis. "Three-Dimensional Bioprinting of Thick Vascularized Tissues." *Proceedings of the National Academy of Sciences of the USA* 113, no. 12 (March 7, 2016): 3179–84. https://doi.org/10.1073/pnas.1521342113.
28. Lydialyle Gibson, "Building Toward a Kidney."
29. "Creating 3-D Tissue and Its Potential for Regeneration." *Harvard Gazette*, March 8, 2016. https://news.harvard.edu/gazette/story/2016/03/creating-3-d-tissue-and-its-potential-for-regeneration.
30. Chen, Ray-Neng, Hsiu-O Ho, and Ming-Thau Sheu. "Characterization of Collagen Matrices Crosslinked Using Microbial Transglutaminase." *Biomaterials* 26, no. 20 (July 2005): 4229–35. https://doi.org/10.1016/j.biomaterials.2004.11.012.
31. David B. Kolesky et al., "Three-Dimensional Bioprinting," 3181.
32. Kang, Hyun-Wook, Sang Jin Lee, In Kap Ko, Carlos Kengla, James J. Yoo, and Anthony Atala. "A 3D Bioprinting System to Produce Human-Scale Tissue Constructs with Structural Integrity." *Nature Biotechnology* 34, no. 3 (February 15, 2016): 312–19. https://doi.org/10.1038/nbt.3413.
33. Ibid., 318.

34. Marrella, Alessandra, Tae Yong Lee, Dong Hoon Lee, Sobha Karuthedom, Denata Syla, Aditya Chawla, Ali Khademhosseini, and Hae Lin Jang. "Engineering Vascularized and Innervated Bone Biomaterials for Improved Skeletal Tissue Regeneration." *Materials Today* 21, no. 4 (May 2018): 362–76. https://doi.org/10.1016/j.mattod.2017.10.005.
35. Johnson, Blake N., and Michael C. McAlpine. "From Print to Patient: 3D-Printed Personalized Nerve Regeneration." *The Biochemist* 38, no. 4 (August 1, 2016): 28–31. https://doi.org/10.1042/bio03804028.

Chapter 5

1. On page 92 of her autobiography, *In Praise of Imperfection*, Levi-Montalcini herself uses the simile "convent cell" rather than "artist's garret." Someone remarked to me that this was "odd" since Levi-Montalcini was Jewish.
2. Levi-Montalcini, Rita, and Luigi Attardi. *In Praise of Imperfection: My Life and Work.* New York: Basic Books, 1988, 89–92.
3. "Rita Levi-Montalcini—Biographical." NobelPrize.org. Accessed December 22, 2019. https://www.nobelprize.org/prizes/medicine/1986/levi-montalcini/biographical.
4. Bernardini, Gene. "The Origins and Development of Racial Anti-Semitism in Fascist Italy." *Journal of Modern History* 49, no. 3 (September 1977): 431–53. https://doi.org/10.1086/241596.
5. Rita Levi-Montalcini, *In Praise of Imperfection*, 88.
6. Ibid., 95.
7. Akter, Farhana. *Tissue Engineering Made Easy*. London: Academic Press, 2016, 10.
8. Abbott, Alison. "Neuroscience: One Hundred Years of Rita." *Nature* 458, no. 7238 (April 2009): 564–67. https://doi.org/10.1038/458564a.
9. Rita Levi-Montalcini, *In Praise of Imperfection*, 92.
10. Holloway, Marguerite. "Profile: Rita Levi-Montalcini." *Scientific American* 268, no. 1 (January 1993): 32–36. https://doi.org/10.1038/scientificamerican0193-32.
11. Yount, Lisa. *Rita Levi-Montalcini: Discoverer of Nerve Growth Factor*. New York: Chelsea House, 2009, 28.
12. Marguerite Holloway, "Profile: Rita Levi-Montalcini," 36.
13. Rita Levi-Montalcini, *In Praise of Imperfection*, 147.
14. Ibid., 151.
15. Lisa Yount, *Rita Levi-Montalcini*, 77.
16. Rita Levi-Montalcini, *In Praise of Imperfection*, 152.
17. Ibid., 153.
18. Shakespeare, William. *A Midsummer Night's Dream* (Act V, Scene 1, Lines 1844–1847), "Open Source Shakespeare: Search Shakespeare's Works, Read the Texts." Accessed December 22, 2019. https://www.opensourceshakespeare.org.
19. "The Nobel Prize in Physiology or Medicine 1986." NobelPrize.org. Accessed December 22, 2019. https://www.nobelprize.org/prizes/medicine/1986/press-release.
20. "About Peripheral Neuropathy—3 Types of Peripheral Nerves." Uchicago.edu, April 16, 2010. http://peripheralneuropathycenter.uchicago.edu/learnaboutpn/aboutpn/symptoms/threenerves.shtml.

21. Norotte, Cyrille, Francoise S. Marga, Laura E. Niklason, and Gabor Forgacs. "Scaffold-Free Vascular Tissue Engineering Using Bioprinting." *Biomaterials* 30, no. 30 (October 2009): 5910–17. https://doi.org/10.1016/j.biomaterials.2009.06.034.

22. Mironov, Vladimir, Thomas Boland, Thomas Trusk, Gabor Forgacs, and Roger R. Markwald. "Organ Printing: Computer-Aided Jet-Based 3D Tissue Engineering." *Trends in Biotechnology* 21, no. 4 (April 2003): 157–61. https://doi.org/10.1016/s0167-7799(03)00033-7.

23. Boland, Thomas, Vladimir Mironov, Anna Gutowska, Elisabeth. A. Roth, and Roger R. Markwald. "Cell and Organ Printing 2: Fusion of Cell Aggregates in Three-Dimensional Gels." *The Anatomical Record* 272A, no. 2 (May 6, 2003): 497–502. https://doi.org/10.1002/ar.a.10059.

24. Jakab, K., A. Neagu, V. Mironov, R. R. Markwald, and G. Forgacs. "Engineering Biological Structures of Prescribed Shape Using Self-Assembling Multicellular Systems." *Proceedings of the National Academy of Sciences of the USA* 101, no. 9 (February 23, 2004): 2864–69. https://doi.org/10.1073/pnas.0400164101.

25. Mironov, Vladimir, Richard P. Visconti, Vladimir Kasyanov, Gabor Forgacs, Christopher J. Drake, and Roger R. Markwald. "Organ Printing: Tissue Spheroids as Building Blocks." *Biomaterials* 30, no. 12 (April 2009): 2164–74. https://doi.org/10.1016/j.biomaterials.2008.12.084.

26. Jakab, Karoly, Cyrille Norotte, Françoise Marga, Keith Murphy, Gordana Vunjak-Novakovic, and Gabor Forgacs. "Tissue Engineering by Self-Assembly and Bio-Printing of Living Cells." *Biofabrication* 2, no. 2 (June 1, 2010): 022001. https://doi.org/10.1088/1758-5082/2/2/022001.

27. Siemionow, Maria, and Grzegorz Brzezicki. "Current Techniques and Concepts in Peripheral Nerve Repair." *International Review of Neurobiology* 87 (2009): 141–72. https://doi.org/10.1016/s0074-7742(09)87008-6.

28. Christopher M. Owens, Françoise Marga, Gabor Forgacs, and Cheryl M. Heesch. "Biofabrication and Testing of a Fully Cellular Nerve Graft." *Biofabrication* 5, no. 4 (November 6, 2013). http://iopscience.iop.org/article/10.1088/1758-5082/5/4/045007.

29. Salzer, J. L., and B. Zalc. "Myelination." *Current Biology* 26 (October 24, 2016): R971–75.

30. Marga, Françoise, Karoly Jakab, Chirag Khatiwala, Benjamin Shepherd, Scott Dorfman, Bradley Hubbard, Stephen Colbert, and Gabor Forgacs. "Toward Engineering Functional Organ Modules by Additive Manufacturing." *Biofabrication* 4 (2012): 022001.

31. Cyrille Norotte et al., "Scaffold-Free Vascular Tissue Engineering," 5916.

32. Stornetta, R. L., S. F. Morrison, D. A. Ruggiero, and D. J. Reis. "Neurons of Rostral Ventrolateral Medulla Mediate Somatic Pressor Reflex." *American Journal of Physiology: Regulatory, Integrative and Comparative Physiology* 256, no. 2 (February 1, 1989): R448–62. https://doi.org/10.1152/ajpregu.1989.256.2.r448.

33. Johnson, Blake N., Karen Z. Lancaster, Gehua Zhen, Junyun He, Maneesh K. Gupta, Yong Lin Kong, Esteban A. Engel, et al. "3D Printed Anatomical Nerve Regeneration Pathways." *Advanced Functional Materials* 25, no. 39 (September 18, 2015): 6205–17. https://doi.org/10.1002/adfm.201501760.

34. Levi-Montalcini, Rita, and Pietro Calissano. "The Nerve-Growth Factor." *Scientific American* 240, no. 6 (June 1979): 68–77. https://doi.org/10.1038/scientificamerican0679-68.

35. Lisa Yount, *Rita Levi-Montalcini*, 97.

36. Labrador, Rafael O., Miquel Butí, and Xavier Navarro. "Influence of Collagen and Laminin Gels Concentration on Nerve Regeneration After Resection and Tube Repair." *Experimental Neurology* 149, no. 1 (January 1998): 243–52. https://doi.org/10.1006/exnr.1997.6650.

37. Weiss, R. "Promising Protein for Parkinson's." *Science* 260, no. 5111 (May 21, 1993): 1072–73. https://doi.org/10.1126/science.8493547.

38. Lin, L., D. Doherty, J. Lile, S. Bektesh, and F. Collins. "GDNF: A Glial Cell Line-Derived Neurotrophic Factor for Midbrain Dopaminergic Neurons." *Science* 260, no. 5111 (May 21, 1993): 1130–32. https://doi.org/10.1126/science.8493557.

39. Bartheld, Christopher S. von, Jami Bahney, and Suzana Herculano-Houzel. "The Search for True Numbers of Neurons and Glial Cells in the Human Brain: A Review of 150 Years of Cell Counting." *Journal of Comparative Neurology* 524, no. 18 (June 16, 2016): 3865–95. https://doi.org/10.1002/cne.24040.

40. Sender, Ron, Shai Fuchs, and Ron Milo. "Revised Estimates for the Number of Human and Bacteria Cells in the Body." *PLoS Biology* 14, no. 8 (August 19, 2016): e1002533. https://doi.org/10.1371/journal.pbio.1002533.

41. Michalski, John-Paul, and Rashmi Kothary. "Oligodendrocytes in a Nutshell." *Frontiers in Cellular Neuroscience* 9 (September 1, 2015). https://doi.org/10.3389/fncel.2015.00340.

42. Ribeiro-Resende, Victor T., Brigitte Koenig, Susanne Nichterwitz, Sven Oberhoffner, and Burkhard Schlosshauer. "Strategies for Inducing the Formation of Bands of Büngner in Peripheral Nerve Regeneration." *Biomaterials* 30, no. 29 (October 2009): 5251–59. https://doi.org/10.1016/j.biomaterials.2009.07.007.

43. Miller, C. "Oriented Schwann Cell Growth on Micropatterned Biodegradable Polymer Substrates." *Biomaterials* 22, no. 11 (June 1, 2001): 1263–69. https://doi.org/10.1016/s0142-9612(00)00278-7.

44. Miller, Cheryl, Srdija Jeftinija, and Surya Mallapragada. "Synergistic Effects of Physical and Chemical Guidance Cues on Neurite Alignment and Outgrowth on Biodegradable Polymer Substrates." *Tissue Engineering* 8, no. 3 (July 2002): 367–78. https://doi.org/10.1089/107632702760184646.

45. Voltaire. *Voltaire's Romances.* New York: Eckler, 1918, 125. https://www.gutenberg.org/files/35595/35595-h/35595-h.htm.

46. Blake N. Johnson, "3D Printed Anatomical Nerve," 6216.

47. Joung, Daeha, Vincent Truong, Colin C. Neitzke, Shuang-Zhuang Guo, Patrick J. Walsh, Joseph R. Monat, Fanben Meng, et al. "3D Printed Stem-Cell Derived Neural Progenitors Generate Spinal Cord Scaffolds." *Advanced Functional Materials* 28, no. 39 (August 9, 2018): 1801850. https://doi.org/10.1002/adfm.201801850.

48. Olson, Jeremy. "University of Minnesota Reports Breakthrough in 3-D Printing for Spinal Cord Repair." *Star Tribune (Minneapolis).* August 13, 2018. https://phys.org/news/2018-08-university-minnesota-breakthrough-d-spinal.html.

49. Führmann, Tobias, Priya N. Anandakumaran, and Molly S. Shoichet. "Combinatorial Therapies After Spinal Cord Injury: How Can Biomaterials Help?" *Advanced Healthcare Materials* 6, no. 10 (March 1, 2017): 1601130. https://doi.org/10.1002/adhm.201601130.

50. McAlpine, Michael. Phone conversation with K.D., February 27, 2018.

51. Parr, A. M., I. Kulbatski, T. Zahir, X. Wang, C. Yue, A. Keating, and C. H. Tator. "Transplanted Adult Spinal Cord-Derived Neural Stem/Progenitor Cells Promote Early Functional Recovery After Rat Spinal Cord Injury." *Neuroscience* 155, no. 3 (August 2008): 760–70. https://doi.org/10.1016/j.neuroscience.2008.05.042.

52. Terzic, Dino, Jacob R. Maxon, Leah Krevitt, Christina Dibartolomeo, Tarini Goyal, Walter C. Low, James R. Dutton, and Ann M. Parr. "Directed Differentiation of Oligodendrocyte Progenitor Cells from Mouse Induced Pluripotent Stem Cells." *Cell Transplantation* 25, no. 2 (February 2016): 411–24. https://doi.org/10.3727/096368915x688137.

53. Presti, David E. *Foundational Concepts in Neuroscience: A Brain–Mind Odyssey.* New York: Norton, 2016, 58.

54. Shakespeare, William. *History of King John* (Act III, Scene 4. Line 1496), "Open Source Shakespeare: Search Shakespeare's Works, Read the Texts." Accessed December 22, 2019. https://www.opensourceshakespeare.org.

55. David E. Presti, *Foundational Concepts in Neuroscience*, 58.

56. Ibid., 60.

57. Inspired by Wodehouse, P. G. *Galahad at Blandings.* London: Penguin, 1999, 116.

58. Meriney, Stephen D., and Erika E. Fanselow. "Ion Channels and Action Potential Generation." In *Synaptic Transmission.* New York: Elsevier, 2019, 35–63. https://doi.org/10.1016/b978-0-12-815320-8.00004-1.

59. Daeha Joung, "3D Printed Stem-Cell Derived Neural Progenitors," 6.

60. "NCI Dictionary of Cancer Terms." National Cancer Institute. Cancer.gov. Accessed December 22, 2019. https://www.cancer.gov/publications/dictionaries/cancer-terms/def/immunocytochemistry.

61. Daeha Joung, "3D Printed Stem-Cell Derived Neural Progenitors," 7.

62. Grienberger, Christine, and Arthur Konnerth. "Imaging Calcium in Neurons." *Neuron* 73, no. 5 (March 2012): 862–85. https://doi.org/10.1016/j.neuron.2012.02.011.

63. Alberts, Bruce, Dennis Bray, Karen Hopkin, Alexander D. Johnson, Julien Lewis, Martin Raff, Keith Roberts, and Peter Walter. *Essential Cell Biology.* New York: Garland, 2014, 548.

64. Murphy, Sean V., and Anthony Atala. "3D Bioprinting of Tissues and Organs." *Nature Biotechnology* 32, no. 8 (2014): 773–85. https://doi.org/10.1038/nbt.2958.

65. Ibid., 782.

Chapter 6

1. "A Life Saving Discovery at MIT MechE." Biomaterials.org. Accessed December 22, 2019. https://www.biomaterials.org/publications/news/life-saving-discovery-mit-meche.

2. Ravage, Barbara. *Burn Unit: Saving Lives After the Flames.* Cambridge, MA: Da Capo Press, 2004, 26–27.

3. Lewis, Anne Marie. "Emergency! Trauma Triad of Death." *Nursing* 30, no. 3 (March 2000): 62–64. https://doi.org/10.1097/00152193-200030030-00028.

4. Sutton, S. B. "Tried in the Fire: New Treatments for Massive Burns." *Harvard Magazine,* September 1984. https://harvardmagazine.com/sites/default/files/Burns_John-Burke.pdf.

5. "A Life-Saving Discovery at MIT MechE."

6. Ibid.

7. Yannas, Ioannis V. "Hesitant Steps from the Artificial Skin to Organ Regeneration." *Regenerative Biomaterials* 5, no. 4 (June 26, 2018). https://doi.org/10.1093/rb/rby012.

8. S. B. Sutton, "Tried in the Fire," 63.

9. Ibid., 60.

10. Carey, Bjorn. "Printable Skin: 'Inkjet' Breakthrough Makes Human Tissue." *Live Science*, February 1, 2005. https://www.livescience.com/118-printable-skin-inkjet-breakthrough-human-tissue.html.

11. Saunders, Rachel E., Julie E. Gough, and Brian Derby. "Delivery of Human Fibroblast Cells by Piezoelectric Drop-on-Demand Inkjet Printing." *Biomaterials* 29, no. 2 (January 2008): 193–203. https://doi.org/10.1016/j.biomaterials.2007.09.032.

12. Derby, Brian. "Bioprinting: Inkjet Printing Proteins and Hybrid Cell-Containing Materials and Structures." *Journal of Materials Chemistry* 18, no. 47 (2008): 5717–21. https://doi.org/10.1039/b807560c.

13. Derby, Brian. Phone conversation with K.D., May 4, 2018.

14. Ng, Wei Long, Jovina Tan Zhi Qi, Wai Yee Yeong, and May Win Naing. "Proof-of-Concept: 3D Bioprinting of Pigmented Human Skin Constructs." *Biofabrication* 10, no. 2 (January 23, 2018): 025005. https://doi.org/10.1088/1758-5090/aa9e1e.

15. Cubo, Nieves, Marta Garcia, Juan F. del Cañizo, Diego Velasco, and Jose L. Jorcano. "3D Bioprinting of Functional Human Skin: Production and in Vivo Analysis." *Biofabrication* 9, no. 1 (December 5, 2016): 015006. https://doi.org/10.1088/1758-5090/9/1/015006.

16. Pourchet, Léa J., Amélie Thepot, Marion Albouy, Edwin J. Courtial, Aurélie Boher, Loïc J. Blum, and Christophe A. Marquette. "Human Skin 3D Bioprinting Using Scaffold-Free Approach." *Advanced Healthcare Materials* 6, no. 4 (December 15, 2016): 1601101. https://doi.org/10.1002/adhm.201601101.

17. Ng, Wei Long, Shuai Wang, Wai Yee Yeong, and May Win Naing. "Skin Bioprinting: Impending Reality or Fantasy?" *Trends in Biotechnology* 34, no. 9 (September 2016): 689–99. https://doi.org/10.1016/j.tibtech.2016.04.006.

18. "Definition of corneous." Merriam-Webster, 2019. https://www.merriam-webster.com/dictionary/corneous.

19. Tobin, Desmond J. "Biochemistry of Human Skin—Our Brain on the Outside." *Chemical Society Reviews* 35, no. 1 (2006): 52–67. https://doi.org/10.1039/b505793k.

20. Wei Long Ng et al., "Skin Bioprinting," 694.

21. Mescher, Anthony L., and Carlos Uchôa. *Junqueira's Basic Histology: Text and Atlas.* 13th ed. New York: McGraw-Hill, 2013, 1.

22. Brandtzaeg, Per. "The Increasing Power of Immunohistochemistry and Immunocytochemistry." *Journal of Immunological Methods* 216, no. 1–2 (July 1998): 49–67. https://doi.org/10.1016/s0022-1759(98)00070-2.

23. Wei Long Ng et al., "Proof-of-Concept: 3D Bioprinting," 11.

24. Huang, Sha, Bin Yao, Jiangfan Xie, and Xiaobing Fu, "3D Bioprinted Extracellular Matrix Mimics Facilitate Directed Differentiation of Epithelial Progenitors for Sweat Gland Regeneration," *Acta Biomaterialia* 32 (2016): 170–77.

25. Ibid.

26. Liu, Nanbo, Sha Huang, Bin Yao, Jiangfan Xie, Xu Wu, and Xiaobing Fu. "3D Bioprinting Matrices with Controlled Pore Structure and Release Function Guide in Vitro Self-Organization of Sweat Gland." *Scientific Reports* 6, no. 1 (October 3, 2016). https://doi.org/10.1038/srep34410.

27. Jakus, Adam E., Alexandra L. Rutz, Sumanas W. Jordan, Abhishek Kannan, Sean M. Mitchell, Chawon Yun, Katie D. Koube, Sung C. Yoo, Herbert E. Whiteley, Claus-Peter Richter, Robert D. Galiano, Wellington K. Hsu, Stuart R. Stock, Erin L. Hsu, and Ramille

N. Shah. "Hyperelastic 'Bone': A Highly Versatile, Growth Factor–Free, Osteoregenerative, Scalable, and Surgically Friendly Biomaterial." *Science Translational Medicine* 8, no. 358 (September 28, 2016): 358ra127. https://doi.org/10.1126/scitranslmed.aaf7704.

28. Shah, Ramille N. Phone conversation with K.D., March 22, 2018.

29. Adam E. Jakus et al., "Hyperelastic 'Bone,' " 4–5.

30. Ibid., 1.

31. Ibid., 6.

32. Ibid.

33. Vincent, James. "3D-Printed 'hyperelastic Bone' Could Be the Future of Reconstructive Surgery." *The Verge*, September 28, 2016. https://www.theverge.com/2016/9/28/13094642/hyperelastic-bone-graft-substance-unveiled.

34. Adam E. Jakus et al., "Hyperelastic 'Bone,' " 10.

35. Ibid., 9.

36. Ibid., 10.

37. Fox, Maggie. "3-D Printed Bone Material Acts Like Real Bone for Custom-Made Implants." NBC News, September 28, 2016. https://www.nbcnews.com/health/health-news/3-d-printed-bone-material-acts-real-bone-custom-made-n656366.

38. Baggaley, Kate. "New 3D Printed Material Can Help Regenerate Bones." *Popular Science*, September 28, 2016. http://www.popsci.com/new-3d-printed-material-helps-bones-regrow.

39. Shah, Ramille N. Phone conversation with K.D., March 22, 2018.

40. Daly, Andrew C., Gráinne M. Cunniffe, Binulal N. Sathy, Oju Jeon, Eben Alsberg, and Daniel J. Kelly. "3D Bioprinting: 3D Bioprinting of Developmentally Inspired Templates for Whole Bone Organ Engineering." *Advanced Healthcare Materials* 5, no. 18 (September 2016): 2352–52. https://doi.org/10.1002/adhm.201670100.

41. Gobin, Andrea S., Doris A. Taylor, Eric Chau, and Luiz C. Sampaio. *Stem Cell and Gene Therapy for Cardiovascular Disease*. Edited by Emerson C. Perin. San Diego, CA : Academic Press, 2016, 351.

42. Dormehl, Luke. "New Technique Could Be the Break 3D-Printed Bones Have Been Waiting For." *Digital Trends*, September 6, 2016. https://www.digitaltrends.com/cool-tech/3d-printing-bone-implants.

43. Andrew C. Daly et al., "3D Bioprinting of Developmentally Inspired Templates," 2353.

44. Ibid.

45. Luke Dormehl, "New Technique."

46. Ruberte, Jesus, Ana Carretero, Marc Navarro, and Mariana López-Luppo. *Morphological Mouse Phenotyping: Anatomy, Histology and Imaging*. Amsterdam: Academic Press, 2017.

47. Cunniffe, Gráinne M., Tomas Gonzalez-Fernandez, Andrew Daly, Binulal N. Sathy, Oju Jeon, Eben Alsberg, and Daniel J. Kelly. "Three-Dimensional Bioprinting of Polycaprolactone Reinforced Gene Activated Bioinks for Bone Tissue Engineering." *Tissue Engineering Part A* 23, no. 17–18 (September 2017): 891–900. https://doi.org/10.1089/ten.tea.2016.0498.

48. Genetic Science Learning Center. "Approaches to Gene Therapy." *Learn. Genetics*, December 1, 2012. https://learn.genetics.utah.edu/content/genetherapy/approaches.

49. Sipp, Douglas, Pamela G. Robey, and Leigh Turner. "Clear up This Stem-Cell Mess." *Nature* 561, no. 7724 (September 2018): 455–57. https://doi.org/10.1038/d41586-018-06756-9.

50. EuroStemCell. "Mesenchymal Stem Cells: The 'Other' Bone Marrow Stem Cells." Eurostemcell.org, November 5, 2015. https://www.eurostemcell.org/mesenchymal-stem-cells-other-bone-marrow-stem-cells.

51. Huang, Xiao, and Christopher S. Brazel. "On the Importance and Mechanisms of Burst Release in Matrix-Controlled Drug Delivery Systems." *Journal of Controlled Release* 73, no. 2–3 (June 2001): 121–36. https://doi.org/10.1016/s0168-3659(01)00248-6.

52. Silva, Amanda K. Andriola, Cyrille Richard, Michel Bessodes, Daniel Scherman, and Otto-Wilhelm Merten. "Growth Factor Delivery Approaches in Hydrogels." *Biomacromolecules* 10, no. 1 (January 12, 2009): 9–18. https://doi.org/10.1021/bm801103c.

53. Lee, Kangwon, Eduardo A. Silva, and David J. Mooney. "Growth Factor Delivery-Based Tissue Engineering: General Approaches and a Review of Recent Developments." *Journal of the Royal Society Interface* 8, no. 55 (August 18, 2010): 153–70. https://doi.org/10.1098/rsif.2010.0223.

54. Kelly, Danny. Phone conversations with K.D., March 26, 2018, and May 1, 2018.

55. Grasman, Jonathan M., Michelle J. Zayas, Raymond L. Page, and George D. Pins. "Biomimetic Scaffolds for Regeneration of Volumetric Muscle Loss in Skeletal Muscle Injuries." *Acta Biomaterialia* 25 (October 1, 2015): 2–15. https://doi.org/10.1016/j.actbio.2015.07.038.

56. Kim, Ji Hyun, Young-Joon Seol, In Kap Ko, Hyun-Wook Kang, Young Koo Lee, James J. Yoo, Anthony Atala, and Sang Jin Lee. "3D Bioprinted Human Skeletal Muscle Constructs for Muscle Function Restoration." *Scientific Reports* 8, no. 1 (August 17, 2018). https://doi.org/10.1038/s41598-018-29968-5.

57. Lieber, Richard L., and Jan Fridén. "Functional and Clinical Significance of Skeletal Muscle Architecture." *Muscle & Nerve* 23, no. 11 (2000): 1647–66. https://doi.org/10.1002/1097-4598(200011)23:11<1647::aid-mus1>3.0.co;2-m.

58. Huxley, A. F., and R. Niedergerke. "Structural Changes in Muscle During Contraction: Interference Microscopy of Living Muscle Fibres." *Nature* 173, no. 4412 (May 22, 1954): 971–73. https://doi.org/10.1038/173971a0.

59. Huxley, Hugh, and Jean Hanson. "Changes in the Cross-Striations of Muscle During Contraction and Stretch and Their Structural Interpretation." *Nature* 173, no. 4412 (May 22, 1954): 973–76. https://doi.org/10.1038/173973a0.

60. Krans, Jacob L. "The Sliding Filament Theory of Muscle Contraction." *Nature Education* 3, no. 9 (2010): 66. https://www.nature.com/scitable/topicpage/the-sliding-filament-theory-of-muscle-contraction-14567666.

61. Marelli, B., M. A. Brenckle, D. L. Kaplan, and F. G. Omenetto. "Silk Fibroin as Edible Coating for Perishable Food Preservation." *Scientific Reports* 6, no. 1 (May 2016). https://doi.org/10.1038/srep25263.

62. Omenetto, F. G., and D. L. Kaplan. "New Opportunities for an Ancient Material." *Science* 329, no. 5991 (July 29, 2010): 528–31. https://doi.org/10.1126/science.1188936.

63. Dixon, Thomas Anthony, Eliad Cohen, Dana M. Cairns, Maria Rodriguez, Juanita Mathews, Rod R. Jose, and David L. Kaplan. "Bioinspired Three-Dimensional Human Neuromuscular Junction Development in Suspended Hydrogel Arrays." *Tissue Engineering Part C: Methods* 24, no. 6 (June 2018): 346–59. https://doi.org/10.1089/ten.tec.2018.0062.

64. Schmid, S. M., and M. Hollmann. "Bridging the Synaptic Cleft: Lessons from Orphan Glutamate Receptors." *Science Signaling* 3, no. 136 (August 17, 2010): pe28. https://doi.org/10.1126/scisignal.3136pe28.

65. Fernandez, Pablo, and Andreas R. Bausch. "The Compaction of Gels by Cells: A Case of Collective Mechanical Activity." *Integrative Biology* 1, no. 3 (2009): 252–59. https://doi.org/10.1039/b822897c.

66. Jansen, Karin A., Dominique M. Donato, Hayri E. Balcioglu, Thomas Schmidt, Erik H. J. Danen, and Gijsje H. Koenderink. "A Guide to Mechanobiology: Where Biology and Physics Meet." *Biochimica et Biophysica Acta (BBA)—Molecular Cell Research* 1853, no. 11 (November 2015): 3043–52. https://doi.org/10.1016/j.bbamcr.2015.05.007.

67. Wang, Lin, Janet Shansky, and Herman Vandenburgh. "Induced Formation and Maturation of Acetylcholine Receptor Clusters in a Defined 3D Bio-Artificial Muscle." *Molecular Neurobiology* 48, no. 3 (January 31, 2013): 397–403. https://doi.org/10.1007/s12035-013-8412-z.

68. Kaplan, David L. Phone conversation with K.D., March 23, 2018.

Chapter 7

1. Specktor, Brandon. "One Liver for One Heart." *Reader's Digest*, March 1, 2018, 67–68.

2. Mayo Clinic Staff. "Autoimmune Hepatitis—Symptoms and Causes." Mayo Clinic, September 12, 2018. https://www.mayoclinic.org/diseases-conditions/autoimmune-hepatitis/symptoms-causes/syc-20352153.

3. Andrews, Travis M. "He Gave Her His Liver. She Gave Him Her Heart." *The Washington Post*, October 25, 2016. https://www.washingtonpost.com/news/morning-mix/wp/2016/10/25/he-gave-her-his-liver-she-gave-him-her-heart.

4. Ibid.

5. Ibid.

6. Brandon Specktor, "One Liver."

7. Keating, Caitlin. "Liver Donor Marries the Woman Whose Life He Saved: 'It Was the Best Decision I Ever Made.'" *People*, November 1, 2016. https://people.com/human-interest/liver-donor-marries-the-woman-whose-life-he-saved-it-was-the-best-decision-i-ever-made.

8. Travis M. Andrews, "He Gave Her His Liver."

9. Lafferty, Susan DeMar. "Southland Couple Share Special Bond." *Chicago Tribune*, January 11, 2016. http://www.chicagotribune.com/suburbs/daily-southtown/news/ct-sta-frankfort-transplant-couple-engaged-st-0110-20160111-story.html.

10. Shakespeare, William. *Henry V* (Act IV, Scene III, Line 2295), "Open Source Shakespeare: Search Shakespeare's Works, Read the Texts." Accessed December 29, 2019. https://www.opensourceshakespeare.org. (From the Saint Crispin's Day Speech.)

11. Susan DeMar Lafferty, "Southland Couple."

12. Power, Carl, and John E. J. Rasko. "Whither Prometheus' Liver? Greek Myth and the Science of Regeneration." *Annals of Internal Medicine* 149, no. 6 (September 16, 2008): 421. https://doi.org/10.7326/0003-4819-149-6-200809160-00009.

13. Higgins, G. M., and R. M. Anderson. "Experimental Pathology of the Liver: I. Restoration of the Liver of the White Rat Following Partial Surgical Removal." *Archives of Pathology* 12 (1931): 186–202.

14. Taub, Rebecca. "Liver Regeneration: From Myth to Mechanism." *Nature Reviews Molecular Cell Biology* 5, no. 10 (October 2004): 836–47. https://doi.org/10.1038/nrm1489.

15. Michalopoulos, G. K. "Liver Regeneration." *Science* 276, no. 5309 (April 4, 1997): 60–66. https://doi.org/10.1126/science.276.5309.60.

16. "Liver: Anatomy and Functions." Johns Hopkins Medicine, 2019. https://www.hopkins-medicine.org/health/conditions-and-diseases/liver-anatomy-and-functions.

17. Angier, Natalie. "The Liver: A 'Blob' That Runs the Body." *The New York Times*, June 12, 2017. https://www.nytimes.com/2017/06/12/health/liver-bodily-function.html.

18. Schaffer, Amanda. "A Renaissance Woman for the Nano Age." *MIT Technology Review*, February 23, 2016. https://www.technologyreview.com/s/546221/a-renaissance-woman-for-the-nano-age.

19. Dunn, J. C., M. L. Yarmush, H. G. Koebe, and R. G. Tompkins. "Hepatocyte Function and Extracellular Matrix Geometry: Long-Term Culture in a Sandwich Configuration." *FASEB Journal* 3, no. 2 (February 1989): 174–77. https://doi.org/10.1096/fasebj.3.2.2914628.

20. Dunn, James C. Y., Ronald G. Tompkins, and Martin L. Yarmush. "Long-Term in Vitro Function of Adult Hepatocytes in a Collagen Sandwich Configuration." *Biotechnology Progress* 7, no. 3 (May 1991): 237–45. https://doi.org/10.1021/bp00009a007.

21. Dunn, J. C., R. G. Tompkins, and M. L. Yarmush. "Hepatocytes in Collagen Sandwich: Evidence for Transcriptional and Translational Regulation." *Journal of Cell Biology* 116, no. 4 (February 15, 1992): 1043–53. https://doi.org/10.1083/jcb.116.4.1043.

22. Bhatia, S. N., M. Toner, R. G. Tompkins, and M. L. Yarmush. "Selective Adhesion of Hepatocytes on Patterned Surfaces." *Annals of the New York Academy of Sciences* 745, no. 1 (December 17, 2006): 187–209. https://doi.org/10.1111/j.1749-6632.1994.tb44373.x.

23. Ibid.

24. Khetani, Salman R., and Sangeeta N. Bhatia. "Microscale Culture of Human Liver Cells for Drug Development." *Nature Biotechnology* 26, no. 1 (November 18, 2007): 120–26. https://doi.org/10.1038/nbt1361.

25. Orfandis, Nicholas. "Overview of Blood Vessel Disorders of the Liver." *Merck Manual Consumer Version*, December 2013. https://www.merckmanuals.com/home/liver-and-gallbladder-disorders/blood-vessel-disorders-of-the-liver/overview-of-blood-vessel-disorders-of-the-liver.

26. Netter, Frank Henry, and Arthur F. Dalley. *Atlas of Human Anatomy*. Teterboro, NJ: Icon Learning Systems, 2001. Plate 270 and Plate 272.

27. Hegyi, Peter, Jozsef Maléth, Julian R. Walters, Alan F. Hofmann, and Stephen J. Keely. "Guts and Gall: Bile Acids in Regulation of Intestinal Epithelial Function in Health and Disease." *Physiological Reviews* 98, no. 4 (October 1, 2018): 1983–2023. https://doi.org/10.1152/physrev.00054.2017.

28. "Liver: Anatomy and Functions," Johns Hopkins Medicine.

29. Natalie Angier, "The Liver: A 'Blob' That Runs the Body."

30. Ibid.

31. Ibid.

32. Faulkner-Jones, Alan, Catherine Fyfe, Dirk-Jan Cornelissen, John Gardner, Jason King, Aidan Courtney, and Wenmiao Shu. "Bioprinting of Human Pluripotent Stem Cells and Their Directed Differentiation into Hepatocyte-Like Cells for the Generation of Mini-Livers in 3D." *Biofabrication* 7, no. 4 (October 21, 2015): 044102. https://doi.org/10.1088/1758-5090/7/4/044102.

33. Faulkner-Jones, Alan, Sebastian Greenhough, Jason A. King, John Gardner, Aidan Courtney, and Wenmiao Shu. "Development of a Valve-Based Cell Printer for the

Formation of Human Embryonic Stem Cell Spheroid Aggregates." *Biofabrication* 5, no. 1 (February 4, 2013): 015013. https://doi.org/10.1088/1758-5082/5/1/015013.

34. National Institutes of Health. "Appendix E: Stem Cell Markers." June 17, 2001. https://stemcells.nih.gov/info/2001report/appendixE.htm.

35. Al-Sabah, A., Z. M. Jessop, I. S. Whitaker, and C. Thornton. "Cell Preparation for 3D Bioprinting." In *3D Bioprinting for Reconstructive Surgery*. Edited by Daniel J. Thomas, Zita M. Jessop, and Iain S. Whitaker, 75–88. New York: Elsevier, 2018. https://doi.org/10.1016/b978-0-08-101103-4.00006-5.

36. Shapiro, Howard M. *Practical Flow Cytometry*. New York: Wiley-Liss, 2003.

37. Goodsell, D. S. "Serum Albumin." *RCSB Protein Data Bank*, January 1, 2003. https://doi.org/10.2210/rcsb_pdb/mom_2003_1.

38. Adams, Nathan. "Albumin." Chemistry World, June 24, 2014. https://www.chemistry-world.com/podcasts/albumin/7482.article.

39. David S. Goodsell, "Serum Albumin."

40. Ma, Xuanyi, Xin Qu, Wei Zhu, Yi-Shuan Li, Suli Yuan, Hong Zhang, Justin Liu, Pengrui Wang, Cheuk Sun Edwin Lai, Fabian Zanella, Gen-Sheng Feng, Farah Sheikh, Shu Chien, and Shaochen Chen. "Deterministically Patterned Biomimetic Human IPSC-Derived Hepatic Model via Rapid 3D Bioprinting." *Proceedings of the National Academy of Sciences of the USA* 113, no. 8 (February 8, 2016): 2206–11. https://doi.org/10.1073/pnas.1524510113.

41. Lee, Benjamin. "Introduction to ±12 Degree Orthogonal Digital Micromirror Devices (DMDs)." Texas Instruments, February 2018. http://www.ti.com/lit/an/dlpa008b/dlpa008b.pdf.

42. Shalu Suri, Li-Hsin Han, Wande Zhang, Ankur Singh, Shaochen Chen, and Christine E. Schmidt. "Solid Freeform Fabrication of Designer Scaffolds of Hyaluronic Acid for Nerve Tissue Engineering." *Biomed Microdevices* 13 (2011), 985.

43. Xuanyi Ma et al., "Deterministically Patterned," Supporting Information, 2.

44. American Society of Mechanical Engineers. "Digital Micromirror Device." Accessed December 29, 2019. https://www.asme.org/about-asme/engineering-history/landmarks/243-digital-micromirror-device.

45. "Why Is the Texas Instruments Digital Micromirror Device (DMD) so Reliable?" Accessed December 29, 2019. http://www.ti.com/pdfs/dlpdmd/135_Myth.pdf.

46. Xuanyi Ma et al., "Deterministically Patterned," 2208.

47. Tsuji, Wakako. "Adipose-Derived Stem Cells: Implications in Tissue Regeneration." *World Journal of Stem Cells* 6, no. 3 (July 26, 2014): 312. https://doi.org/10.4252/wjsc.v6.i3.312.

48. Haurowitz, Felix, and Daniel E. Koshland. "Protein—Enzymes." In *Encyclopædia Britannica,* December 26, 2019. https://www.britannica.com/science/protein/Enzymes.

49. Chen, Shaochen. E-mails to K.D., February 9, 2018.

50. Yanagi, Yusuke, Koichi Nakayama, Tomoaki Taguchi, Shin Enosawa, Tadashi Tamura, Koichiro Yoshimaru, Toshiharu Matsuura, et al. "In Vivo and Ex Vivo Methods of Growing a Liver Bud Through Tissue Connection." *Scientific Reports* 7, no. 1 (October 26, 2017). https://doi.org/10.1038/s41598-017-14542-2.

51. Dhawan, Anil. "Clinical Human Hepatocyte Transplantation: Current Status and Challenges." *Liver Transplantation* 21, no. S1 (November 10, 2015): S39–44. https://doi.org/10.1002/lt.24226.

52. Takebe, Takanori, Keisuke Sekine, Masahiro Enomura, Hiroyuki Koike, Masaki Kimura, Takunori Ogaeri, Ran-Ran Zhang, Yasuharu Ueno, Yun-Wen Zheng, Naoto Koike, Shinsuke Aoyama, Yasuhisa Adachi, and Hideki Taniguchi. "Vascularized and Functional Human Liver from an IPSC-Derived Organ Bud Transplant." *Nature* 499, no. 7459 (July 25, 2013): 481–84. https://doi.org/10.1038/nature12271.

53. Takanori Takebe, Masahiro Enomura, Emi Yoshizawa, Masaki Kimura, Hiroyuki Koike, Yasuharu Ueno, Takahisa Matsuzaki, Takashi Yamazaki, Takafumi Toyohara, Kenji Osafune, Hiromitsu Nakauchi, Hiroshi Y. Yoshikawa, and Hideki Taniguchi. "Vascularized and Complex Organ Buds from Diverse Tissues via Mesenchymal Cell-Driven Condensation." *Cell Stem Cell* 16, no. 5 (May 7, 2015): 556–65. https://doi.org/10.1016/j.stem.2015.03.004.

54. Dusinska, Maria, Elise Rundén-Pran, Jürgen Schnekenburger, and Jun Kanno. "Toxicity Tests: In Vitro and In Vivo." In *Adverse Effects of Engineered Nanomaterials: Exposure, Toxicology, and Impact on Human Health*. Edited by Bengt Fadeel, Antonio Pietroiusti, and Anna A. Shvedova, 68. London: Elsevier, 2017.

55. Stevens, Kelly R., Margaret A. Scull, Vyas Ramanan, Chelsea L. Fortin, Ritika R. Chaturvedi, Kristin A. Knouse, Jing W. Xiao, Canny Fung, Teodelinda Mirabella, Amanda X. Chen, Margaret G. McCue, Michael T. Yang, Heather E. Fleming, Kwanghun Chung, Ype P. de Jong, Christopher S. Chen, Charles M. Rice, and Sangeeta N. Bhatia. "In Situ Expansion of Engineered Human Liver Tissue in a Mouse Model of Chronic Liver Disease." *Science Translational Medicine* 9, no. 399 (July 19, 2017): eaah5505. https://doi.org/10.1126/scitranslmed.aah5505.

56. Stevens, K. R., M. D. Ungrin, R. E. Schwartz, S. Ng, B. Carvalho, K. S. Christine, R. R. Chaturvedi, et al. "InVERT Molding for Scalable Control of Tissue Microarchitecture." *Nature Communications* 4, no. 1 (May 14, 2013). https://doi.org/10.1038/ncomms2853.

57. Azvolinsky, Anna. "Engineered Human Liver Tissue Grows in Mice." *The Scientist Magazine*, July 19, 2017. https://www.the-scientist.com/daily-news/engineered-human-liver-tissue-grows-in-mice-31198.

58. Miller, Jordan S., Kelly R. Stevens, Michael T. Yang, Brendon M. Baker, Duc-Huy T. Nguyen, Daniel M. Cohen, Esteban Toro, Alice A. Chen, Peter A. Galie, Xiang Yu, Ritika Chaturvedi, Sangeeta N. Bhatia, and Christopher S. Chen. "Rapid Casting of Patterned Vascular Networks for Perfusable Engineered Three-Dimensional Tissues." *Nature Materials* 11, no. 9 (July 1, 2012): 768–74. https://doi.org/10.1038/nmat3357.

59. "Organovo and Tarveda Therapeutics Announce Definitive Merger Agreement." Businesswire, December 16, 2019. https://www.businesswire.com/news/home/20191216005228/en.

60. Sistare, Frank D., William B. Mattes, and Edward L. LeCluyse. "The Promise of New Technologies to Reduce, Refine, or Replace Animal Use While Reducing Risks of Drug Induced Liver Injury in Pharmaceutical Development." *ILAR Journal* 57, no. 2 (December 1, 2016): 186–211. https://doi.org/10.1093/ilar/ilw025.

61. Natalie Angier, "The Liver: A 'Blob' That Runs the Body."

62. Frank D. Sistare et al., "The Promise of New Technologies," 186.

63. Nutton, Vivian. *Ancient Medicine*. London: Routledge, 2012, 238. https://doi.org/10.4324/9780203081297.

64. Jouanna, Jacques. "The Legacy of the Hippocratic Treatise *The Nature of Man*: The Theory of the Four Humours." *Greek Medicine from Hippocrates to Galen*, July 25, 2012, 335–59. https://doi.org/10.1163/9789004232549_017.

Chapter 8

1. Acierno, Louis J., and L. Timothy Worrell. "James Bryan Herrick." Edited by J. Willis Hurst and W. Bruce Fye. *Clinical Cardiology* 23, no. 3 (March 2000): 230–32. https://doi.org/10.1002/clc.4960230322.
2. Association of American Physicians. *Proceedings of the Association of American Physicians.* Vol. 27. Philadelphia, PA: Dornan, 1912. https://hdl.handle.net/2027/uva.3470150456.
3. Ross, R. S. "A Parlous State of Storm and Stress: The Life and Times of James B. Herrick." *Circulation* 67, no. 5 (May 1983): 955–59. https://doi.org/10.1161/01.cir.67.5.955.
4. Means, James Howard. *The Association of American Physicians: Its First Seventy-Five Years.* New York, McGraw-Hill, 1961, 108.
5. Jones, David S. "Visions of a Cure: Visualization, Clinical Trials, and Controversies in Cardiac Therapeutics, 1968–1998." *Isis* 91, no. 3 (September 2000): 504–41. https://doi.org/10.1086/384853.
6. Ibid., 508.
7. Richard S. Ross, "A Parlous State," 957.
8. Herrick, James B. "An Intimate Account of My Early Experience with Coronary Thrombosis." *American Heart Journal* 27, no. 1 (January 1944): 1–18. https://doi.org/10.1016/s0002-8703(44)90602-2.
9. Herrick, James B. "Thrombosis of the Coronary Arteries." *Journal of the American Medical Association* 72, no. 6 (February 8, 1919): 387. https://doi.org/10.1001/jama.1919.02610060001001.
10. Fye, W. Bruce. "A History of the Origin, Evolution, and Impact of Electrocardiography." *American Journal of Cardiology* 73, no. 13 (May 1994): 937–49. https://doi.org/10.1016/0002-9149(94)90135-x.
11. James B. Herrick, "Thrombosis of the Coronary Arteries," 389.
12. W. Bruce Fye, "A History of the Origin," 943.
13. Herrick, J. B. *A Short History of Cardiology.* Springfield, IL: Charles C. Thomas, 1942, 229.
14. James B. Herrick, "An Intimate Account," 15.
15. Morber, Jenny. "Custom Organs, Printed to Order." Pbs.org, March 18, 2015. https://www.pbs.org/wgbh/nova/article/3d-printed-organs.
16. Ibid.
17. Wodehouse, Pelham Grenville. *Something Fresh.* London: Arrow Books, 2008, 15.
18. Beans, Carolyn. "Inner Workings: The Race to Patch the Human Heart." *Proceedings of the National Academy of Sciences of the USA* 115, no. 26 (June 26, 2018): 6518–20. https://doi.org/10.1073/pnas.1808317115.
19. Gao, Ling, Molly E. Kupfer, Jangwook P. Jung, Libang Yang, Patrick Zhang, Yong Da Sie, Quyen Tran, Visar Ajeti, Brian T. Freeman, Vladimir G. Fast, Paul J. Campagnola, Brenda M. Ogle, and Jianyi Zhang. "Myocardial Tissue Engineering with Cells Derived from Human-Induced Pluripotent Stem Cells and a Native-Like, High-Resolution, 3-Dimensionally Printed Scaffold." *Circulation Research* 120, no. 8 (April 14, 2017): 1318–25. https://doi.org/10.1161/circresaha.116.310277.
20. Aamodt, Joseph M., and David W. Grainger. "Extracellular Matrix-Based Biomaterial Scaffolds and the Host Response." *Biomaterials* 86 (April 2016): 68–82. https://doi.org/10.1016/j.biomaterials.2016.02.003.

21. Karp, Gerald. *Cell and Molecular Biology: Concepts and Experiments*. Hoboken, NJ: Wiley, 2010, 236.

22. Roberts, Billy J. "Direct Normal Solar Irradiance." National Renewable Energy Laboratory, February 22, 2018. https://www.nrel.gov/gis/assets/pdfs/solar_dni_2018_01.pdf.

23. "Coronary." Lexico Dictionaries. Accessed January 12, 2020. https://www.lexico.com/en/definition/coronary.

24. Awada, Hassan K., Mintai P. Hwang, and Yadong Wang. "Towards Comprehensive Cardiac Repair and Regeneration After Myocardial Infarction: Aspects to Consider and Proteins to Deliver." *Biomaterials* 82 (March 2016): 94–112. https://doi.org/10.1016/j.biomaterials.2015.12.025.

25. Prabhu, Sumanth D., and Nikolaos G. Frangogiannis. "The Biological Basis for Cardiac Repair After Myocardial Infarction." *Circulation Research* 119, no. 1 (June 24, 2016): 91–112. https://doi.org/10.1161/circresaha.116.303577.

26. Hassan K. Awada et al., "Towards Comprehensive Cardiac Repair," 100.

27. Purslow, P. P. *Collagen: Structure and Mechanics*. Edited by Peter Fratzl. New York: Springer, 2008, 325.

28. Hassan K. Awada et al., "Towards Comprehensive Cardiac Repair," 100.

29. Ling Gao et al., "Myocardial Tissue Engineering," 1322.

30. Ogle, Brenda. Phone conversation with K.D., April 11, 2018.

31. Montgomery, Miles, Samad Ahadian, Locke Davenport Huyer, Mauro Lo Rito, Robert A. Civitarese, Rachel D. Vanderlaan, Jun Wu, Lewis A. Reis, Abdul Momen, Saeed Akbari, Aric Pahnke, Ren-Ke Li, Christopher A. Caldarone, and Milica Radisic. "Flexible Shape-Memory Scaffold for Minimally Invasive Delivery of Functional Tissues." *Nature Materials* 16, no. 10 (August 14, 2017): 1038–46. https://doi.org/10.1038/nmat4956.

32. Zhang, Jianyi, Wuqiang Zhu, Milica Radisic, and Gordana Vunjak-Novakovic. "Can We Engineer a Human Cardiac Patch for Therapy?" *Circulation Research* 123, no. 2 (July 6, 2018): 244–65. https://doi.org/10.1161/circresaha.118.311213.

33. Ogle, Brenda. E-mail with K.D., December 19, 2018.

34. Nguyen, Patricia K., Evgenios Neofytou, June-Wha Rhee, and Joseph C. Wu. "Potential Strategies to Address the Major Clinical Barriers Facing Stem Cell Regenerative Therapy for Cardiovascular Disease." *JAMA Cardiology* 1, no. 8 (November 1, 2016): 953. https://doi.org/10.1001/jamacardio.2016.2750.

35. Ringer, Sydney. "Concerning the Influence Exerted by Each of the Constituents of the Blood on the Contraction of the Ventricle." *Journal of Physiology* 3, no. 5–6 (August 1, 1882): 380–93. https://doi.org/10.1113/jphysiol.1882.sp000111.

36. Ringer, Sydney. "A Further Contribution Regarding the Influence of the Different Constituents of the Blood on the Contraction of the Heart." *Journal of Physiology* 4, no. 1 (January 1, 1883): 29–42. https://doi.org/10.1113/jphysiol.1883.sp000120.

37. Cheng, Heping, and W. J. Lederer. "Calcium Sparks." *Physiological Reviews* 88, no. 4 (October 2008): 1491–545. https://doi.org/10.1152/physrev.00030.2007.

38. "EKG vs. Echo—Heart Health Can Depend on Both." Washington University Physicians, July 15, 2015. https://wuphysicians.wustl.edu/your-health-update-newsletter/post/3618/ekg-vs-echo-heart-health-can-depend-on-both.

39. Ibid.

40. Jang, Jinah, Hun-Jun Park, Seok-Won Kim, Heejin Kim, Ju Young Park, Soo Jin Na, Hyeon Ji Kim, et al. "3D Printed Complex Tissue Construct Using Stem Cell-Laden

Decellularized Extracellular Matrix Bioinks for Cardiac Repair." *Biomaterials* 112 (January 2017): 264–74. https://doi.org/10.1016/j.biomaterials.2016.10.026.

41. Milanowski, Ann. "25 Amazing Facts About Your Heart." Cleveland Clinic, July 30, 2019. https://health.clevelandclinic.org/25-amazing-facts-about-your-heart.

42. Folse, Roland, and Eugene Braunwald. "Determination of Fraction of Left Ventricular Volume Ejected per Beat and of Ventricular End-Diastolic and Residual Volumes." *Circulation* 25, no. 4 (April 1962): 674–85. https://doi.org/10.1161/01.cir.25.4.674.

43. Cikes, Maja, and Scott D. Solomon. "Beyond Ejection Fraction: An Integrative Approach for Assessment of Cardiac Structure and Function in Heart Failure." *European Heart Journal* 37, no. 21 (September 28, 2015): 1642–50. https://doi.org/10.1093/eurheartj/ehv510.

44. Laskowski, Edward R. "What's a Normal Resting Heart Rate?" Mayo Clinic, August 29, 2018. https://www.mayoclinic.org/healthy-lifestyle/fitness/expert-answers/heart-rate/faq-20057979.

45. Wang, Zhan, Sang Jin Lee, Heng-Jie Cheng, James J. Yoo, and Anthony Atala. "3D Bioprinted Functional and Contractile Cardiac Tissue Constructs." *Acta Biomaterialia* 70 (April 2018): 48–56. https://doi.org/10.1016/j.actbio.2018.02.007.

46. The Editors of Encyclopaedia Britannica. "Systole: Heart Function." In *Encyclopædia Britannica*, April 16, 2015. https://www.britannica.com/science/systole-heart-function.

47. The Editors of Encyclopaedia Britannica. "Diastole: Heart Function." In *Encyclopædia Britannica*, December 2, 2011. https://www.britannica.com/science/diastole-heart-function.

48. Ye, Fei, Fangping Yuan, Xiaohong Li, Nigel Cooper, Joseph P. Tinney, and Bradley B. Keller. "Gene Expression Profiles in Engineered Cardiac Tissues Respond to Mechanical Loading and Inhibition of Tyrosine Kinases." *Physiological Reports* 1, no. 5 (October 2013). https://doi.org/10.1002/phy2.78.

49. Sheehy, Sean P., Anna Grosberg, and Kevin Kit Parker. "The Contribution of Cellular Mechanotransduction to Cardiomyocyte Form and Function." *Biomechanics and Modeling in Mechanobiology* 11, no. 8 (July 7, 2012): 1227–39. https://doi.org/10.1007/s10237-012-0419-2.

50. Scannell, Jack W., Alex Blanckley, Helen Boldon, and Brian Warrington. "Diagnosing the Decline in Pharmaceutical R&D Efficiency." *Nature Reviews Drug Discovery* 11, no. 3 (March 1, 2012): 191–200. https://doi.org/10.1038/nrd3681.

51. "Why Do You Breathe?" British Lung Foundation, February 5, 2018. https://www.blf.org.uk/support-for-you/how-your-lungs-work/why-do-we-breathe.

Chapter 9

1. Roueché, Berton. "The Fog." In *American Earth: Environmental Writing Since Thoreau*. Edited by Bill McKibben. New York: The Library of America, 2008, 295–312.

2. Kiester, Edwin, Jr. "A Darkness in Donora." *Smithsonian Magazine*, November 1999. https://www.smithsonianmag.com/history/a-darkness-in-donora-174128118.

3. Berton Roueché, "The Fog," 311.

4. Bhatia, Sangeeta N., and Donald E. Ingber. "Microfluidic Organs-on-Chips." *Nature Biotechnology* 32, no. 8 (2014): 760–72. https://doi.org/10.1038/nbt.2989.

5. "Science on Tap." Wyss Institute, October 23, 2017. https://wyss.harvard.edu/media-post/science-on-tap. Comment is at 2:07 during the video presentation.

6. Ingber, Don. "Spanning the Divide." *SciArt Magazine*, October 2018. https://www.sciart-magazine.com/spanning-the-divide.html.

7. Snelson, Kenneth. "Snelson on the Tensegrity Invention." *International Journal of Space Structures* 11, no. 1–2 (April 1996): 43–48. https://doi.org/10.1177/026635119601-207.

8. Fuller, Richard Buckminster. Tensile-Integrity Structures. United States Patent Office, issued November 13, 1962. US Patent Number 3,063,521.

9. Ingber, Donald E. E-mail to K.D., August 27, 2019.

10. Ingber, Donald E. "Tensegrity-Based Mechanosensing from Macro to Micro." *Progress in Biophysics and Molecular Biology* 97, no. 2–3 (June 2008): 163–79. https://doi.org/10.1016/j.pbiomolbio.2008.02.005.

11. Ingber, Donald E. "Cellular Mechanotransduction: Putting All the Pieces Together Again." *FASEB Journal* 20, no. 7 (May 2006): 811–27. https://doi.org/10.1096/fj.05-5424rev.

12. Ulrich, Tom. "The Backstory Behind Organs-on-Chips—Vector." Boston Children's Hospital, October 8, 2014. https://vector.childrenshospital.org/2014/10/the-backstory-behind-organs-on-chips.

13. Wang, N., J. Butler, and D. Ingber. "Mechanotransduction Across the Cell Surface and Through the Cytoskeleton." *Science* 260, no. 5111 (May 21, 1993): 1124–27. https://doi.org/10.1126/science.7684161.

14. Tom Ulrich, "The Backstory."

15. Ingber, Donald E. "Developmentally Inspired Human 'Organs on Chips.'" *Development* 145, no. 16 (May 18, 2018): dev156125. https://doi.org/10.1242/dev.156125.

16. Xia, Younan, and George Whitesides. "Soft Lithography." *Angewandte Chemie International Edition* 37, no. 5 (December 17, 1998): 550–75. https://doi.org/10.1002/(sici)1521-3773(19980316)37:5<550::aid-anie550>3.0.co;2-g.

17. Singhvi, R., A. Kumar, G. Lopez, G. Stephanopoulos, D. Wang, G. Whitesides, and D. Ingber. "Engineering Cell Shape and Function." *Science* 264, no. 5159 (April 29, 1994): 696–98. https://doi.org/10.1126/science.8171320.

18. Chen, Christopher S., Milan Mrksich, Sui Huang, George M. Whitesides, and Donald E. Ingber. "Geometric Control of Cell Life and Death." *Science* 276, no. 5317 (May 30, 1997): 1425–28. https://doi.org/10.1126/science.276.5317.1425.

19. Tom Ulrich, "The Backstory."

20. Christopher S. Chen et al., "Geometric Control."

21. Squires, Todd M., and Stephen R. Quake. "Microfluidics: Fluid Physics at the Nanoliter Scale." *Reviews of Modern Physics* 77, no. 3 (October 6, 2005): 977–1026. https://doi.org/10.1103/revmodphys.77.977.

22. Eric Hanson Hewlett Packard Laboratories. "How an Inkjet Printer Works." Imaging.org, 2015. https://www.imaging.org/site/IST/Resources/Imaging_Tutorials/How_an_Ink_Jet_Printer_Works/IST/Resources/Tutorials/Inkjet_Printer.aspx.

23. Perlman, Howard. "Drip Calculator: How Much Water Does a Leaking Faucet Waste?" USGS Water Science School. Accessed January 5, 2020. https://water.usgs.gov/edu/activity-drip.html.

24. Castillo-León, Jaime. "Microfluidics and Lab-on-a-Chip Devices: History and Challenges." Edited by Jaime Castillo-León and Winnie E. Svendsen. *Lab-on-a-Chip Devices and*

Micro-Total Analysis Systems, October 14, 2014, 1–15. https://doi.org/10.1007/978-3-319-08687-3_1.

25. Tom Ulrich, "The Backstory."

26. Ibid.

27. Takayama, Shuichi, Emanuele Ostuni, Philip LeDuc, Keiji Naruse, Donald E. Ingber, and George M. Whitesides. "Subcellular Positioning of Small Molecules." *Nature* 411, no. 6841 (June 2001): 1016. https://doi.org/10.1038/35082637.

28. Huh, D., H. Fujioka, Y.-C. Tung, N. Futai, R. Paine, J. B. Grotberg, and S. Takayama. "Acoustically Detectable Cellular-Level Lung Injury Induced by Fluid Mechanical Stresses in Microfluidic Airway Systems." *Proceedings of the National Academy of Sciences of the USA* 104, no. 48 (November 15, 2007): 18886–91. https://doi.org/10.1073/pnas.0610868104.

29. Ibid., 18886.

30. Moore, Nicole Casal. "Lung-on-a-Chip Leads to New Insights on Pulmonary Diseases." University of Michigan News, November 21, 2007. https://news.umich.edu/lung-on-a-chip-leads-to-new-insights-on-pulmonary-diseases.

31. Ibid.

32. Donald E. Ingber, "Developmentally Inspired."

33. Ibid.

34. Huh, Dongeun, Benjamin D. Matthews, Akiko Mammoto, Martín Montoya-Zavala, Hong Yuan Hsin, and Donald E. Ingber. "Reconstituting Organ-Level Lung Functions on a Chip." *Science* 328, no. 5986 (2010): 1662–68. https://doi.org/10.1126/science.1188302.

35. Tom Ulrich, "The Backstory."

36. Dongeun Huh et al., "Reconstituting Organ-Level Lung Functions," 1665.

37. Tom Ulrich, "The Backstory."

38. Huh, Dongeun, Daniel C. Leslie, Benjamin D. Matthews, Jacob P. Fraser, Samuel Jurek, Geraldine A. Hamilton, Kevin S. Thorneloe, Michael Allen McAlexander, and Donald E. Ingber. "A Human Disease Model of Drug Toxicity-Induced Pulmonary Edema in a Lung-on-a-Chip Microdevice." *Science Translational Medicine* 4, no. 159 (November 7, 2012): 159ra147. https://doi.org/10.1126/scitranslmed.3004249.

39. Kim, Hyun Jung, Dongeun Huh, Geraldine Hamilton, and Donald E. Ingber. "Human Gut-on-a-Chip Inhabited by Microbial Flora That Experiences Intestinal Peristalsis-Like Motions and Flow." *Lab on a Chip* 12, no. 12 (2012): 2165. https://doi.org/10.1039/c2lc40074j.

40. Tom Ulrich, "The Backstory."

41. Skardal, Aleksander, Sean V. Murphy, Mahesh Devarasetty, Ivy Mead, Hyun-Wook Kang, Young-Joon Seol, Yu Shrike Zhang, Su-Ryon Shin, Liang Zhao, Julio Aleman, Adam R. Hall, Thomas D. Shupe, Andre Kleensang, Mehmet R. Dokmeci, Sang Jin Lee, John D. Jackson, James J. Yoo, Thomas Hartung, Ali Khademhosseini, Shay Soker, Colin E. Bishop, and Anthony Atala. "Multi-Tissue Interactions in an Integrated Three-Tissue Organ-on-a-Chip Platform." *Scientific Reports* 7, no. 1 (August 18, 2017). https://doi.org/10.1038/s41598-017-08879-x.

42. Srinivasan, Balaji, Aditya Reddy Kolli, Mandy Brigitte Esch, Hasan Erbil Abaci, Michael L. Shuler, and James J. Hickman. "TEER Measurement Techniques for in Vitro Barrier Model Systems." *Journal of Laboratory Automation* 20, no. 2 (April 2015): 107–26. https://doi.org/10.1177/2211068214561025.

43. Paxton, Steve, Michelle Peckham, and Adele Knibbs. "The Leeds Histology Guide." University of Leeds, 2003. https://www.histology.leeds.ac.uk/respiratory/basic_structure.php.

44. Paxton, Steve, Michelle Peckham, and Adele Knibbs. "The Leeds Histology Guide." Leeds. ac.uk, 2003. https://www.histology.leeds.ac.uk/respiratory/respiratory.php.

45. "Alveoli." Johns Hopkins School of Medicine, 1995. http://oac.med.jhmi.edu/res_phys/encyclopedia/alveoli/alveoli.html.

46. Dezube, Rebecca. "Exchanging Oxygen and Carbon Dioxide." Merck Manual Consumer Version, June 2019. https://www.merckmanuals.com/home/lung-and-airway-disorders/biology-of-the-lungs-and-airways/exchanging-oxygen-and-carbon-dioxide.

47. Ochs, Matthias, Jens R. Nyengaard, Anja Jung, Lars Knudsen, Marion Voigt, Thorsten Wahlers, Joachim Richter, and Hans Jørgen G. Gundersen. "The Number of Alveoli in the Human Lung." *American Journal of Respiratory and Critical Care Medicine* 169, no. 1 (January 2004): 120–24. https://doi.org/10.1164/rccm.200308-1107oc.

48. "Beta Blockers." Mayo Clinic, August 16, 2019. https://www.mayoclinic.org/diseases-conditions/high-blood-pressure/in-depth/beta-blockers/art-20044522.

49. Inspired by Wodehouse, P. G. *Service with a Smile*. New York: Norton, 2013, 96.

50. The Editors of Encyclopaedia Britannica. "Interleukin." In *Encyclopædia Britannica*, August 21, 2019. https://www.britannica.com/science/interleukin.

51. Skardal, Aleksander. Phone conversation with K.D., January 16, 2019.

52. "Printers to Produce Life-Saving Organs." Phys.org, December 12, 2005. https://phys.org/news/2005-12-printers-life-saving.html.

53. Collins, Lois M. "U. Makes a Healing 'Bio-Paper.'" *Deseret News*, November 5, 2005. https://www.deseretnews.com/article/635158922/U-makes-a-healing-bio-paper.html.

54. Park, Ju Young, Hyunryul Ryu, Byungjun Lee, Dong-Heon Ha, Minjun Ahn, Suryong Kim, Jae Yun Kim, Noo Li Jeon, and Dong-Woo Cho. "Development of a Functional Airway-on-a-Chip by 3D Cell Printing." *Biofabrication* 11, no. 1 (October 30, 2018): 015002. https://doi.org/10.1088/1758-5090/aae545.

55. The Editors of Encyclopaedia Britannica. "Cytokine." In *Encyclopædia Britannica*, November 21, 2018. https://www.britannica.com/science/cytokine.

56. Balaji Srinivasan et al., "TEER Measurement Techniques."

57. Ibid., 109.

58. Ibid., 108.

59. Colloff, Matthew. *Dust Mites*. Collingwood, Victoria, Australia: CSIRO, 2009, 5.

60. Hull, Chuck. E-mail to K.D., July 19, 2019.

61. Horváth, Lenke, Yuki Umehara, Corinne Jud, Fabian Blank, Alke Petri-Fink, and Barbara Rothen-Rutishauser. "Engineering an in Vitro Air–Blood Barrier by 3D Bioprinting." *Scientific Reports* 5, no. 1 (January 22, 2015). https://doi.org/10.1038/srep07974.

62. Ibid., Supplementary Information Figure S1.

63. Ibid., 4.

64. "A Lush Legacy." Lush Cosmetics. Accessed January 5, 2020. https://www.lushusa.com/stories/article_a-lush-legacy.html.

65. "2017 Lush Prize Winners." Lush Cosmetics. Accessed January 5, 2020. https://www.lushusa.com/stories/article_meet-lush-prize-winners-2017.html.

66. Ingber, Donald. "Human Organs-on-Chips May One Day Replace Animal Testing." Science Nordic, March 13, 2017. https://sciencenordic.com/biology-denmark-medicine/human-organs-on-chips-may-one-day-replace-animal-testing/1443448.

67. "BIC Museum of Microscopy—Overview." Cell Biology, Neurobiology and Biophysics, Utrecht University, January 31, 2013. https://cellbiology.science.uu.nl/facilities/biology-imaging-center-bic/bic-museum-of-microscopy-overview.

Chapter 10

1. Bellomo, G. "A Short History of 'Glomerulus.'" *Clinical Kidney Journal* 6, no. 2 (March 27, 2013): 250–51. https://doi.org/10.1093/ckj/sft022.
2. Weening, Jan J., and J. Charles Jennette. "Historical Milestones in Renal Pathology." *Virchows Archiv* 461, no. 1 (June 3, 2012): 3–11. https://doi.org/10.1007/s00428-012-1254-7.
3. Fine, Leon G. *Renal Physiology : People and Ideas.* Edited by Carl W. Gottschalk, Robert W. Berliner, and Gerhard H. Giebisch. Bethesda, MD: American Physiological Society, 1987, 20–21.
4. Wilson, L. G. "The Development of the Knowledge of Kidney Function in Relation to Structure—Malpighi to Bowman." *Bulletin of the History of Medicine* 34, no. 2 (March 1960): 175–81. https://www.jstor.org/stable/44446678.
5. Jamison, Rex L. "Resolving an 80-Yr-Old Controversy: The Beginning of the Modern Era of Renal Physiology." *Advances in Physiology Education* 38, no. 4 (December 2014): 286–95. https://doi.org/10.1152/advan.00105.2014.
6. Leon G. Fine, "Evolution of Renal Physiology," 22.
7. Ibid., 23.
8. L. G. Wilson, "The Development of the Knowledge of Kidney Function."
9. Rex L. Jamison, "Resolving an 80-Yr-Old Controversy."
10. Schmidt, Carl F. *Alfred Newton Richards, 1876–1966: A Biographical Memoir.* Vol. 42. Washington, DC: National Academy of Sciences, 1971. http://www.nasonline.org/publications/biographical-memoirs/memoir-pdfs/richards-alfred-n.pdf.
11. Rex L. Jamison, "Resolving an 80-Yr-Old Controversy."
12. Carl F. Schmidt, "Alfred Newton Richards," 293.
13. Wearn, J. T. "Section II: On Working with Dr. Richards in Philadelphia." *Annals of Internal Medicine* 71, no. 5, part 2 (November 1, 1969): 43–58. https://doi.org/10.7326/0003-4819-71-5-43.
14. Carl F. Schmidt, "Alfred Newton Richards," 293.
15. Ibid.
16. Rex L. Jamison, "Resolving an 80-Yr-Old Controversy."
17. "Your Kidneys & How They Work." National Institute of Diabetes and Digestive and Kidney Diseases, June 2018. https://www.niddk.nih.gov/health-information/kidney-disease/kidneys-how-they-work.
18. Geggel, Laura. "How Much Blood Is in the Human Body?" Live Science, March 3, 2016. https://www.livescience.com/32213-how-much-blood-is-in-the-human-body.html.
19. The Editors of Encyclopedia Britannica. "Nephron." In *Encyclopædia Britannica*, July 3, 2015. https://www.britannica.com/science/nephron.
20. Davies, Kelvin J. A. "Adaptive Homeostasis." *Molecular Aspects of Medicine* 49 (June 2016): 1–7. https://doi.org/10.1016/j.mam.2016.04.007.
21. Cannon, Walter B. *Ses Amis, Ses Collègues, Ses Élèves.* Edited by Auguste Pettit. Paris: Les Editions Médicales, 1926, 91.

22. Guyton, Arthur C., John E. Hall, and Reed Elsevier. *Textbook of Medical Physiology*. 11th ed. Philadelphia, PA: Elsevier Saunders, 2007, 4.

23. Cannon, Walter B. *The Wisdom of the Body*. New York: Norton, 1963.

24. "History of the Merck Manuals." Merck Manual Professional Edition. Accessed January 5, 2020. https://www.merckmanuals.com/professional/resourcespages/history.

25. Lewis, James, III. "Overview of Electrolytes." Merck Manual Consumer Version, September 2018. https://www.merckmanuals.com/home/hormonal-and-metabolic-disorders/electrolyte-balance/overview-of-electrolytes.

26. "Mineral." Merriam-Webster. Accessed January 5, 2020. https://www.merriam-webster.com/dictionary/mineral.

27. Healthwise Staff. "Minerals: Their Functions and Sources." Michigan Medicine, November 7, 2018. https://www.uofmhealth.org/health-library/ta3912.

28. "Salt and Sodium." The Nutrition Source, July 18, 2013. https://www.hsph.harvard.edu/nutritionsource/salt-and-sodium.

29. Murphy, Sherry L., Jiaquan Xu, Kenneth D. Kochanek, and Elizabeth Arias. "Mortality in the United States, 2017." NCHS Data Brief no. 328. Centers for Disease Control and Prevention, November 2018. https://www.cdc.gov/nchs/products/databriefs/db328.htm.

30. "The Top 10 Causes of Death." World Health Organization, May 24, 2018. https://www.who.int/news-room/fact-sheets/detail/the-top-10-causes-of-death.

31. Homan, Kimberly A., David B. Kolesky, Mark A. Skylar-Scott, Jessica Herrmann, Humphrey Obuobi, Annie Moisan, and Jennifer A. Lewis. "Bioprinting of 3D Convoluted Renal Proximal Tubules on Perfusable Chips." *Scientific Reports* 6, no. 1 (October 11, 2016). https://doi.org/10.1038/srep34845.

32. Tiong, Ho Yee, Peng Huang, Sijing Xiong, Yao Li, Anantharaman Vathsala, and Daniele Zink. "Drug-Induced Nephrotoxicity: Clinical Impact and Preclinical in Vitro Models." *Molecular Pharmaceutics* 11, no. 7 (March 3, 2014): 1933–48. https://doi.org/10.1021/mp400720w.

33. Redfern, Will S., Russ Bialecki, Lorna Ewart, Tim G. Hammond, Lew Kinter, Silvana Lindgren, Chris E. Pollard, Mike Rolf, and Jean-Pierre Valentin. "Impact and Prevalence of Safety Pharmacology-Related Toxicities Throughout the Pharmaceutical Life Cycle." *Journal of Pharmacological and Toxicological Methods* 62, no. 2 (September 2010): e29. https://doi.org/10.1016/j.vascn.2010.11.098.

34. St. Johnston, Daniel, and Julie Ahringer. "Cell Polarity in Eggs and Epithelia: Parallels and Diversity." *Cell* 141, no. 5 (May 2010): 757–74. https://doi.org/10.1016/j.cell.2010.05.011.

35. Ibid.

36. Lackie, John. *A Dictionary of Biomedicine*. New York: Oxford University Press, 2010. https://doi.org/10.1093/acref/9780199549351.001.0001.

37. Lin, Neil Y. C., Kimberly A. Homan, Sanlin S. Robinson, David B. Kolesky, Nathan Duarte, Annie Moisan, and Jennifer A. Lewis. "Renal Reabsorption in 3D Vascularized Proximal Tubule Models." *Proceedings of the National Academy of Sciences of the USA* 116, no. 12 (March 4, 2019): 5399–404. https://doi.org/10.1073/pnas.1815208116.

38. Rabbani, Harris Sajjad, Dani Or, Ying Liu, Ching-Yao Lai, Nancy B. Lu, Sujit S. Datta, Howard A. Stone, and Nima Shokri. "Suppressing Viscous Fingering in Structured Porous Media." *Proceedings of the National Academy of Sciences of the USA* 115, no. 19 (April 23, 2018): 4833–38. https://doi.org/10.1073/pnas.1800729115.

39. Marelli-Berg, F. M., M. Clement, C. Mauro, and G. Caligiuri. "An Immunologist's Guide to CD31 Function in T-Cells." *Journal of Cell Science* 126, no. 11 (June 1, 2013): 2343–52. https://doi.org/10.1242/jcs.124099.

40. Adams, Nathan. "Albumin." Chemistry World, June 24, 2014. https://www.chemistry-world.com/podcasts/albumin/7482.article.

41. Wasserman, David H. "Four Grams of Glucose." *American Journal of Physiology: Endocrinology and Metabolism* 296, no. 1 (January 2009): E11–21. https://doi.org/10.1152/ajpendo.90563.2008.

42. Ibid., E11.

43. Ibid.

44. Mather, Amanda, and Carol Pollock. "Glucose Handling by the Kidney." *Kidney International* 79 (March 2011): S1–6. https://doi.org/10.1038/ki.2010.509.

45. Ibid., S2.

46. Ibid., S1.

47. Richards, A. N., B. B. Westfall, and P. A. Bott. "Renal Excretion of Inulin, Creatinine and Xylose in Normal Dogs." *Experimental Biology and Medicine* 32, no. 1 (October 1, 1934): 73–75. https://doi.org/10.3181/00379727-32-7564p.

48. Plosker, Greg L. "Dapagliflozin." *Drugs* 72, no. 17 (December 2012): 2289–312. https://doi.org/10.2165/11209910-000000000-00000.

49. Smalley, Eric. "NYU's Nadrian Seeman." Technology Research News, May 2005. http://www.trnmag.com/Stories/2005/050405/View_Nadrian_Seeman_050405.html.

50. Ibid.

Chapter 11

1. Plimpton, George. "E. L. Doctorow, The Art of Fiction No. 94." *The Paris Review, No. 101*, December 1986. https://www.theparisreview.org/interviews/2718/e-l-doctorow-the-art-of-fiction-no-94-e-l-doctorow.

2. Racaniello, Vincent. "Poliovirus." *Neurotropic Viral Infections.* Edited by Carol Shoshkes Reiss. 2016, 1–26. https://doi.org/10.1007/978-3-319-33133-1_1.

3. Oshinsky, David M. *Polio: An American Story*. Oxford, UK: Oxford University Press, 2006, 8–9.

4. Ibid., 11.

5. Smallman-Raynor, M. R., and A. D. Cliff. *Poliomyelitis: Emergence to Eradication*. Oxford, UK: Oxford University Press, 2006, 93–96.

6. Gould, Tony. *A Summer Plague: Polio and Its Survivors*. New Haven, CT: Yale University Press, 1997, 28.

7. M. R. Smallman-Raynor and A. D. Cliff, "Poliomyelitis," 150.

8. "Philip Drinker, Polio, and That 'Damn Machine.'" Harvard T.H. Chan School of Public Health, April 24, 2013. https://www.hsph.harvard.edu/news/centennial-philip-drinker-polio.

9. Drinker, Philip. "The Use of a New Apparatus for the Prolonged Administration of Artificial Respiration." *Journal of the American Medical Association* 92, no. 20 (May 18, 1929): 1658–60. https://doi.org/10.1001/jama.1929.02700460014005.

10. Waxman, Olivia B. "The Inspiring Depression-Era Story of How the 'March of Dimes' Got Its Name." *Time*, January 3, 2018. https://time.com/5062520/march-of-dimes-history.

11. Enders, J. F., T. H. Weller, and F. C. Robbins. "Cultivation of the Lansing Strain of Poliomyelitis Virus in Cultures of Various Human Embryonic Tissues." *Science* 109, no. 2822 (January 28, 1949): 85–87. https://doi.org/10.1126/science.109.2822.85.

12. Tony Gould, *A Summer Plague,* 124.

13. Bodian, David, Isabel M. Morgan, and Howard A. Howe. "Differentiation of Types of Poliomyelitis Viruses: III. The Grouping of Fourteen Strains into Three Basic Immunological Types." *American Journal of Epidemiology* 49, no. 2 (March 1949): 234–47. https://doi.org/10.1093/oxfordjournals.aje.a119273.

14. David M. Oshinsky, *Polio: An American Story*, 121.

15. "A Science Odyssey: People and Discoveries: Jonas Salk." Pbs.org, 1998. https://www.pbs.org/wgbh/aso/databank/entries/bmsalk.html.

16. David M. Oshinsky, *Polio: An American Story*, 199.

17. Ibid., 207.

18. Ibid., 268.

19. Pottol, Kristy S. "Armed Forces Institute of Regenerative Medicine Annual Report 2013." Armed Forces Institute of Regenerative Medicine, 2013. https://www.afirm.mil/assets/documents/annual_report_2013.pdf.

20. Crown, Ellen. "USAMRDC: AFIRM II Kick off Gathers Army, Academic Regenerative Medicine Experts." Army.mil, June 1, 2019. https://mrdc.amedd.army.mil/index.cfm/media/articles/2013/AFIRM_II_kick_off_gathers_experts.

21. Lindbergh, Stacey. "Medical Technology Enterprise Consortium Status Report." Medical Technology Enterprise Consortium, August 2017. https://mtec-sc.org/wp-content/uploads/2016/12/MTEC-2017-Status-Report-v7.pdf.

22. "Hansjörg Wyss Gives $125M to Create Institute." *Harvard Gazette*, October 9, 2008. https://news.harvard.edu/gazette/story/2008/10/hansjorg-wyss-gives-125m-to-create-institute.

23. "Hansjörg Wyss Doubles His Gift." *Harvard Gazette*, May 21, 2013. https://news.harvard.edu/gazette/story/2013/05/wyss-gift.

24. Shaw, Jonathan. "Catalyzing Bioengineering." *Harvard Magazine*, August 8, 2019. https://www.harvardmagazine.com/2019/09/wyss-institute-gift.

25. Sinha, Gunjan. "Cell Presses." *Nature Biotechnology* 32, no. 8 (August 2014): 716–19. https://doi.org/10.1038/nbt.2983.

26. Murphy, Sean V, and Anthony Atala. "3D Bioprinting of Tissues and Organs." *Nature Biotechnology* 32, no. 8 (2014): 773–85. https://doi.org/10.1038/nbt.2958.

27. Baber, David. *Television Game Show Hosts: Biographies of 32 Stars.* Jefferson, NC: Mcfarland, 2009, 33.

28. Gibson, Lydialyle. "Building Toward a Kidney." *Harvard Magazine*, December 7, 2016. https://harvardmagazine.com/2017/01/building-toward-a-kidney.

29. Ibid.

30. Kang, Hyun-Wook, Sang Jin Lee, In Kap Ko, Carlos Kengla, James J. Yoo, and Anthony Atala. "A 3D Bioprinting System to Produce Human-Scale Tissue Constructs with Structural Integrity." *Nature Biotechnology* 34, no. 3 (February 15, 2016): 312–19. https://doi.org/10.1038/nbt.3413.

31. Harris, Margaret. "Print Me an Organ." *Physics World*, June 4, 2018. https://physicsworld.com/a/print-me-an-organ.

32. Ibid.

33. Miller, Jordan S. "The Billion Cell Construct: Will Three-Dimensional Printing Get Us There?" *PLoS Biology* 12, no. 6 (June 17, 2014): e1001882. https://doi.org/10.1371/journal.pbio.1001882.

34. Baserga, Renato. *The Biology of Cell Reproduction.* Cambridge, MA: Harvard University Press, 1985. "It is possible to calculate, approximately, the number of cells in an animal by measuring the total amount of DNA. For instance, a mouse 25 gm in weight contains a total of 20 mg of DNA. Since diploid somatic cells (in mammals) have 6×10^{-12} gm of DNA, one can calculate that there are, in mouse, about 3×10^9 cells." Dividing the estimated number of cells by 25 gm gives a typical mammalian cell density of around 1.2×10^8 cells/gm. Assuming roughly 1 gm/mL gives about 1.2×10^8 cells/mL for the density per unit volume. Lastly, according to Jordan S. Miller, "The Billion Cell Construct," human cells and mouse cells are about the same size.

35. Grosskopf, Abigail K., Ryan L. Truby, Hyoungsoo Kim, Antonio Perazzo, Jennifer A. Lewis, and Howard A. Stone. "Viscoplastic Matrix Materials for Embedded 3D Printing." *ACS Applied Materials & Interfaces* 10, no. 27 (March 2018): 23353–61. https://doi.org/10.1021/acsami.7b19818.

36. Skylar-Scott, Mark A., Sebastien G. M. Uzel, Lucy L. Nam, John H. Ahrens, Ryan L. Truby, Sarita Damaraju, and Jennifer A. Lewis. "Biomanufacturing of Organ-Specific Tissues with High Cellular Density and Embedded Vascular Channels." *Science Advances* 5, no. 9 (September 2019): eaaw2459. https://doi.org/10.1126/sciadv.aaw2459.

37. Lin, Ruei-Zeng, and Hwan-You Chang. "Recent Advances in Three-Dimensional Multicellular Spheroid Culture for Biomedical Research." *Biotechnology Journal* 3, no. 9–10 (October 2008): 1285–85. https://doi.org/10.1002/biot.1285.

38. Fennema, Eelco, Nicolas Rivron, Jeroen Rouwkema, Clemens van Blitterswijk, and Jan de Boer. "Spheroid Culture as a Tool for Creating 3D Complex Tissues." *Trends in Biotechnology* 31, no. 2 (February 2013): 108–15. https://doi.org/10.1016/j.tibtech.2012.12.003.

39. Lee, Dongkyoung, Shiva Pathak, and Jee-Heon Jeong. "Design and Manufacture of 3D Cell Culture Plate for Mass Production of Cell-Spheroids." *Scientific Reports* 9, no. 1 (September 27, 2019). https://doi.org/10.1038/s41598-019-50186-0.

40. Lancaster, M. A., and J. A. Knoblich. "Organogenesis in a Dish: Modeling Development and Disease Using Organoid Technologies." *Science* 345, no. 6194 (July 17, 2014): 1247125. https://doi.org/10.1126/science.1247125.

41. Mark A. Skylar-Scott et al., "Biomanufacturing of Organ-Specific Tissues."

42. Madeline A. Lancaster and J. A. Knoblich, "Organogenesis in a Dish," 1247125-2.

43. Abbasi, Adeel, Francis DeRoos, José Artur Paiva, J. M. Pereira, Brian G. Harbrecht, Donald P. Levine, Patricia D. Brown, et al. "Cardiac Contractility." In *Encyclopedia of Intensive Care Medicine.* Edited by J. L. Vincent and J. B. Hall. New York: Springer 2012, 460–62. https://doi.org/10.1007/978-3-642-00418-6_192.

44. Abraham, T. P., and R. A. Nishimura. "Myocardial Strain: Can We Finally Measure Contractility?" *Journal of the American College of Cardiology* 37, no. 3 (March 2001): 731–34. https://doi.org/10.1016/s0735-1097(00)01173-6.

45. Mirsky, Israel, and William W. Parmley. "Assessment of Passive Elastic Stiffness for Isolated Heart Muscle and the Intact Heart." *Circulation Research* 33, no. 2 (August 1973): 233–43. https://doi.org/10.1161/01.res.33.2.233.

46. T. P. Abraham and R. A. Nishimura, "Myocardial Strain," 731.

47. Bussadori, C., A. Moreo, M. Di Donato, B. De Chiara, D. Negura, E. Dall'Aglio, E. Lobiati, M. Chessa, C. Arcidiacono, J. S. Dua, F. Mauri, and M. Carminati. "A New 2D-Based Method for Myocardial Velocity Strain and Strain Rate Quantification in a Normal Adult and Paediatric Population: Assessment of Reference Values." *Cardiovascular Ultrasound* 7, no. 1 (February 13, 2009). https://doi.org/10.1186/1476-7120-7-8.

48. Lee, Naomi. "The Lancet Technology: 3D Printing for Instruments, Models, and Organs?" *Lancet* 388, no. 10052 (October 2016): 1368. https://doi.org/10.1016/s0140-6736(16)31735-4.

49. Ibid.

50. Powley, Tanya. "Printing Whole Organs Remains a Long Way Off." *Financial Times*, June 11, 2015. https://www.ft.com/content/2fb2fb5a-ffa5-11e4-bc30-00144feabdc0.

51. Volk, Steve. "The Doctor and the Salamander." *Discover*, March 2015.

52. Schultheiss, Dirk, Rainer M. Engel, Ranice W. Crosby, Gary P. Lees, Michael C. Truss, and Udo Jonas. "Max Brödel (1870–1941) and Medical Illustration in Urology." *Journal of Urology* 164, no. 4 (October 2000): 1137–42. https://doi.org/10.1016/s0022-5347(05)67128-5.

53. Macchi, Veronica, Edgardo Picardi, Antonino Inferrera, Andrea Porzionato, Alessandro Crestani, Giacomo Novara, Raffaele De Caro, and Vincenzo Ficarra. "Anatomic and Radiologic Study of Renal Avascular Plane (Brödel's Line) and Its Potential Relevance on Percutaneous and Surgical Approaches to the Kidney." *Journal of Endourology* 32, no. 2 (February 2018): 154–59. https://doi.org/10.1089/end.2017.0689.

54. Steve Volk, "The Doctor and the Salamander," 37.

55. Gompertz, Benjamin. "On the Nature of the Function Expressive of the Law of Human Mortality, and on a New Mode of Determining the Value of Life Contingencies." *Philosophical Transactions of the Royal Society of London* 115 (1825): 513–83. https://www.jstor.org/stable/107756.

56. Kirkwood, Thomas B. L. "Deciphering Death: A Commentary on Gompertz (1825) 'On the Nature of the Function Expressive of the Law of Human Mortality, and on a New Mode of Determining the Value of Life Contingencies.'" *Philosophical Transactions of the Royal Society B: Biological Sciences* 370, no. 1666 (April 19, 2015): 20140379. https://doi.org/10.1098/rstb.2014.0379.

57. Olshansky, S. Jay, and Bruce A. Carnes. "Ever Since Gompertz." *Demography* 34, no. 1 (February 1997): 1–15. https://doi.org/10.2307/2061656.

58. Strehler, B. L., and A. S. Mildvan. "General Theory of Mortality and Aging." *Science* 132, no. 3418 (July 1, 1960): 14–21. https://doi.org/10.1126/science.132.3418.14.

59. Fries, James F. "The Compression of Morbidity." *The Milbank Memorial Fund Quarterly: Health and Society* 61, no. 3 (1983): 397–419. https://doi.org/10.2307/3349864.

60. Khan, Sadiya S., Benjamin D. Singer, and Douglas E. Vaughan. "Molecular and Physiological Manifestations and Measurement of Aging in Humans." *Aging Cell* 16, no. 4 (May 23, 2017): 624–33. https://doi.org/10.1111/acel.12601.

61. James F. Fries, "The Compression of Morbidity," 399.

62. Sadiya S. Khan et al., "Molecular and Physiological Manifestations."

63. Bernard L. Strehler and A. S. Mildvan, "General Theory of Mortality and Aging."

64. James F. Fries, "The Compression of Morbidity," 399.

65. Atala, Anthony. Phone conversation with K.D., May 30, 2018.

66. Steve Volk, "The Doctor and the Salamander," 37.

67. Atala, Anthony, and Gabor Forgacs. "Three-Dimensional Bioprinting in Regenerative Medicine: Reality, Hype, and Future." *Stem Cells Translational Medicine* 8, no. 8 (May 6, 2019): 744–45. https://doi.org/10.1002/sctm.19-0089.

68. Ibid., 744.

69. Scharnhorst, Gary. *Mark Twain: The Complete Interviews*. Tuscaloosa, AL: University of Alabama Press, 2006, 317.

70. The Editors of Encyclopaedia Britannica. "Sir Peter B. Medawar." In *Encyclopædia Britannica*, September 28, 2019. https://www.britannica.com/biography/Peter-Medawar.

71. Medawar, P. B. *Induction and Intuition in Scientific Thought*. London: Routledge, 2008, 1.

72. Anthony Atala and Gabor Forgacs. "Three-Dimensional Bioprinting," 745.

73. Atala, Anthony, Stuart B. Bauer, Shay Soker, James J. Yoo, and Alan B. Retik. "Tissue-Engineered Autologous Bladders for Patients Needing Cystoplasty." *Lancet* 367, no. 9518 (April 2006): 1241–46. https://doi.org/10.1016/s0140-6736(06)68438-9.

74. Hakenberg, Oliver Walter. "Re: Tissue-Engineered Autologous Bladders for Patients Needing Cystoplasty." *European Urology* 50, no. 2 (August 2006): 382–83. https://doi.org/10.1016/j.eururo.2006.05.030.

Epilogue

1. Nancy. Conversation with K.D., August 2, 2019.

Glossary

Acellular Containing no cells.

Acetylcholine A chemical synthesized in certain neurons (nerve cells), acetylcholine is a neurotransmitter, a chemical that sends signals to other cells, for example, at neuro-muscular junctions. It was the first neuro-transmitter to be so identified.

Actin A filamentous protein that partici-pates in muscle contraction in concert with the protein myosin.

Additive manufacturing The process of cre-ating an object by building it one layer at a time. The term includes 3D printing.

Agarose A biocompatible polymer derived from seaweed that has been used as a hy-drogel and scaffold material.

Albumin The major protein in human blood plasma. Albumin is water-soluble and able to bind to molecules that aren't water-soluble, such as fatty acids, and deliver them to cells.

Alginate A biocompatible polymer derived from seaweed that has been used as a hy-drogel and scaffold material.

Alveolus A hollow air sac in lung tissue (the plural form is alveoli) in which blood gives up its carbon dioxide cellular waste gas in exchange for oxygen. (Cells use oxygen to derive energy from nutrients by the process known as cellular respiration.)

Amino acid The building blocks of proteins. The key elements of an amino acid are carbon, hydrogen, oxygen, and nitrogen with other elements present in side chains (a chemical group attached to the main chain/backbone) of certain amino acids. Although hundreds of naturally occurring amino acids have been identified, only 22 amino acids combine to form a vast number of biologically impor-tant proteins. Of these 22 amino acids, 9 are called "essential" for humans because they cannot be produced from other compounds by the human body and thus must be ingested as food.

Amniotic fluid A source of stem cells con-tained within the amniotic sac (or amnion) that is contained within the placenta, the temporary organ that connects a baby to its mother by means of an umbilical cord.

Aneural Lacking nerves.

Angiogenic An angiogenic growth factor is a molecule that promotes the formation of blood vessels.

Anisotropic An anisotropic object is one that has a physical property exhibiting dif-ferent values when measured in different directions (e.g., wood is stronger along the grain than across the grain).

Antibody A protein produced by the body's immune system in response to the presence of a substance called an antigen.

Antigen Any substance that causes the body's immune system to produce antibodies against it. Antigens may have entered the body from the outside or may have been gen-erated within the body. Antigens appearing on the surface of cells can be used as markers to help identify and classify cells.

Antihistamine Medications used to reduce the effects of histamines—chemical sub-stances produced by the body during an al-lergic reaction.

Apical surface In the context of epi-thelial cells, the surface that faces the lumen (opening) of a tube or the external environment, as opposed to the basal surface that faces the basement membrane (the structure separating epithelial cells from the underlying connective tissue).

Apoptosis A genetically determined process of cell self-destruction, sometimes called programmed cell death. It is a process that eliminates damaged or unneeded cells. If apoptosis is halted, for example, by genetic mutation, the result may be uncontrolled cell growth and tumor formation.

Arrhythmia An irregular or abnormal heart beat rhythm.

Arteriole A very small artery.

Articular cartilage A highly specialized connective tissue that functions to provide a smooth, lubricated surface for articulation—joint movement—and to facilitate the transmission of loads with minimal friction. It's found in your knees, hips, shoulders, and elsewhere. Unlike most tissues, articular cartilage does not have nerves or blood vessels and is composed of a dense extracellular matrix with a sparse distribution of cells called chondrocytes. Articular cartilage is a form of hyaline cartilage that's found on the articular surface of bones—the surfaces present at joints—and so it's given a distinct name to reflect that.

Aspirate Draw out (or draw in) by suction.

Atherosclerosis Arterial disease characterized by fatty deposits on the vessels' inner walls.

Avascular Lacking blood vessels.

Axon An axon (typically one axon per neuron) is a long projection of a neuron/nerve cell that carries electrical signals away from the cell body.

Barrier function In the context of endothelial and epithelial cells, barrier function describes permeability, the property of regulating the free movement of molecules between tissues and/or interstitial compartments. For example, the permeability of endothelial and epithelial cells permits solutes and small molecules to pass through but limits the passage of larger molecules.

Basal surface In the context of epithelial cells, the surface that faces the underlying connective tissue, as opposed to the apical surface that faces the lumen (opening) of a tube or the external environment.

Basement membrane The basement membrane is a specialized form of extracellular matrix that comprises the supporting layer for epithelial and endothelial monolayers, thus separating these monolayers from the underlying connective tissue.

Bile duct Bile ducts are drainage tubes that carry bile from the liver—where bile is produced—to the gallbladder and from the gallbladder to the small intestine. *Bile* is a substance that aids in the digestion and absorption of fats and fat-soluble vitamins.

Biocompatible Not harmful or toxic to living cells and tissues.

Bioconstruct A fabricated cellular or tissue construction.

Bioink One proposed definition is a formulation of cells suitable for processing by an automated biofabrication technology that may also contain biologically active components and biomaterials. However, others maintain that bioinks need not contain cells—or even biological materials. Rather, they argue the term should include materials that may be devoid of cells but that serve a role in the development of a functional tissue, where these roles are broadly described as structural, functional, or sacrificial. For example, in this point of view, a so-called sacrificial bioink—sometimes called a fugitive bioink—can be used to define channels mimicking the vascular system. The sacrificial bioink is then removed to arrive at open channels that can be perfused with nutrients and oxygen to enhance the viability and function of nearby cells within the tissue.

Bioluminescence Light emitted by living organisms through chemical reactions in their bodies.

Biomimetic Mimicking a naturally occurring component or method.

Bioprinting Describes computer-aided transfer processes for patterning and assembling—by direct spatial placement and arrangement—of cells and cell-compatible material. This is often performed layer-by-layer using a 3D printer specially designed to be suitable for living cells and other biological material.

Bioreactor A chamber that creates an environment to mimic that of the target tissue, regulating a combination of biochemical, mechanical, and electrical variables to facilitate dynamic interactions between tissue and environment. Bioreactors are often used subsequent to the actual bioprinting of a construct (often referred to as post-processing or post-printing) to aid in maturation and growth.

Black Death A plague in mid-fourteenth-century Europe attributed to the pathogen *Yersinia pestis* bacteria, the cause of bubonic plague. The Black Death is estimated to have killed 25 million people.

Bleomycin A medicine used to treat several types of cancer.

Blood plasma The liquid portion of blood. Plasma is largely water but also contains critical solutes, including several electrolytes and other substances.

Brödel's line A two-dimensional plane (rather than a one-dimensional line) that denotes a section of the kidney that is relatively avascular and thus allows for an incision with minimal blood loss. Medical illustrator Max Brödel delineated this feature.

Brønsted–Lowry acids and bases A Brønsted–Lowry acid is a substance that is able to donate a hydrogen ion, H^+ (this is the same as a proton), and hence increases the concentration of H^+ when it dissolves in a solvent. (The higher the concentration of hydrogen ions produced by an acid placed in a solvent, the higher is the acidity of the solution.) A Brønsted–Lowry base is a substance that is able to accept a hydrogen ion, H^+.

Brush border This describes closely packed finger-like projections (known as microvilli) that cover the surface of the epithelium in several tissues in the body, including the proximal convoluted tubules of the kidneys.

Bud In general, an organ develops from a condensed tissue mass prior to vascularization. This condensed tissue mass appears during the early stages of organogenesis (a phase of embryonic development) and is termed an organ bud.

Build platform A shelf in a bioprinter on which the bioprinted construct is assembled during the production process.

Burst release profile High initial release rate of materials seen at the beginning of controlled-release processes such as drug delivery.

Calcium imaging A microscopy technique that uses the flux (flow) of fluorescent calcium ions to visualize the local concentration of calcium ions in cells or tissues.

Calcium transient A brief flow of calcium ions (Ca^{2+}) that signals a contraction of muscle tissue. Calcium transients trigger the conversion of an electrical stimulus to a mechanical response.

Capillary bed A web or network of capillaries, the body's smallest blood vessels, form a capillary bed. Capillary beds receive blood from small arteries called arterioles. The blood passes through the network of capillaries and then into small veins called venules. While blood is passing through the network, the thin walls of the capillaries allow nutrients, gases such as oxygen and carbon dioxide, water, and wastes to be exchanged by means of diffusion between the blood and the surrounding tissues of the body.

Cardiac patch A swath of engineered tissue designed to replace heart muscle damaged during a heart attack.

Cardiomyocyte Muscle cells that comprise the heart muscle.

Casting Forming an object by pouring molten material into a mold.

Cavernoma A mass of abnormal blood vessels.

Cell adhesion molecule Cell surface proteins that mediate cell–cell and cell–extracellular matrix binding.

Cell aggregate Cells that are loosely grouped together form what are known as cell aggregates (by contrast, cells that are tightly joined are called tissues).

Cell membrane Cell membranes serve as semipermeable barriers and are composed of a double layer of lipids (a lipid bilayer) with various sugar molecules and proteins residing on the membrane's surface. The proteins serve as receptor sites for messenger molecules. Many of the proteins traverse the entire lipid bilayer (these are called transmembrane proteins) and facilitate molecular communication between the outside and the inside of the cell.

Cell polarity The intrinsic asymmetry in many types of cells that have distinct anterior and posterior ends that differ in their shape, structure, function, or organization of cellular components.

Cell signaling Cells receive and process signals that originate outside their membranous borders. Membrane proteins in cells bind to signaling molecules and subsequently transmit the signal through a series of molecular interactions to internal signaling pathways.

Cell surface markers Proteins on the surface of cells that serve as chemical indicators of specific cell types, including both the cell's lineage and its stage in the differentiation process.

Cellular respiration Cells use oxygen to derive energy from nutrients by the process known as cellular respiration.

Central nervous system That part of the body's nervous system consisting of the brain and spinal cord.

Centrifuge An instrument that spins tubes of material at high speeds. The spinning produces a sustained centrifugal force so that material of higher density is forced to the bottom of the tubes while material of lower density is displaced and moves to the top of the tubes.

Chondrocyte The sole cell type of healthy, normal cartilage.

Chromosome Thread-like structures containing DNA and found in the nucleus of a cell.

Collagen A fibrous protein that is a major component of the extracellular matrix and connective tissues.

Colocalized Located very close together.

Compaction (of gels by cells) A phenomenon in which cells distributed within an extracellular matrix gel induce forces on the fibrous material of the gel causing the gel to become more tightly packed.

Computer-aided design (CAD) The use of computer hardware and software to create an optimized two- or three-dimensional engineering design.

Confluent Denotes the situation in which a monolayer of cells covers the entirety of a surface.

Confocal microscopy A type of fluorescence microscopy in which a screen with a pinhole is used so that only a tiny portion of the sample is illuminated by laser light at a given time (the laser is scanned across the entire sample to get a full image). A confocal microscope gives improved resolution compared to a conventional microscope because the pinhole screen blocks out-of-focus fluorescent light. (Because the focal point of the objective lens of the microscope forms an image at the position of the pinhole, these two points are said to be "conjugate points." Since the pinhole is **conj**ugate to the **focal** point of the lens, it is described as a **confocal** pinhole.)

Contractility Cardiac contractility is the inherent strength of the heart's contraction during systole.

Cord blood Refers to umbilical cord blood from a newborn that contains stem cells from the baby's blood.

Corneocyte A keratinocyte in its last stage of differentiation. They are found on our skin's very outermost surface.

Coronary arteries Blood vessels that supply blood to the heart muscle. They are branches of the aorta (the main blood supplier to the body).

Creatinine A cellular waste product produced by muscle metabolism.

Cross-link A bond that links one polymer chain to another or one part of a polymer chain to another part of the same chain. *To cross-link* is the process of forming such bonds.

Cytoarchitecture The typical pattern of cellular arrangement within a particular tissue or organ.

Cytokines A broad class of protein signaling molecules that act through cell surface receptors. When secreted by the immune system, they mediate and regulate immunity and inflammation.

Cytokine storm A dysregulation of the body's innate immune system characterized by elevated circulating cytokine levels, acute systemic inflammatory symptoms, and secondary organ dysfunction (i.e., secondary to unchecked inflammation rather than primary organ insult owing to disease, infection, or trauma) beyond that which could be attributed to a normal response to a pathogen, if a pathogen is present. In the absence of a pathogen, a cytokine storm can occur owing to a genetic disorder, for example.

Cytoplasm A thick solution that includes all the material inside the cell but outside the nucleus of the cell. The cytoplasm contains all the cell's organelles as well as other material; it is given structure by the cytoskeleton.

Cytoskeleton A structure made of filamentous proteins that helps cells maintain their shape and internal organization and provides mechanical support that enables cells to carry out essential functions. The cytoskeleton can be thought of as the bones and muscles of our cells.

Decellularized extracellular matrix The extracellular matrix devoid of the cells that inhabit it.

Denature To modify the molecular structure of a protein compared to its highly ordered structure in its natural state.

Dendrite A tree-like projection from the body of a neuron (nerve cell) that receives signals from other neurons.

Dermis A layer of skin below the outermost epidermis that is predominantly extracellular matrix with a low density of fibroblast cells.

Developmental biology The study of the processes by which multicellular organisms differentiate, grow, and develop, as controlled by their genes.

Diastole The part of the beating cycle of the heart when the heart muscle relaxes and allows the chambers to fill with blood.

Differential adhesion hypothesis The proposal put forth by biologist Malcolm Steinberg to explain cell sorting during embryogenesis in which liquid-like tissue spreading and cell segregation arise from differences in the adhesiveness (the stickiness) of different types of tissues.

Differentiation The specialization of cells to become distinct cell types.

Diffusion The movement of a substance from an area of higher concentration to an area of lower concentration.

Diffusivity Describes how fast one substance can diffuse through another substance; that is, the rate of diffusion.

Digital mask A digital mask is a virtual representation of a static physical mask. A digital mask is not static but, rather, it's dynamically generated by the on/off operation of each mirror in a computer-programmable array of micromirrors on a digital micromirror device.

Digital micromirror device A large array of very small aluminum mirrors, each of which can rotate individually and which can be in one of two possible rotational positions. In one position—the *on* position—a mirror will reflect light onto a designated surface. In the other position—the *off* position—the reflected light will not reach the designated surface. This enables the display of a digitized pattern of reflected light on the surface.

DNA An abbreviation for deoxyribonucleic acid, a molecule that codes genetic information for the transmission of inherited traits.

Echocardiogram A test using ultrasound (high-frequency sound waves) to visualize the heart.

Ejection fraction A measurement of how much blood the left ventricle of the heart pumps out (ejects) with each contraction of the heart muscle.

Elastic cartilage One of the three types of cartilage, it's found in the external ear and elsewhere. It is a structural tissue but for non-load-bearing body parts. It contains widely dispersed fine collagen fibers, providing strength.

Elasticity Characterizes the response of a material to a force that deforms it.

Elastomer A rubbery material composed of polymers that can recover their original shape after being stretched.

Electrocardiogram A test that measures the heart's electrical activity. It's abbreviated as ECG or EKG.

Electrolyte Minerals dissolved in the body's fluids that create electrically charged ions— that is, atoms with a surplus or deficit of electrons.

Electromechanical coupling Cardiac electromechanical coupling is the process that enables electrical excitation of heart cells to result in synchronous mechanical contraction of the tissue as a whole.

Embedded printing 3D printing in which the print nozzle moves within a soft material that surrounds and supports the ink being printed.

Embryoid body Three-dimensional aggregates of pluripotent stem cells that self-organize to develop tissues.

End-diastole The volume of blood in the left ventricle after it fills with blood prior to its contraction.

Endocytosis This describes how cells envelop macromolecules and particles from the surrounding medium by enclosing them in a small portion of the cell's lipid bilayer membrane. The membrane first invaginates and then pinches off inside the cell to form a vesicle containing the ingested substance.

Endothelial cell The cells that line the interior surface of blood vessels. A sheet of such cells is collectively referred to as the endothelium.

Endothelium A monolayer of endothelial cells that lines the inner walls of blood vessels (and lymphatic vessels).

End-systole The volume of blood in the left ventricle after it finishes contracting and ejecting blood.

Engraftment The word *engraft* is often used instead of *graft*, and engraftment is the process in which transplanted tissue integrates with host tissue. The integration process involves blood vessels and nerves connecting the transplanted and host tissues.

Enzyme A protein that accelerates the rate of a chemical reaction but without being altered in the process.

Epidermis The skin's outermost layer containing densely packed cells called keratinocytes. The epidermis also contains cells called melanocytes at the junction between the epidermis and the skin layer beneath the epidermis called the dermis.

Epinephrine Also called adrenaline, it is a hormone secreted mainly by the adrenal glands. It increases cardiac output and raises glucose levels in the blood. Epinephrine is also synthesized in the laboratory and used as a medication for several conditions, including anaphylaxis (a serious allergic reaction) and cardiac arrest.

Epithelial cell Cells that cover body surfaces, epithelial cells attach to one another forming an uninterrupted layer of cells separating the underlying tissues from the outside world. Examples are the epidermis of the skin and the linings of respiratory, urinary, and digestive tracts.

Epithelium Layers of epithelial cells that line hollow organs and glands and form the epidermal layer of the skin.

Eroom's law An observation stating that the number of new drugs approved per billion U.S. dollars spent on commercial drug research and development (dollars are inflation-adjusted) has halved approximately every 9 years since 1950. This is Moore's law spelled backwards to emphasize the contrast between drug development and the exponential advances in technologies such as the computer chip industry. Eroom's Law suggests that pharmaceutical research and development efficiency has declined fairly steadily.

Ex vivo Literally *outside the living body*, this is a procedure in which experiments are done with whole tissue slices with minimal alteration of the natural environment of the tissue.

Exocytosis The release of cellular substances contained within cell vesicles. This occurs by fusion of the membrane of the vesicle with the cell's membrane and subsequent release of the vesicle's contents to the exterior of the cell.

Extracellular matrix The non-cellular component of tissues and organs that provides physical scaffolding for cells and initiates critical biochemical and biomechanical cues necessary for tissue structural development, differentiation, and homeostasis. The extracellular matrix is made by the cells themselves.

Extrusion bioprinter Extrusion-based bioprinting extrudes (pushes out) continuous filaments of the relevant bioink.

Fatty acids The lipids that comprise the lipid bilayer membrane of cells have two contrasting ends, one of which is a fatty acid. Fatty acids are molecular chains that make up the portion of the lipid molecules that do not like to be in contact with water. As a result, the fatty acid chains face one another within the lipid bilayer to avoid the water outside the cell and within the cell's interior.

Fibrin An insoluble protein converted from fibrinogen by the clotting enzyme thrombin. Fibrin forms a fibrous mesh that impedes blood flow.

Fibrinogen A protein in blood that is converted to fibrin by the enzyme thrombin and forms a fibrin-based blood clot.

Fibroblasts The most common cell in connective tissue, fibroblasts produce a precursor to collagen that is an integral part of the extracellular matrix.

Fibrocartilage Cartilage that contains bundles of collagen in areas that require great support. Fibrocartilage forms the intervertebral discs in the spinal cord. Fibrocartilage is also found in the meniscus of the knee joint (along with articular cartilage). Damage to the meniscal cartilage of the knee can expose the articular cartilage to higher stresses and breakdown.

Fibronectin A protein of the extracellular matrix that binds to many other proteins including integrins on the cell surface, serving as a cell adhesion molecule connecting cells to the extracellular matrix.

Fibrotic scarring Fibrotic scarring describes the replacement of previously healthy tissue with tissue formed by an excess of connective tissue fibers.

Flow cytometry A technology used to measure and analyze multiple physical characteristics of cells as they flow in a fluid stream through a laser beam. The laser light that is scattered off the cells in several directions is captured and analyzed by software to uncover cellular properties such as size and internal complexity.

Fluid compartment It's convenient to speak of body fluids in terms of specific fluid compartments, with the compartments separated by a form of physical barrier. Three main compartments are fluid within cells, fluid in the space around cells, and blood plasma. To maintain normal function, the body must keep fluid levels from varying too much in each of these areas.

Fluorescence A physical phenomenon in which a chemical compound emits light of a particular color very soon after being illuminated by light of another color. Fluorescent molecules are used as reporters or tracers, and fluorescent tags can be attached to molecules that don't fluoresce themselves. Fluorescence can be used to determine the locations of certain structures within the cell, the presence of specific membrane constituents on the cell's exterior (for the identification of cell type), or to verify a certain process within the cell such as enzyme activity.

Fluorescent protein A family of proteins that emits a bright fluorescent color when exposed to certain wavelengths of light. Fluorescent proteins are very useful as reporter molecules that signal where a particular gene has been expressed—for example, in a living organism or in a bioprinted construct.

Fractional shortening A measurement made with an echocardiogram to determine the shortening of the left ventricle diameter between end-diastole and end-systole. The measurement helps to evaluate the heart's muscular contractility.

Free radicals Also called reactive oxygen species, free radicals are a type of unstable molecule containing oxygen. They chemically react readily with molecules in a cell and can cause damage to DNA, RNA, and proteins.

Fugitive ink Also known as sacrificial ink, these materials can be printed in the form of conduits or channels within a construct and then removed by simple means—such as lowering of temperature—to leave behind open passages for the creation of vascular networks for nutrient diffusion, for example.

Gait duty cycle A gait refers to a particular sequence of lifting and placing the feet during legged locomotion—walking or running for instance. Each repetition of the sequence is called a gait cycle. The gait duty cycle is what you get when you divide the stand time by the total amount of time the limb is either standing or swinging. Stand time is the duration of time that the limb is in contact with the ground, and swing time is the duration of time that the limb is not in contact with the ground.

Gap junctions Aggregates of intercellular channels that permit direct cell–cell transfer of ions and small molecules.

Gelatin A protein made from the connective tissue of animals. Gelatin results from the denaturing of collagen, so the three-dimensional shape of the collagen is lost and what remains are protein fragments of various molecular weights.

GelMa An abbreviation for gelatin methacrylate, a commonly used cross-linkable hydrogel that possesses some essential properties of native extracellular matrix.

Gene expression The process by which molecular products such as proteins develop from our DNA. Gene expression can also create molecular products other than proteins that help the cell to assemble proteins.

Gene therapy An experimental technique designed to introduce genetic material into cells to compensate for abnormal genes or to make a beneficial protein. A new gene is inserted directly into a cell via a carrier called a vector that is genetically engineered to deliver the gene.

Genotype Refers to the genetic makeup of an organism—that is, it describes an organism's complete set of genes.

Germ cell layer Refers to the three cellular layers—endoderm, mesoderm, and ectoderm—that will form all tissues of the embryo and thus give rise to all cells of the adult body.

Ghost heart A decellularized heart in which the remaining extracellular matrix appears as translucent tissue, devoid of cells.

Glass In science, glass refers to many different amorphous substances with different origins, as opposed to everyday life in which glass (e.g., a window) is an amorphous substance specifically derived from silicon dioxide.

Glial cell Also referred to as *glia*, glial cells can be found surrounding neurons in both the central nervous system and the peripheral nervous system. Glial cells form an insulating sheath of myelin that encircles—albeit with strategically placed gaps—the axons of neurons to speed the transmission of electrical impulses. Glial cells also provide physical support for neurons.

Glomerulus The blood-filtering unit of the kidney, the glomerulus is a specialized bundle of capillaries found at the beginning of each nephron. A thin, membranous, double-walled capsule (called Bowman's capsule) surrounds the glomerulus, and the filtrate from the capillaries passes through this capsule to get to the proximal convoluted tubule.

Glucose A sugar molecule that powers many processes in both animal and plant cells.

Glucose transporter Each cell in our body gathers the glucose it needs from the blood using membrane-bound proteins called glucose transporters. The transporter protein has an opening facing the outside of the cell, and it picks up a molecule of glucose. Then it shifts shape and opens toward the inside, releasing the glucose into the cell.

Glycogen Glycogen consists of many connected glucose molecules. It's stored in several places in the body; primarily the liver and skeletal muscles, but also in smaller amounts in the kidneys, the brain, and elsewhere.

Golgi apparatus An organelle within cells consisting of a group of flattened disc-like structures, the Golgi apparatus functions as a factory to process (modify, package, and transport) proteins and sort them for transport to their ultimate destinations—for example, to the cell membrane or secretion outside of the cell. It's sometimes called the Golgi complex or simply the Golgi.

Gradient The rate at which a quantity—for example, temperature or concentration—changes over distance or time.

Growth factor A substance, often a protein, that can stimulate cell growth, proliferation, and differentiation.

Hepatocyte The main cell type of the liver.

Histamine A chemical released by mast cells in the body's immune system. Histamines are released as an inflammatory response to injury or infection and as an allergic response to an encounter with an antigen.

Histology The microscopic study of the structure of cells and tissues to find the ways in which individual components are structurally and functionally related.

Homeostasis Maintenance of nearly constant conditions in the internal environment of the human body.

Hooke's law For relatively small distances, the force needed to extend or compress a spring is proportional to that distance.

Humors An ancient belief that the so-called four bodily humors (black bile, phlegm, yellow bile, and blood) and the balance and interactions of these four humors regulate the behavior of the human body.

Hyaline cartilage One of the three types of cartilage, hyaline cartilage is the most abundant in the human body. Hyaline cartilage is found lining bones in joints, where it's given the special name of articular cartilage. It's also found elsewhere, such as at the ends of the ribs, in the tracheal rings of the windpipe that connects the throat to the lungs, and in the nose (in the nasal septum separating the nasal cavity into the two nostrils).

Hydrogel A highly hydrated polymeric network with a structure similar to a cell's natural extracellular matrix, used to encapsulate cells and other biological material.

Hydroxyapatite A calcium mineral that is the main inorganic component of bone.

Hyperglycemia Hyperglycemia denotes an excess of glucose in the bloodstream. It's commonly called high blood sugar and is a signature of diabetes as well as a known risk factor for vascular disease.

Immunocytochemistry A lab test that uses antibodies to test for certain antigens in a sample of cells. It's performed with intact cells but with their surrounding extracellular matrix removed.

Immunofluorescence The labeling of antibodies or antigens with fluorescent dyes to demonstrate the presence of a particular antigen or antibody in a tissue.

Immunohistochemistry A lab test that uses antibodies as a way to identify the location of antigen proteins in thin tissue sections within the context of their familiar extracellular matrix.

In situ In its (original) place; in position.

In vitro In a test tube, culture dish, etc.— outside a living body under artificial conditions.

In vivo Within a living organism.

Induced pluripotent stem cells Adult cells that have been genetically reprogrammed to an embryonic stem cell-like state.

Induction The ability of a cell or tissue to influence the fate of nearby cells or tissues by chemical signals during development. An example is embryonic induction in which one tissue influences the differentiation of another tissue.

Inkjet bioprinter Inkjet-based bioprinting uses thermal or acoustic forces to eject drops of bioink onto a substrate.

Innervation The process of providing nerves to tissues.

Integrin A transmembrane protein that forms a link between the cytoskeleton inside cells and the extracellular matrix outside cells.

Interleukin A type of protein that mediates communication between cells. Interleukins are particularly important in stimulating immune responses such as inflammation; they are not stored within cells but are rapidly secreted in response to a stimulus such as an infectious agent. Interleukins are part of a larger group of cell signaling molecules called cytokines.

Inulin A sugar found in many plant roots, it is of great value in kidney studies. Inulin isn't reabsorbed by the proximal tubules, so it serves as a hallmark for barrier function integrity and an indicator of selective reabsorption.

Ion An atom or molecule with a net electrical charge, either positive or negative.

Iron lung The iron lung is a negative pressure ventilator, which means that the machine alternately decreases air pressure around the thorax and the abdomen (to stimulate spontaneous inhalation) and then returns the surrounding air pressure to normal atmospheric pressure. It was commonly used to enable breathing in people with polio-induced respiratory paralysis.

Kenzan method This is a scaffold-free printing method, essentially a spheroid-assembling method. Medical-grade metal needles robotically impale preformed spheroids, providing them with spatial organization and a controlled opportunity to interact, fuse, and secrete extracellular matrix.

Keratin A structural protein found in skin (the epidermal layer), scales, hair, feathers, nails, claws, horns, and hooves.

Keratinocytes Cells found in the epidermis that play a prominent role in the barrier function of the skin.

Kidney wafer A slice of healthy kidney tissue—a kidney wafer—might be used as a graft to augment a failing kidney.

Laser-assisted bioprinter Laser-assisted bioprinting typically consists of a pulsed laser beam to transfer material from a target onto a receiving substrate. The target is an optically transparent material with a coating that acts as a laser absorption layer. When the absorption layer is covered with a biological material, the absorbed laser pulse ejects biomaterial from the target and propels it through the air toward the receiving substrate.

Limb bud In an embryo, a limb bud is a small protuberance from which a limb develops.

Lipid bilayer Lipid bilayers form the structure of the cell's outer membrane. All of the lipid molecules in cell membranes are composed of two distinct ends. One end is referred to as polar, which means that it has both an electrically positive and an electrically negative portion. Overall, the polar end may carry a net electrical charge or it may not. Regardless, the polar end of a lipid molecule likes to be in contact with water. The other end of the lipid molecule is composed of two chains of fatty acids. The fatty acid chains do not like to be in contact with water. The two lipid layers (the lipid bilayer) that form the cell membrane have their polar ends facing toward the outside of the cell and toward the interior of the cell. The fatty acid ends of each lipid molecule face each other. So the fatty acid ends are sandwiched between the polar ends.

Lobules Literally small lobes, they represent the organizational structure of the liver. Lobules are idealized as hexagonal arrangements of hepatocytes radiating outward from a central vein.

Lumen The cavity or opening of a tubular structure such as a blood vessel.

Marker expression Cell surface markers are proteins that appear on the surface of cells (serving as receptors for signaling molecules to identify a particular type of cell) and gene expression is the process by which products such as proteins develop from our DNA. So, *marker expression* refers to the presence (and prevalence) of cell surface marker proteins.

Matrigel A trade name for a commonly used hydrogel that contains many components typically found in the extracellular matrix.

Mechanobiology A field that studies how cells sense and use mechanical information provided by the extracellular matrix environment—whether their native matrix or otherwise—and how mechanical forces influence cellular growth, development, differentiation, and disease.

Mechanotransduction Processes through which cells sense and respond to mechanical stimuli by converting them to biochemical signals that produce specific cellular responses.

Melanocyte A specialized skin cell found in the bottom layer of the epidermis that produces the protective skin-darkening pigment melanin.

Mesenchymal stem cells Putative adult stem cells present in tissues such as umbilical cord, bone marrow, and fat tissue. They are said to differentiate into a variety of tissues, including bone, cartilage, muscle, fat, and connective tissue. However, there is considerable concern in the scientific community on the use/misuse of the term: Sipp, Douglas, Pamela G. Robey, and Leigh Turner. "Clear up This Stem-Cell Mess." *Nature* 561, no. 7724 (September 2018): 455–57. https://doi.org/10.1038/d41586-018-06756-9.

Messenger RNA (mRNA) Information stored in DNA found in the nucleus of cells is transcribed (copied) into another molecule called messenger RNA (mRNA). Messenger RNA then travels out of the cell nucleus and into the cytoplasm of the cell where two other forms of RNA use the information in the mRNA to synthesize proteins.

Metabolite A product of metabolic breakdown (or a substance necessary for a metabolic reaction).

Microchannels Channels with diameters of several hundred microns that allow nutrients and oxygen to diffuse into bioprinted structures to keep them alive while they develop a system of blood vessels.

Microfluidics A field dedicated to miniaturized plumbing and fluid manipulation using channels measuring from tens to hundreds of microns in diameter. By combining cells with microfluidic channels, one can manipulate a cell's microenvironment.

Micron One millionth of a meter, also referred to as a micrometer. In English units, 1 micron is about 39 millionths of an inch. To provide some sense of scale, the thickness of a human hair is very variable but is approximately 60–120 microns wide.

Microreactors Small versions of bioreactors.

Microvascular Relating to the part of the circulatory system made up of minute vessels such as the smallest capillaries that average 0.003–0.004 millimeters (3–4 microns) in diameter.

Microvilli Cellular membrane protrusions that increase the surface area of the cell from which they project.

Mineral Minerals are solid crystalline elements or compounds—inorganic substances that occur in nature. Some minerals function as electrolytes.

Mineralization of bone In bone mineralization, osteoblasts (bone-forming cells) produce crystals of the inorganic material calcium phosphate and deposit them in an organized fashion in the surrounding extracellular matrix. Calcium and phosphorus are the principal minerals found in bone.

Moore's law An observation made by Gordon Moore, founder of computer chip manufacturer Intel Corporation, that the number of transistors in an integrated circuit doubles approximately every two years.

Morphology Refers to the size, shape, and structure of a cell or organoid.

Motor nerve (motor neuron) A nerve in the central nervous system that sends signals to the muscles of the body.

Movable type Movable type means that individual type pieces with a single letter or character can be easily assembled and relocated to print any combination of characters that is desired.

Multi-nozzle print head A bioprinter that uses more than one print head/nozzle.

Multiphoton-excited photochemistry Whereas fluorescence is usually excited in a material by the absorption of a single photon of light, the absorption of two or more photons results in a sharper image having more contrast. Multiphoton-excited photochemistry is a fluorescence imaging technique that is similar to confocal microscopy. Multiphoton-excited photochemistry has an advantage in imaging optically thick specimens.

Myelin An insulating sheath that encircles (with strategically placed gaps) the axons of nerve cells to speed the transmission of electrical impulses.

Myoblast An embryonic precursor that differentiates into a muscle cell.

Myocardial infarction A heart attack.

Myofiber A bundle of so-called myofibrils that forms the basic structural element of muscle tissue.

Myofibril The smallest functional unit in skeletal muscle.

Myosin A filamentous protein that, in coordination with another filamentous protein named actin, is responsible for muscle contraction.

Myotube A myotube is a step in the formation of muscle tissue and is formed by the fusion of multiple precursors called myoblasts. This fusion results in the myotube having multiple nuclei.

Necrosis Cell death that, unlike apoptosis, is not programmed but is the result of injury, infection, cancer, and so on.

Necrotic core In bioprinting, this means that there are a large number of dead cells in the interior of a bioprinted construct.

Negative pressure ventilator Also called an iron lung, a negative pressure ventilator is a machine that alternately decreases air pressure around the thorax and the abdomen (to stimulate spontaneous inhalation) and then returns the surrounding air pressure to normal atmospheric pressure. It was commonly used to enable breathing in people with polio-induced respiratory paralysis.

Nephron The basic functional and structural unit of the kidney, the nephron is responsible for filtering excess fluid and waste products out of the blood, reabsorbing the useful material (glucose, for example) from the filtrate, and excreting the non-useful material in the form of urine.

Nerve growth factor The first neurotrophic factor discovered, nerve growth factor facilitates the growth and organization of the nervous system during embryonic development. This protein remains active in the adult organism in many ways, including maintenance of phenotypic and functional characteristics of neurons and immune cells.

Nerve impulse An electrical signal generated by a nerve cell that travels along its axon. Prior to a nerve impulse there is a positive charge outside the cell compared to inside the cell: that is, an electrical gradient across the cell membrane. A nerve impulse is a sudden reversal of the electrical charge across the cell membrane of the nerve cell. The impulse begins when the nerve cell receives a chemical signal from another cell that causes sodium–potassium pumps in the cell membrane to open. This allows the excess of positive ions that were outside the cell to rush back into the cell, causing a flip-flop: Now the inside of the cell has a positive charge compared to the outside of the cell. This reversal of the electrical gradient ripples down the axon as an electric current—that is, a nerve impulse.

Neurite A projection from the cell body of a neuron that can be either an axon or a dendrite.

Neurofilaments Protein polymers found in the cytoplasm of neurons.

Neuromuscular junction The chemical synapse between motor neurons and skeletal muscle fibers.

Neuron Also called a nerve cell, a neuron is an electrically excitable cell in the nervous system. It functions to process and transmit information. A neuron is typically composed of a cell body, an axon, and a dendritic tree (multiple dendrites). Neurons communicate with each other at junctions called synapses, where an axon from one neuron relays a signal to the dendritic tree of another neuron. The signal transmitted at the synapse can be in the form of chemical or electrical transmission. Most signal transmission is chemical and mediated by molecules called neurotransmitters.

Neurotransmitter A chemical messenger released by neurons that enables transmission of electrical signals across synapses from one neuron to another neuron.

Neurotrophic factors Molecules that enhance the growth and survival potential of neurons. They play important roles in both development—where they act as guidance cues for developing neurons—and in the mature nervous system.

Nozzle The portion of a bioprinter that dispenses material to build a bioconstruct.

Oligodendrocytes Glial cells that produce myelin for the neurons of the central nervous system.

Organ Multiple tissue types that act together to participate in specific functions.

Organ bud In an embryo, an organ bud is a condensed tissue mass—prior to blood perfusion—that will develop into an organ.

Organelle A small, specialized structure within the cytoplasm of a cell that performs specific functions. One example of an organelle is the Golgi apparatus.

Organogenesis In tissue engineering (which encompasses bioprinting), organogenesis refers to the design and development of engineered tissue. (The word "organogenesis" also describes a phase of embryonic development.)

Organoid A three-dimensional, multicellular, in vitro tissue construct that mimics some features of its corresponding in vivo organ, such that it can be used to study cellular and molecular aspects of that organ.

Organ-on-a-chip A human organ-on-a-chip is a microfluidic cell culture device. It has separate parenchymal—that is, functional, organ-specific tissue—and vascular compartments. Living human cells line these compartments. The device mimics the multicellular architecture, tissue–tissue interfaces, and the relevant physical microenvironment of key functional units of living organs and provides dynamic vascular perfusion in vitro. These devices are called *chips* because they were originally fabricated using methods adapted from those used for manufacturing computer microchips.

Organ reserve The ability of an organ to successfully return to its original physiological level of activity, its normal state of homeostatic balance, after being stressed.

Osteoinduction The initial step in the formation of bone.

Parenchyma The functional part of an organ, in contrast to the stroma that refers to the structural and connective tissue of the organ.

Partial organs A small functioning part of a full organ.

Perfusate The generic name for a perfused fluid.

Perfusion The passage of fluid through a circulatory system. In bioprinting the fluid is typically a liquid culture medium or blood.

Peripheral nervous system The nervous system excluding the brain and spinal column—that is, all the nerves in different locations in our body exclusive of the brain and spinal column.

Peristaltic Refers to coordinated waves of contraction and relaxation.

pH A scale used to determine the extent to which an aqueous (or other) solution is acidic or basic. pH is a measure of hydrogen ion concentration that translates the values of the concentration of hydrogen ions into numbers between 0 and 14. A solution with a pH of less than 7 is considered acidic and a solution with a pH greater than 7 is considered basic. Pure water, which is neither acidic nor basic, has a pH of 7.

Phase III clinical trials U.S. Food and Drug Administration phase III clinical trials are done to test efficacy and monitor adverse reactions. They usually involve several hundred to several thousand participants and test how well a new drug works compared with existing medications for the same condition. Phase III trials are larger and longer in duration than phase II or phase I clinical trials. Phase III studies are sometimes known as pivotal studies.

Phenotype The observable physical properties of an organism as determined by its genotype (the set of genes the organism carries) and by environmental influences on the organism's genes.

Photocurable A material made of components—polymers, for example—that can be hardened or cross-linked by either visible or ultraviolet light.

Photolithography In the semiconductor industry, photolithography is used to create a three-dimensional pattern on a surface. The experimenter coats the surface with a light-sensitive polymer called a photoresist and applies a patterned mask to the surface; only unmasked areas of the polymer are exposed to light. Chemicals known as developers dissolve either the photoresist that was exposed or the photoresist that was not exposed, depending on the choice of photoresist. The end result is a transfer of the geometry of the patterned mask to form a three-dimensional relief image on the surface.

Photopolymerization A technique that uses visible or ultraviolet light to cause polymers to become cross-linked—that is, bonded together.

Plasmid A small (typically), circular DNA molecule found in bacteria and used in molecular biology as a vector—that is, a segment of DNA that carries genetic material into a cell.

Pluripotent Describes the ability of a cell to develop into the three primary germ cell layers of the early embryo and therefore into all cells of the adult body. Embryonic stem cells and induced pluripotent stem cells are characterized by their pluripotency.

Polarized epithelium Epithelial cells are an example of a cell type with distinct anterior and posterior ends. Epithelial cells have an apical surface and a basal surface to establish a barrier function and to provide directionality for vectorial transport and other functions.

Poliomyelitis A virus-borne disease often called by the shortened name polio. The poliovirus enters the body through the mouth, multiplying along the way to the digestive tract, where it multiplies even further. In paralytic polio, the virus leaves the digestive tract, enters the bloodstream, and then attacks nerve cells. In severe cases, the throat and chest may be paralyzed, and death may result if the patient isn't assisted by some means of artificial breathing support.

Polymer A large molecule composed of many repeating identical subunits.

Post-processing The maturation (incubation) of a printed construct in a bioreactor following bioprinting.

Prestressed A type of tensegrity structure in which, even before any external force is applied, all the structural members are already in tension or compression.

Printability The printability of a bioink describes whether the ink can be accurately and precisely deposited where and when you want it to be.

Progenitor cells Early descendants of stem cells but more constrained in their differentiation potential and their capacity for self-renewal. Note that there is difficulty in the literature in precisely defining progenitor cells since their phenotype and differentiation capabilities are highly dependent on culture conditions, environmental cues, and experimental methodologies.

Programmable Able to change physical properties such as shape or optical characteristics based on a user's input or autonomous sensing.

Propranolol Known as a beta blocker, propranolol is a medication that is used to treat high blood pressure and arrhythmias.

Protein Proteins are large molecules made by the combined effort of DNA and three types of RNA found in cells. Proteins are said to be *expressed* by portions of DNA called genes. Proteins are the major macromolecular molecules in our body. They are the building blocks of cells and execute most cell functions. Among their vast repertoire, they act as channels or pumps to control passage of small molecules into and out of the cell, they carry messages from one cell to another or act as signal transducers that relay sets of signals inward from the cell membrane to the cell nucleus, they act as tiny molecular machines with moving parts, and they serve as antibodies and enzymes. Proteins are made up of long chains of amino acids.

Protein pump A transmembrane protein that drives the active transport of ions or small molecules across the cell membrane (across the lipid bilayer).

Proximal convoluted tubule That portion of the nephron that reabsorbs the majority of the constituents of the glomerular filtrate that have nutritional significance—for example, glucose.

Pulmonary Relating to the lungs.

Reabsorption Transport of important nutrients such as amino acids, glucose, and vitamin transport proteins out of the glomerular filtrate takes place in the proximal convoluted tubule of the kidney. Capillaries that run next to the proximal convoluted tubule reabsorb these nutrients. One speaks of *reabsorption* rather than absorption because the nutrients were initially absorbed by the digestive tract following ingestion of food.

Reactive oxygen species Also called free radicals, reactive oxygen species are a type of unstable molecule containing oxygen. They chemically react readily with molecules in a cell and can cause damage to DNA, RNA, and proteins.

Recombinant protein A protein that is encoded by a gene from recombinant DNA. Recombinant DNA is formed by techniques that combine DNA from different species to produce new genetic combinations that are not found in nature.

Renal Relating to the kidneys.

Rheology The study of the flow and deformation of materials.

RNA An abbreviation for ribonucleic acid. There are many types of RNA. Three types are most well-known. One is messenger RNA (mRNA) that carries genetic code from the DNA in the nucleus of a cell to the cytoplasm of that cell. The other two types are transfer RNA (tRNA) and ribosomal RNA (rRNA) both of which are present in the cytoplasm of cells; they act together to synthesize proteins from messenger RNA (a process referred to as *translation*). Transfer RNA carries amino acids to ribosomal RNA which, along with other associated molecules, directs the steps of protein synthesis, linking amino acids together to make a protein molecule.

Sacrificial ink Also known as fugitive ink, it can serve as a temporary support structure and can then be removed by lowering or raising the temperature or by the ink's solubility in water, depending on the specific ink.

Saltatory conduction Many nerve axons are wrapped with myelin, but there are breaks in the wrapping. As a nerve impulse travels down the axon, it is said to jump from the location of one myelin break to the next. This mode of conduction is called saltatory from a Latin word meaning "to leap." Saltatory conduction is a faster way to travel down an axon than traveling down an axon without a myelin sheath.

Scaffold Materials that provide mechanical strength and stiffness and on which cells can attach, migrate, proliferate, and differentiate.

Scalable In the context of bioprinting strategies, a method is scalable if it can be readily adapted to form larger size or higher density bioconstructs.

Schwann cell A glial cell in the peripheral nervous system that produces the insulating myelin sheath around the axons of nerve cells.

Seeding (of cells) Seeding cells means spreading a defined amount (volume or cell number) of cells onto a surface.

Selective reabsorption In the kidney, selective reabsorption is a process in which particular molecules of value are reabsorbed from the glomerular filtrate and returned to the blood.

Sense receptors We have sense receptors for vision, taste, smell, hearing, and balance that are in our eyes, mouth, nose, and ears (for both sound and balance), respectively. We have sense receptors in our skin and distributed throughout our body for senses like touch, temperature, and pain. Most sense receptors are specialized epithelial cells. Each type of receptor must transduce (convert) its sensory input into an electrical signal so that the input can be carried by nerves from the peripheral nervous system to the central nervous system for processing.

Sensory nerves Nerves that carry sensory information from a receptor such as those in the eye or ear to a central location in the nervous system, i.e., the spinal cord or brain.

Shear stress Describes internal forces that act parallel to a cross-sectional area of an object causing layers or planes internal to the object to slide past one another. In bioprinting, cells experience shear stress as they flow out of the nozzle, manifested as sliding forces or internal frictional forces.

Shear thinning Behavior in which the viscosity decreases as the shear rate increases—that is, as the rate at which the material's internal layers move past each other.

Signal receptor molecules Proteins (generally transmembrane proteins) on a cell surface that act as receptors by binding to signaling molecules from another cell and then initiate a physiological response in the target cell. The binding outside the cell gives rise to signaling within the cell through a cascade of molecular events.

Signal transduction The process in which a signal provided by a chemical messenger is translated via a series of molecular events into a cellular response.

Silicone A polymeric, rubber-like material.

Silk fibroin A structural protein made by silkworms and spiders.

Sodium–potassium pump A transmembrane protein that generates a gradient of ions, pumping sodium ions out of the cell and potassium ions into the cell.

Soft lithography As opposed to photolithography used in computer chip manufacturing, soft lithography is a suite of methods that share the common feature that they use a patterned elastomer as the mask, stamp, or mold.

Solubilization The action of chemical reagents on a substance to produce a structural breakdown into a liquid form that can then be dissolved in a solution.

Solute A substance dissolved in another substance, called the solvent, to create a solution.

Solution A mixture of two (or more) substances in which a substance called the solute is dissolved in another substance called the solvent.

Solvent A substance that dissolves a solute to form a solution.

Spatial light modulator An object that modulates a beam of light, usually the light's intensity. Modulation means the controlled variation of some property of the light, such as its amplitude (the intensity of a light wave is proportional to the square of its amplitude) or its frequency.

Spatial resolution The minimum distance at which two objects can be distinguished from one another.

Sphericity How an object compares to a perfect sphere.

Spheroid An aggregated, mutually adherent population of cells adopting a spherical or sphere-like shape.

Stem cells Unspecialized cells capable of replenishing themselves through cell division, sometimes after long periods of inactivity. Under certain physiologic or experimental conditions, they can be induced to become tissue- or organ-specific cells.

Stepper motor An electric motor that divides a full rotation into a number of equal steps.

Stereolithography An additive manufacturing method that creates a three-dimensional object by photocuring liquid polymers layer by layer using ultraviolet light. Chuck Hull invented stereolithography; it was the first 3D printing method.

Strain With respect to the heart muscle, strain is a dimensionless quantity that represents the percentage change in dimension of the heart's left ventricle from a resting state to a state after a force has been applied. This allows assessment of the stiffness of the heart's left ventricle. Healthy heart tissue should be flexible enough to extend in response to demands placed on it; stiffening can contribute to (and be a measure of) heart dysfunction.

Stroma The structural and connective tissue of an organ, as opposed to the parenchyma, which refers to the functional part of the organ.

Synapse The site of transmission of electric nerve impulses between two neurons, usually located between the axon terminal of one neuron and the dendritic tree of another neuron. The transmission is typically accomplished by the release of a chemical messenger called a neurotransmitter.

Synovial joints The most common type of joint in the body, a key characteristic of a synovial joint is the presence of a joint cavity filled with synovial fluid. Such a cavity is not present in fibrous joints (e.g., skull bones) or cartilaginous joints (e.g., intervertebral discs). Examples of synovial joints are knees, hips, shoulders, wrists, and elbows.

Systole The part of the beating cycle of the heart when the heart muscle contracts to pump blood out.

3D printing An additive manufacturing process—constructing one layer at a time—that builds three-dimensional solid objects from a digital file.

TEER An acronym for transepithelial/transendothelial electrical resistance, TEER is a method used for quantitatively measuring and monitoring over time the integrity of the in vitro barrier function of tissues.

Tensegrity A combined word from *tension* and *integrity* coined by Buckminster Fuller and denoting an architectural system composed of opposing tension and compression elements that balance each other, thereby creating a resting tension or "prestress" that stabilizes the entire structure.

Thrombin An enzyme that catalyzes the conversion of fibrinogen to fibrin to facilitate the clotting of blood.

Thrombus A thrombus is commonly known as a blood clot. It's a mix of fibrin and blood platelets often containing red and white blood cells. A thrombus can occur within a blood vessel or a chamber of the heart. If it occurs in a blood vessel, it will be adjacent to the endothelium. If it occurs in a chamber of the heart, it will be adjacent to the endocardium (the membrane that lines the inside of the chambers of the heart and forms the surface of the heart valves).

Tissue A group of structurally and functionally similar cells and their extracellular matrix material.

Tissue engineering A field that applies the principles of biology and engineering to the development of functional substitutes for damaged tissue.

Trachea Commonly referred to as the windpipe, the trachea is a tube carrying air from the mouth and nose to the lungs.

Transcription factor A protein involved in the process of converting (transcribing) DNA into RNA.

Transfection Introducing a piece of foreign DNA via a vector.

Transglutaminase An enzyme that catalyzes the cross-linking of proteins. It has been called a protein stapler, and in the food industry it is often referred to as meat glue.

Transmembrane protein Many proteins are permanently embedded in the cell membrane. Those embedded proteins that go completely through the cell membrane—so that part of the protein is outside the cell and part is within the cell—are known as transmembrane proteins.

Ultrafiltrate In the kidney, this is the fluid of water, glucose, ions, and amino acids—but virtually free of blood cells and large proteins—that has passed through the capillary walls into the Bowman's capsule that surrounds them. The process is referred to as ultrafiltration, whereas the fluid is variously referred to as glomerular filtrate or ultrafiltrate.

Ultraviolet light Electromagnetic radiation having a wavelength shorter than that of the violet end of the visible electromagnetic spectrum but longer than that of X-rays.

Upregulation An increase in the number of a particular receptor on the surface of cells thus increasing the cellular response to the corresponding signaling messenger.

Urea A major end product (waste product) in the metabolic breakdown of proteins.

Vascular endothelial growth factor An angiogenic growth factor that induces proliferation, migration, and permeability of endothelial cells (and thus stimulation of blood vessels).

Vascularization The process of providing blood vessels to tissues.

Vector A segment of DNA that carries genetic material into a cell.

Vectorial transport The process by which polarized epithelial cells enforce *one-way* directional transport of ions and molecules in and out of the cell.

Venule A very small vein.

Vesicle A small structure within or outside of a cell that contains liquid (or other contents) surrounded by a lipid bilayer.

Viability The number of living cells in a sample divided by the initial number of cells in the sample gives the cell viability.

Viscoelasticity A property of materials that exhibit both viscous and elastic characteristics under stress and deformation, thus showing attributes of both the liquid and the solid phase.

Viscosity A measure of the resistance to deformation or flow.

Viscous fingering The emergence of a finger-like invasion pattern at the interface of two fluids when a fluid of lower viscosity displaces a resident viscous fluid in a porous medium.

Well plates Flat plastic plates with multiple shallow concavities called *wells*. The multiple wells are often used as an array of small test tubes.

Index